ESTUARINE RESEARCH, MONITORING, and RESOURCE PROTECTION

Marine Science Series

The CRC Marine Science Series is dedicated to providing state-of-the-art coverage of important topics in marine biology, marine chemistry, marine geology, and physical oceanography. The series includes volumes that focus on the synthesis of recent advances in marine science.

CRC MARINE SCIENCE SERIES

SERIES EDITOR

Michael J. Kennish, Ph.D.

PUBLISHED TITLES

Artificial Reef Evaluation with Application to Natural Marine Habitats, William Seaman, Jr.
Chemical Oceanography, Second Edition, Frank J. Millero
Coastal Ecosystem Processes, Daniel M. Alongi
Ecology of Estuaries: Anthropogenic Effects, Michael J. Kennish
Ecology of Marine Bivalves: An Ecosystem Approach, Richard F. Dame
Ecology of Marine Invertebrate Larvae, Larry McEdward
Ecology of Seashores, George A. Knox
Environmental Oceanography, Second Edition, Tom Beer
Estuary Restoration and Maintenance: The National Estuary Program, Michael J. Kennish
Eutrophication Processes in Coastal Systems: Origin and Succession of Plankton Blooms and Effects on Secondary Production in Gulf Coast Estuaries, Robert J. Livingston
Handbook of Marine Mineral Deposits, David S. Cronan
Handbook for Restoring Tidal Wetlands, Joy B. Zedler
Intertidal Deposits: River Mouths, Tidal Flats, and Coastal Lagoons, Doeke Eisma
Marine Chemical Ecology, James B. McClintock and Bill J. Baker
Morphodynamics of Inner Continental Shelves, L. Donelson Wright
Ocean Pollution: Effects on Living Resources and Humans, Carl J. Sindermann
Physical Oceanographic Processes of the Great Barrier Reef, Eric Wolanski
The Physiology of Fishes, Second Edition, David H. Evans
Pollution Impacts on Marine Biotic Communities, Michael J. Kennish
Practical Handbook of Estuarine and Marine Pollution, Michael J. Kennish
Practical Handbook of Marine Science, Third Edition, Michael J. Kennish
Seagrasses: Monitoring, Ecology, Physiology, and Management, Stephen A. Bortone
Trophic Organization in Coastal Systems, Robert J. Livingston

ESTUARINE RESEARCH, MONITORING, and RESOURCE PROTECTION

Edited by
Michael J. Kennish

Institute of Marine and Coastal Sciences
Rutgers University
New Brunswick, New Jersey

WITHDRAWN

CRC PRESS

Boca Raton   London   New York   Washington, D.C.

The cover design was created by Scott M. Haag of the Center for Remote Sensing and Spatial Analysis at Rutgers University. It is a Landsat image of the Jacques Cousteau National Estuarine Research Reserve and surrounding coastal bays and watersheds of New Jersey. The original satellite image is from the U.S. Geological Survey EROS Data Center, Sioux Falls, South Dakota (http://idcm.usgs.gov/).

Library of Congress Cataloging-in-Publication Data

Estuarine research, monitoring, and resource protection / edited by Michael J. Kennish.
    p. cm. -- (Marine science series)
  Includes bibliographical references and index.
  ISBN 0-8493-1960-9
  1. National Estuarine Research Reserve System. 2. Estuarine ecology--United States--Case studies. 3. Environmental monitoring--United States--Case studies. I. Kennish, Michael J. II. Series.

QH76.E86 2003
577.7'86'0973--dc21   2003053062

This book contains information obtained from authentic and highly regarded sources. Reprinted material is quoted with permission, and sources are indicated. A wide variety of references are listed. Reasonable efforts have been made to publish reliable data and information, but the author and the publisher cannot assume responsibility for the validity of all materials or for the consequences of their use.

Neither this book nor any part may be reproduced or transmitted in any form or by any means, electronic or mechanical, including photocopying, microfilming, and recording, or by any information storage or retrieval system, without prior permission in writing from the publisher.

The consent of CRC Press LLC does not extend to copying for general distribution, for promotion, for creating new works, or for resale. Specific permission must be obtained in writing from CRC Press LLC for such copying.

Direct all inquiries to CRC Press LLC, 2000 N.W. Corporate Blvd., Boca Raton, Florida 33431.

Trademark Notice: Product or corporate names may be trademarks or registered trademarks, and are used only for identification and explanation, without intent to infringe.

Visit the CRC Press Web site at www.crcpress.com

© 2004 by CRC Press LLC

No claim to original U.S. Government works
International Standard Book Number 0-8493-1960-9
Library of Congress Card Number 2003053062
Printed in the United States of America 1 2 3 4 5 6 7 8 9 0
Printed on acid-free paper

Dedication

*This book is dedicated to
the National Estuarine Research Reserve System.*

Preface

Estuarine Research, Monitoring, and Resource Protection is principally designed as a reference volume for estuarine and watershed scientists, resource managers, decision makers, and other professionals who deal with coastal zone issues. Information contained in this volume will be useful to individuals conducting either basic or applied research on estuaries. It will also be valuable to administrators engaged in coastal resource management programs.

This book is an outgrowth of my work as research coordinator of the Jacques Cousteau National Estuarine Research Reserve (JCNERR) in New Jersey. I thank my colleagues at the JCNERR who comprise a remarkably cohesive and competent group of researchers, administrators, and support staff. They include Michael P. DeLuca (reserve manager), Scott Haag (GIS coordinator), Josephine Kozic (volunteer coordinator), Janice McDonnell (assistant manager), Eric Simms (education coordinator), and Lisa Weiss (watershed coordinator). These individuals are dedicated professionals who have played major roles in the successful development and expansion of the program site.

Estuarine Research, Monitoring, and Resource Protection provides an overview of the National Estuarine Research Reserve System (NERRS). I would like to thank many members of the NERRS program who have supplied data, site profile reports, and other information vital to the production of the volume. At the Estuarine Reserves Division of the National Oceanic and Atmospheric Administration (NOAA), I thank Laurie McGilvray (chief), Maurice Crawford (research coordinator), and Erica Seiden (program specialist). At NERR program sites, I thank Betty Wenner (research coordinator) and Saundra Upchurch (reserve biologist) of the Ashepoo–Combahee–Edisto (ACE) Basin NERR, Lee Edmiston (research coordinator) of the Apalachicola NERR, Julie Bortz (research coordinator) of the Chesapeake Bay (Maryland) NERR, Willy Reay (reserve manager) and Ken Moore (research coordinator) of the Chesapeake Bay (Virginia) NERR, Bob Scarborough (research coordinator) of the Delaware NERR, Kerstin Wasson (research coordinator) of the Elkorn Slough NERR, Brian Smith (research coordinator) of the Great Bay NERR, Rick Gleeson (research coordinator) of the Guana Tolomato Matanzas (GTM) NERR, Chuck Nieder (research coordinator) of the Hudson River NERR, Carmen Gonzalez (reserve manager) of the Jobos Bay NERR, Carl Schoch (research coordinator) of the Kachemak Bay NERR, Kenny Reposa (research coordinator) of the Narragansett Bay NERR, Steve Ross (research coordinator) of the North Carolina NERR, Chris Buzzelli (research coordinator) of the North Inlet-Winyah Bay NERR, Dave Klarer (research coordinator) of the Old Woman Creek NERR, Doug Bulthuis (research coordinator) of the Padilla Bay NERR, Mike Shirley (research coordinator) of the Rookery Bay NERR, Dorset Hurley (research coordinator) of the Sapelo Island NERR, Steve Rumrill (research coordinator) of the South Slough NERR, Jeff Crooks (research coordinator) of the Tijuana River NERR, Chris Weidman (research coordinator) of the Waquoit Bay NERR, Scott Phipps (research coordinator) of the Weeks Bay NERR, and Michele Dionne (research coordinator) of the Wells NERR. Special thanks to Tammy Small, Manager of the Centralized Data Management Office, for providing water quality data on NERRS program sites. Dwight Trueblood, Co-Director of the Cooperative Institute for Coastal and Estuarine Environmental Technology (CICEET), is likewise thanked for his involvement in the NERRS Program. Special gratitude is extended to the Waquoit Bay NERR, Delaware NERR, ACE Basin NERR, Weeks Bay NERR, and Tijuana River NERR, whose profile reports constituted valuable sources of information for this publication.

I would also like to acknowledge the work of Ken Able (Rutgers University) and his staff on the JCNERR system in New Jersey, Skip Livingston (Florida State University) on the Apalachicola NERR in Florida, Ivan Valiela (Boston University Marine Program) on the Waquoit Bay NERR, and Joy Zedler on the Tijuana River NERR in Southern California. These investigators have produced extensive databases on important estuarine systems in the NERRS program.

I am especially grateful to the editorial and production personnel of CRC Press who are responsible for publishing this book. In particular, I express appreciation to John B. Sulzycki, senior editor, and Christine Andreasen, production editor, of the editorial and production departments, respectively.

This is Publication Number 2003–17 of the Institute of Marine and Coastal Sciences, Rutgers University, and Contribution Number 100-23 of the Jacques Cousteau National Estuarine Research Reserve. Work on this volume was conducted under an award from the Estuarine Reserves Division, Office of Ocean and Coastal Resource Management, National Ocean Service, National Oceanic and Atmospheric Administration.

Editor

Michael J. Kennish, Ph.D., is a research professor in the Institute of Marine and Coastal Sciences at Rutgers University, New Brunswick, New Jersey, and the research coordinator of the Jacques Cousteau National Estuarine Research Reserve in Tuckerton, New Jersey. He holds B.A., M.S., and Ph. D. degrees in geology from Rutgers University. Dr. Kennish's professional affiliations include the American Fisheries Society (Mid-Atlantic Chapter), American Geophysical Union, American Institute of Physics, Estuarine Research Federation, New Jersey Academy of Science, and Sigma Xi.

Dr. Kennish has conducted biological and geological research on coastal and deep-sea environments for more than 25 years. While maintaining a wide range of research interests in marine ecology and marine geology, Dr. Kennish has been most actively involved with studies of marine pollution and other anthropogenic impacts on estuarine and marine ecosystems as well as biological and geological investigations of deep-sea hydrothermal vents and seafloor spreading centers. He is the author or editor of 11 books dealing with various aspects of estuarine and marine science. In addition to these books, Dr. Kennish has published more than 130 research articles and book chapters and presented papers at numerous conferences. His biogeographical profile appears in *Who's Who in Frontiers of Science and Technology, Who's Who Among Rising Young Americans, Who's Who in Science and Engineering,* and *American Men and Women of Science.*

Introduction

Estuaries rank among the most productive aquatic ecosystems on earth. They also rank among the most heavily impacted by human activities. Kennish (2002a) recently assessed the environmental state of estuaries and predicted their condition by the year 2025. He identified ten principal anthropogenic stressors on estuaries that, taken together, can mediate significant changes in the structure, function, and controls of these vital coastal ecotones (Table 1). Tier I anthropogenic stressors (i.e., habitat loss and alteration, eutrophication, organic loading, and fisheries overexploitation) are the most serious, having the potential to generate global-scale impacts.

Anthropogenic impacts can be differentiated into three major groups, including those that degrade water quality (e.g., pathogens, nutrients, chemical contaminants, and sewage wastes), result in the loss or alteration of habitat (e.g., wetland reclamation, shoreline development, and dredging), and act as biotic stressors (e.g., overfishing and introduced/invasive species). Nearly all U.S. estuaries are affected in some way by anthropogenic activities, and the scientific literature is replete with reference to human-induced alteration of these coastal systems (for a review see Kennish, 1992, 1997, 2001a). It is critically important to understand these anthropogenic impacts in order to formulate sound management decisions regarding the protection of coastal resources.

Estuaries are particularly susceptible to anthropogenic stressors because of rapid population growth and development in coastal watersheds nationwide. Demographic trends indicate that the coastal zone will continue to be the target of heavy human settlement during the 21st century (Kennish, 2002a). Hence, human activities potentially impacting estuaries will likely become more pervasive in the years ahead.

Kennish (2002a) has shown that an array of estuarine impacts will accompany coastal watershed development during the next 25 years. Among the most severe will be habitat loss and alteration associated with large-scale modifications of coastal watersheds (e.g., deforestation and construction, marsh diking and ditching, and channelization and impoundments), estuarine shorelines (e.g., bulkheads, revetments, retaining walls, and lagoons), and estuarine basins (e.g., dredging and dredged material disposal, channel and inlet stabilization, harbor and marina development, and mariculture and commercial fishing activities). Nutrient enrichment and inputs of oxygen-depleting substances will accelerate as impervious surfaces and hydrological modifications increase in watershed areas. Eutrophication is expected to become more widespread, with greater incidences of hypoxia and anoxia, particularly in shallow coastal bays with limited circulation and flushing. Bricker et al. (1999) recorded moderate to high eutrophic conditions in more than 80 estuaries in conterminous U.S. waters, mostly located along the Atlantic and Gulf of Mexico coasts. They also projected that eutrophic conditions will worsen in 86 U.S. estuaries by 2020. Nutrient overenrichment is thus a serious concern.

TABLE 1
Ranking of Future Anthropogenic Threats to Estuarine Environments Based on Assessment of Published Literature[a]

Stressor	Principal Impacts
1. Habitat loss and alteration	Elimination of usable habitat for estuarine biota
2. Eutrophication	Exotic and toxic algal blooms; hypoxia and anoxia of estuarine waters; increased benthic invertebrate mortality; fish kills; altered community structure; shading; reduced seagrass biomass; degraded water quality
3. Sewage	Elevated human pathogens; organic loading; increased eutrophication; degraded water and sediment quality; deoxygenated estuarine waters; reduced biodiversity
4. Fisheries overexploitation	Depletion or collapse of fish and shellfish stocks; altered food webs; changes in the structure, function, and controls of estuarine ecosystems

5. Chemical contaminants
  Higher priority
    Synthetic organic compounds
  Lower priority
    Oil (PAHs)
    Metals
    Radionuclides
	Adverse effects on estuarine organisms including tissue inflammation and degeneration, neoplasm formation, genetic derangement, aberrant growth and reproduction, neurological and respiratory dysfunction, digestive disorders, and behavioral abnormalities; reduced population abundance; sediment toxicity

6. Freshwater diversions	Altered hydrological, salinity, and temperature regimes; changes in abundance, distribution, and species composition of estuarine organisms
7. Introduced invasive species	Changes in species composition and distribution; shifts in trophic structure; reduced biodiversity; introduction of detrimental pathogens
8. Sea level rise	Shoreline retreat; loss of wetlands habitat; widening of estuary mouth; altered tidal prism and salinity regime; changes in biotic community structure
9. Subsidence	Modification of shoreline habitat; degraded wetlands; accelerated fringe erosion; expansion of open water habitat
10. Debris/litter (plastics)	Habitat degradation; increased mortality of estuarine organisms due to entanglement in debris and subsequent starvation and suffocation

[a] For example, McIntyre, 1992, 1995; Windom, 1992; Yap, 1992; Jones, 1994; Kennish, 1997, 1998, 2000, 2001a, b; Goldberg, 1995, 1998.

Source: Kennish, M.J. 2002. *Environmental Conservation* 29: 78–107.

Other serious stressors are overfishing, which will threaten some fish and shellfish stocks and alter estuarine food webs (Sissenwine and Rosenberg, 1996), and chemical contaminants (especially synthetic organic compounds), which will continue to be most problematic in urban industrialized estuaries (Kennish, 2002b).

Altered stream hydrology coupled to freshwater diversions will also be a problem, and these modifications could affect broad geographic regions. Introduced/invasive species, coastal subsidence, and sediment input/turbidity will likewise impact many estuarine systems. All of these stressors can cause shifts in the structure of estuarine biotic communities or the degradation of valuable estuarine habitat.

Kennish (2002a, p. 102) stated, "As the coastal population increases during the next two decades, anthropogenic impacts on estuaries will likely escalate unless effective management strategies are formulated. Best management practices must be initiated to protect freshwater and coastal wetlands, to minimize input of toxic agents, nutrients, and disease vectors to receiving water bodies, to mollify physical alterations of river–estuary systems that could lead to adverse changes involving nutrient transfer and salinity distribution, and to maintain adequate freshwater inflow to sustain natural productivity and the important nursery function of the systems (Livingston, 2001). It will also be advantageous to limit shoreline development, reduce invasive species, and prevent overfishing. These measures may entail adapting strict management guidelines."

More monitoring and research are needed to identify impacts in the estuarine basins themselves and to develop remedial measures to revitalize altered habitat. In particular, ecosystem level research is necessary to fundamentally understand the natural and anthropogenic processes operating in these coastal environments. Assessment programs must specifically delineate water quality and habitat conditions. Improved nonpoint source pollution controls are required to ameliorate water and sediment quality impacts. Alternative landscaping (e.g., replacing lawns with ground covers, shrubs, trees, and other natural vegetation), modified agricultural practices (e.g., application of new methods to reduce erosion, runoff, and sedimentation), and structural controls (e.g., constructed wetlands, detention facilities, and filtration basins) can significantly mitigate stormwater runoff and contaminant mobilization in adjoining watersheds. In addition, proper restoration efforts should be instituted to return degraded habitat to more natural conditions (NOAA/NOS, 1999). However, these efforts are typically labor intensive, time consuming, and costly. Moreover, they often fall short in terms of the recovery goals of the impacted habitat.

Several federal government programs are providing valuable data for assessing environmental conditions in U.S. estuaries and coastal watersheds. These include the National Estuarine Research Reserve System (NERRS), National Estuary Program (NEP), Coastal Zone Management (CZM) Program, National Status and Trends (NS&T) Program, National Coastal Assessment Program, Environmental Monitoring and Assessment Program, National Marine Fisheries Service National Habitat Program, U.S. Fish and Wildlife Service Coastal Program, and National Wetlands Inventory. Of these programs, NERRS is unique because it consists of a network of 25 protected sites that yield information on national estuarine trends of local or regional concern vital to promoting informed resource management. This network of protected areas represents a federal, state, and community partnership in which environmental monitoring and research as well as a comprehensive program of education and outreach strengthen understanding, appreciation, and stewardship of estuaries, coastal habitats, and associated watersheds. NERRS encompasses more than a million hectares of estuarine, wetland, and upland habitats in all biogeographical regions of the U.S.

NERRS sites are essentially coastal ecosystems used as demonstration sites for long-term research and monitoring and resource protection, as well as education and interpretation. The objective of this book is to examine in detail the NERRS program, focusing on environmental research, monitoring, and restoration components. The NERRS sites generally represent pristine and undisturbed areas that can serve as reference locations to assess other estuarine systems impacted by anthropogenic activities. One of the principal reasons for creating the NERRS program was to improve the management of estuarine resources by providing an integrated mechanism for the detection and measurement of local, regional, and national trends in estuarine conditions. Increasing and competing demands for coastal resources require a coordinated program such as NERRS to improve coastal zone management. Research and education programs of NERRS can guide estuarine and watershed management for sustained support of coastal resources.

This initial volume of *Estuarine Research, Monitoring, and Resource Protection* describes the workings of the NERRS program — its organization, goals, and management strategies. It does not provide a critique of the program aims and achievements, which will be the focus of a later volume. The second volume will assess how the NERRS program has succeeded overall in achieving technical and management objectives.

Chapter 1 of *Estuarine Research, Monitoring, and Resource Protection* is a comprehensive treatment of the principal components of the NERRS program. Chapters 2 to 7 concentrate on the physical, chemical, and biological characterization of selected NERRS sites, as follows:

- Chapter 2: Waquoit Bay National Estuarine Research Reserve
- Chapter 3: Jacques Cousteau National Estuarine Research Reserve
- Chapter 4: Delaware National Estuarine Research Reserve
- Chapter 5: Ashepoo–Combahee–Edisto (ACE) Basin National Estuarine Research Reserve
- Chapter 6: Weeks Bay National Estuarine Research Reserve
- Chapter 7: Tijuana River National Estuarine Research Reserve

These case studies offer a cross section of NERRS sites on the Atlantic, Pacific, and Gulf of Mexico coasts and therefore give broad coverage of the program.

It is important to specify that the success of the NERRS program depends on the unselfish cooperation of government agencies, academic institutions, public interest groups, concerned citizens, and the general public. These entities must all work together to ensure protection of the water quality, habitat, and resources in the system of estuarine and coastal watersheds comprising the NERRS program. The case studies of reserve sites reported in this book demonstrate how critical it is to maintain the ecological integrity of our coastal environments.

REFERENCES

Bricker, S.B., C.G. Clement, D.E. Pirhalla, S.P. Orlando, and D.R.G. Farrow. 1999. National Estuarine Eutrophication Assessment: Effects of Nutrient Enrichment in the Nation's Estuaries. Technical Report, National Oceanic and Atmospheric Administration, National Ocean Service, Special Projects Office and the National Centers for Coastal Ocean Science, Silver Spring, MD.

Goldberg, E.D. 1995. Emerging problems in the coastal zone for the twenty-first century. *Marine Pollution Bulletin* 31: 152–158.

Goldberg, E.D. 1998. Marine pollution — an alternative view. *Marine Pollution Bulletin* 36: 112–113.

Jones, G. 1994. Global warming, sea level change and the impact on estuaries. *Marine Pollution Bulletin* 28: 7–14.

Kennish, M.J. 1992. *Ecology of Estuaries: Anthropogenic Effects*. CRC Press, Boca Raton, FL.

Kennish, M.J. (Ed.). 1997. *Practical Handbook of Estuarine and Marine Pollution*. CRC Press, Boca Raton, FL.

Kennish, M.J. 1998. *Pollution Impacts on Marine Biotic Communities*. CRC Press, Boca Raton, FL.

Kennish, M.J. (Ed.). 2000. *Estuary Restoration and Maintenance: The National Estuary Program*. CRC Press, Boca Raton, FL.

Kennish, M.J. (Ed.). 2001a. *Practical Handbook of Marine Science,* 3rd ed. CRC Press, Boca Raton, FL.

Kennish, M.J. 2001b. Coastal salt marsh systems: a review of anthropogenic impacts. *Journal of Coastal Research* 17: 731–748.

Kennish, M.J. 2002a. Environmental threats and environmental future of estuaries. *Environmental Conservation* 29: 78–107.

Kennish, M.J. 2002b. Sediment contaminant concentrations in estuarine and coastal marine environments: potential for remobilization by boats and personal watercraft. *Journal of Coastal Research* Special Issue 37, pp. 151–178.

Livingston, R.J. 2001. *Eutrophication Processes in Coastal Systems*. CRC Press, Boca Raton, FL.

Livingston, R.J. 2003. *Trophic Organization in Coastal Systems*. CRC Press, Boca Raton, FL.

McIntyre, A.D. 1992. The current state of the oceans. *Marine Pollution Bulletin* 25: 1–4.

McIntyre, A.D. 1995. Human impact on the oceans: the 1990s and beyond. *Marine Pollution Bulletin* 31: 147–151.

NOAA/NOS. 1999. National Ocean Service Strategic Plan, 1999–2004. Technical Report, NOAA/National Ocean Service, Silver Spring, MD.

Sissenwine, M.P. and A.A. Rosenberg. 1996. Marine fisheries at a critical juncture. In: Pirie, R.G. (Ed.). *Oceanography: Contemporary Readings in Ocean Sciences*. 3rd ed. Oxford University Press, New York, pp. 293–302.

Windom, H.L. 1992. Contamination of the marine environment from land-based sources. *Marine Pollution Bulletin* 25: 32–36.

Yap, H.T. 1992. Marine environmental problems: experiences of developing regions. *Marine Pollution Bulletin* 25: 37–40.

Contents

Chapter 1 National Estuarine Research Reserve System:
Program Components 1

Introduction 1
NERRS Mission 6
NERRS Program Components 9
    Monitoring and Research: System-Wide Monitoring Program 9
        SWMP Development 10
        SWMP Components 12
    Special High-Priority Initiatives 23
        Habitat Restoration 23
        Invasive Species 24
    Education and Outreach 25
Summary and Conclusions 28
References 29

Case Study 1

Chapter 2 Waquoit Bay National Estuarine Research Reserve 35

Introduction 35
Watershed 36
    Upland Pitch Pine/Oak Forests 36
    Sandplain Grasslands 37
    Vernal Pools and Coastal Plain Pond Shores 38
    Riparian Habitats 38
    Freshwater Wetlands 38
    Salt Marshes 39
    Mudflats and Sandflats 39
    Beaches and Dunes 41
Estuary 42
    Tidal Creeks and Channels 42
    Waquoit Bay 43
        Environment 43
        Organisms 43
Anthropogenic Impacts 49
    Eutrophication 49
Summary and Conclusions 51
References 53

Case Study 2

Chapter 3 Jacques Cousteau National Estuarine Research Reserve 59
    Introduction 59
    Environmental Setting 61
    Mullica River–Great Bay Estuary 62
        Water Quality 62
    Watershed Biotic Communities 65
        Plant Communities 65
            Salt Marshes 65
            Brackish Tidal Marshes 66
            Freshwater Marshes 69
            Lowland Plant Communities 71
            Upland Plant Communities 80
            Barrier Island Plant Communities 81
        Animal Communities 82
            Amphibians and Reptiles 82
            Mammals 85
            Birds 86
            Fish 92
    Estuarine Biotic Communities 94
        Plant Communities 94
            Benthic Flora 94
            Phytoplankton 96
        Animal Communities 98
            Zooplankton 98
            Benthic Fauna 100
            Finfish 104
    Summary and Conclusions 110
    References 111

Case Study 3

Chapter 4 Delaware National Estuarine Research Reserve 119
    Introduction 119
    Lower St. Jones River Reserve Site 120
        Watershed 120
            Upland Vegetation 121
            Wetland Vegetation 121
        Aquatic Habitat 125
            Water Quality 125
        Anthropogenic Impacts 128

Pollution	128
Habitat Alteration	132
Biotic Communities	133
Phytoplankton	133
Zooplankton	134
Benthic Fauna	136
Finfish	138
Amphibians and Reptiles	140
Birds	140
Mammals	145
Upper Blackbird Creek Reserve Site	151
Watershed	151
Upland Vegetation	151
Wetland Vegetation	152
Aquatic Habitat	156
Anthropogenic Impacts	157
Pollution and Habitat Alteration	157
Biotic Communities	157
Phytoplankton	157
Zooplankton	157
Benthic Fauna	158
Finfish	160
Amphibians and Reptiles	160
Birds	160
Mammals	161
Commercially and Recreationally Important Species	162
Summary and Conclusions	163
References	165

Case Study 4

Chapter 5 Ashepoo–Combahee–Edisto (ACE) Basin National Estuarine Research Reserve 171

Introduction	171
Watershed	173
Plant Communities	173
Animal Communities	175
Amphibians and Reptiles	176
Mammals	183
Birds	186
Insects	189
Estuary	190
Physical-Chemical Characteristics	190

   Biotic Communities 192
    Phytoplankton 192
    Zooplankton 194
    Benthic Invertebrates 195
    Fish 197
  Coastal Marine Waters 200
   Animal Communities 200
    Fish 200
    Reptiles 201
    Mammals 202
    Birds 202
  Endangered and Threatened Species 202
  Anthropogenic Impacts 203
  Summary and Conclusions 205
  References 206

Case Study 5

Chapter 6   Weeks Bay National Estuarine Research Reserve 217

 Introduction 217
 Weeks Bay 217
  Physical Description 217
 Watershed 220
  Plant Communities 220
   Upland Habitats 220
   Wetland Habitats 220
  Animal Communities 220
   Herpetofauna 220
   Mammals 223
   Birds 223
 Estuary 223
  Plant Communities 223
   Phytoplankton and Microphytobenthos 223
  Animal Communities 225
   Zooplankton 225
   Benthic Fauna 225
   Fish 226
 Anthropogenic Impacts 227
 Summary and Conclusions 228
 References 229

Case Study 6

Chapter 7 Tijuana River National Estuarine Research Reserve 235
    Introduction 235
    Watershed 237
        Habitat 238
            Salt Marsh 238
            Salt Pannes 240
            Brackish Marsh 240
            Riparian Habitat 241
            Wetland–Upland Transition 242
            Dunes and Beach Habitat 243
            Intertidal Flats 244
    Estuary 244
        Aquatic Habitat: Tidal Creeks and Channels 244
            Plants 245
            Benthic Invertebrates 245
            Fish 248
            Birds 252
    Anthropogenic Impacts 252
    Summary and Conclusions 256
    References 259

Index 261

1 National Estuarine Research Reserve System: Program Components

INTRODUCTION

Estuaries are recognized worldwide as critically important coastal environments with exceptional biotic production. They rank among the most vital ecological systems on earth, providing valuable resources for the world economy (Alongi, 1998; Kennish, 2001a). Many commercially and recreationally important finfish and shellfish species depend on estuaries for survival. Estuarine-dependent species comprise more than 90% of the total fisheries landings in the Gulf of Mexico alone (Kennish, 2000). Furthermore, estuarine- and wetland-dependent species account for about 75% of the total U.S. annual seafood harvest of more than 4 million metric tons (Weber, 1995). Estuarine and coastal marine fisheries return more than $23 billion annually to the U.S. economy (Kennish, 2000).

Aside from their significance to world fisheries, estuaries support several other multi-billion dollar commercial and recreational interests. Among the most notable are tourism, shipping, marine transportation, marine biotechnology, oil and gas recovery, mineral exploration, and electric power generation. Estuaries generate employment opportunities directly or indirectly for millions of people in the U.S. and abroad.

Because of their great commercial and recreational importance, estuaries are often utilized excessively by a burgeoning coastal population. Approximately 60% of the world population now resides near the coasts (Goldberg, 1994). In the U.S., nearly 140 million people (~53% of the total population) live along the coastal zone in close proximity to estuaries (Cohen et al., 1997; NOAA, 1998). Statistical trends indicate that the coastal population is expected to approach 6 billion people worldwide by the year 2025 (Kennish, 2002).

Increasing coastal watershed development, urbanization, and industrialization during the past century resulted in habitat alteration, pollution, and overuse of many estuarine systems. Various anthropogenic stressors (e.g., species introductions, overfishing, freshwater diversions, point and nonpoint source pollution inputs, nutrient overenrichment, waste dumping, and wetland reclamation) created a multitude of estuarine problems (Kennish, 1992, 1997). Water quality and habitat degradation accelerated at such an alarming rate that by the 1960s and early 1970s

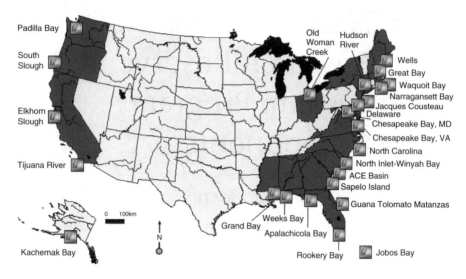

FIGURE 1.1 Map showing the site locations of the National Estuarine Research Reserve System. (From the National Estuarine Research Reserve Program, Silver Spring, MD.)

widespread public outcry led to the enactment of vital state and federal legislation to protect estuarine and marine environments. Particularly noteworthy was passage by Congress of the Coastal Zone Management Act (CZMA) of 1972. This legislation ushered in the National Estuarine Sanctuary Program. More specifically, Section 315 of the CZMA, as amended, authorized the establishment of "estuarine sanctuaries" or "protected research areas, which could include any or all parts of an estuary, adjoining transitional areas and adjacent uplands, set aside to provide scientists and students the opportunity to examine over a period of time the ecological relationships within the area."

An important element of the newly formed National Estuarine Sanctuary Program was the development of a state–federal partnership to establish, manage, and maintain representative estuarine ecosystems and to promote stewardship of coastal resources by engaging local communities and residents, as well as regional groups. Congress designated the National Oceanic and Atmospheric Administration (NOAA) as the federal agency administering the overall program. South Slough, Oregon, became the first estuarine sanctuary in 1974. Congress reauthorized the CZMA in 1985, and at that time, the name of the National Estuarine Sanctuary Program was changed to the National Estuarine Research Reserve System (NERRS, 2002a).

NOAA has designated 25 reserve sites in 21 states and Puerto Rico since 1974 (Figure 1.1, Table 1.1), and during the past three decades, extensive monitoring and research data have been collected at these sites (NERRS, 2002b). These 25 reserve sites represent 15 biogeographical regions and a wide range of estuarine conditions. They are located along the Atlantic Coast, Gulf of Mexico, Pacific Coast, Caribbean Sea, and Great Lakes, covering nearly 500,000 ha of estuarine waters, wetlands, and uplands. This network of protected and coordinated sites has been established for

TABLE 1.1
National Estuarine Research Reserve Sites

ACE (Ashepoo–Combahee–Edisto) Basin NERR, South Carolina
Apalachicola NERR, Florida
Chesapeake Bay NERR, Maryland
Chesapeake Bay NERR, Virginia
Delaware NERR, Delaware
Elkhorn Slough NERR, California
Grand Bay NERR, Mississippi
Great Bay NERR, New Hampshire
GTM (Guana Tolomato Matanzas) NERR, Florida
Hudson River NERR, New York
Jacques Cousteau NERR, New Jersey
Jobos Bay NERR, Puerto Rico
Kachemak Bay NERR, Alaska
Narragansett Bay NERR, Rhode Island
North Carolina NERR, North Carolina
North Inlet-Winyah Bay NERR, South Carolina
Old Woman Creek NERR, Ohio
Padilla Bay NERR, Washington
Rookery Bay NERR, Florida
San Francisco Bay NERR, California[a]
Sapelo Island NERR, Georgia
South Slough NERR, Oregon
St. Lawrence River NERR, New York[a]
Texas NERR, Texas[a]
Tijuana River NERR, California
Waquoit Bay NERR, Massachusetts
Weeks Bay NERR, Alabama
Wells NERR, Maine

[a] Proposed site.

research and monitoring activities, as well as to increase public awareness of the importance of estuarine and coastal resources (NOAA, 2002a). NOAA's Estuarine Reserves Division (ERD) in Silver Spring, Maryland, part of the Office of Ocean and Coastal Resource Management, serves as the management center of NERRS, overseeing operations and budgets, as well as providing administrative support for the reserve sites.

Each reserve, which is managed on a day-to-day basis by a state agency or a university, consists of a discrete area containing key habitat within an estuarine system that is protected by state law from significant ecological change (NERRS, 2002a, b). NERRS strives to improve the health of the nation's coastal habitats by generating information that fosters sound resource management. In so doing, it augments the CZM program. For example, integrated programs of NERRS address specific high-priority resource management concerns such as degraded coastal water quality, loss and alteration of estuarine and watershed habitat, habitat restoration, reduction of biodiversity, and problematic effects of pollution and invasive species. The reserves are essentially demonstration sites where monitoring and research data are used to assess coastal issues of local, regional, and national interest for the purpose of sustaining estuarine systems (NERRS, 2002c).

Coastal states play an integral role in the designation of reserve sites and their operation, acting jointly with NOAA to establish a site program. A partnership exists among NOAA, the coastal state, local communities, and regional groups to address resource management issues in the reserve (NERRS, 2002a–c). Reserve sites may also establish partnerships with other government agencies, conservation organizations, universities, and local school systems (NERRS, 1994a). In addition to providing the base funding for operation, research, and education for the NERRS program, NOAA also sets standards for operating the reserves, supports activities of each reserve, sponsors a graduate research fellowship program, and facilitates decision making at the national level. NOAA base funding may be augmented at individual sites by local and state allocations, as well as grants.

The role of each reserve site is multifaceted. Reserve staff and other investigators gather data through research and long-term monitoring programs at the site. The main goal of these programs is to characterize the natural and anthropogenic processes governing stability and change in the reserve, and in so doing, assist coastal resource managers in making informed decisions (Greene and Trueblood, 1999). To accomplish this goal, it is not only necessary to characterize the current physical, chemical, and biological conditions of the reserve but also to assess changes in conditions through time. It is critical to develop a baseline monitoring program that enables natural resource program managers and other coastal decision makers to detect trends in water quality and habitat loss and alteration. Data collected by the research and monitoring programs should foster greater understanding of the relationship between disturbance and physical, chemical, and biological processes required to sustain biotic communities in the reserve.

A System-wide Monitoring Program (SWMP) has been established by NERRS to identify and track short-term variability and long-term changes in the integrity and biodiversity of site estuaries and their coastal watersheds for the purpose of contributing to effective coastal zone management (NERRS, 2002a). Important components are water quality monitoring, biomonitoring, and the assessment of land use/land cover characteristics within the reserve boundaries. Monitoring data help to define baseline conditions and establish trends for the NERR system of estuaries. Monitoring can delineate the status of resources in the reserves. Monitoring funds from NOAA are available to each reserve as part of an operations grant. NOAA supports basic monitoring programs in NERRS under the Code of Federal Regulations (15CFR Part 921 Subpart G).

The Centralized Data Management Office (CDMO), located at the North Inlet-Winyah Bay NERR (Belle W. Baruch Institute for Marine Biology and Coastal Research) at the University of South Carolina in Georgetown, South Carolina, serves as a databank and principal technical support for the NERR SWMP, archiving and analyzing monitoring data and information products from each reserve (Wenner et al., 2001). Its major function, therefore, is the management of the basic infrastructure and data protocol to support the assimilation and exchange of data, metadata, and information within the framework of NERRS sites, state CZM programs, and related state- and federally funded education, monitoring, and research initiatives (NERRS, 2002a; Sanger et al., 2002). It also serves a critical role in quality control of data for the NERRS program. The CDMO formats NERR SWMP data to meet the standards of the Federal Geographical Data Committee, and the data are disseminated to all users over the World Wide Web (http://cdmo.baruch.sc.edu).

Aside from monitoring and research, environmental education and stewardship are other vital components of the NERRS program (NERRS, 1994a, b). The primary goal of the education component is to improve public awareness of estuarine-related issues and coastal resource problems, as well as to be responsive to concerns of the general public. To accomplish this goal, each reserve offers workshops, field trips, and other educational opportunities that improve understanding of estuaries, watersheds, and coastal resources. Such educational efforts engender public interest and participation. NERRS also develops the reserves as resource education centers that address coastal matters of local, state, regional, national, and global significance. The success of the education programs is enhanced by the ability of NERRS to link research, resource management, stewardship, and education.

One of the major goals of stewardship in the NERRS program is to improve protection of estuarine resources for designated uses such as fishing, shellfish harvesting, swimming, and other recreational activities. Effective sustainable yield management of estuarine and watershed resources is critical to the long-term viability of estuarine and coastal systems. This can be achieved, in part, by integrating resource management programs across local, state, and federal levels of government through coordination and establishment of joint research, education, and stewardship. It would be particularly advantageous to develop a proactive management program at a regional watershed scale that allocates resources to priority issues and concerns, especially through partnerships and sharing of the resources. One objective, for example, would be to protect key land and water areas that are vital ecological units of the reserve.

Reserve staff members conduct education and training programs for community leaders, resource users, schoolchildren, and the general public on the natural and human environment of the reserve, as well as on resource management needs. Through this educational process, they interpret and communicate current scientific findings to these audiences. The reserve staff interacts with various partners and local communities to address water quality problems, habitat loss and alteration, invasive species, and declining resources. The Coastal Training Program (CTP) and decision maker workshops also provide technical training for local government officials and administrators on regional coastal management problems, offering relevant science-based educational programs on a variety of topics. CTP may target development, land use, water supply planning, open space conservation, habitat restoration, estuarine water quality, and other skill-based training modules. The focus of coastal decision maker workshops is to improve decision making related to coastal resource management. CTP, in turn, increases collaboration, facilitates information and technology transfer, and promotes greater understanding of anthropogenic impacts on coastal resources. The NERRS framework, therefore, effectively links science and monitoring programs with education and resource stewardship initiatives to disseminate scientific information for the betterment of coastal watershed and resource management.

The NERRS Strategic Plan developed in 1994 (NERRS, 1995) and revised in 2002 (NERRS, 2002c), together with a series of multi-year action plans initiated in 1996, identifies the primary goals, objectives, and actions of the program. The multi-year action plans, which are revised annually for 3-year periods, have been formulated to assist NOAA in its environmental stewardship mission to sustain healthy coasts. The sixth NERRS Action Plan (NERRS, 2002d), covering the fiscal years 2002, 2003, and 2004, conveys the following long-term goals of the program:

1. Advance the state of knowledge about the requirements for sustainable estuarine ecosystems and the interactions of humans with those ecosystems
2. Improve decisions affecting estuarine and coastal resources
3. Move the operations and the infrastructure of NERRS forward

It is necessary to update the NERRS Action Plan on a regular basis to take into account new advances and developments in the program.

NERRS also performs a needs assessment to identify the common issues, capacity needs, and data uses in the system. Results of a needs assessment initiative undertaken by NOAA's ERD and Coastal Services Center in the summer of 2002 reveal that upland land cover and benthic or subtidal habitats (e.g., habitat mapping) are the two most common data needs of the reserve sites (Schuyler et al., 2002). Other common data needs include topography and bathymetry, invasive species, habitat change, erosion, and water quality. The need for change detection analysis is also deemed to be important by the reserve sites. Management needs are broader and more diverse, dealing with policy and planning and acquisition and restoration, as well as education and research. In addition, needs assessment helps to delineate the remote sensing technology requirements of the reserves. Remote sensing and Geographic Information System (GIS) applications are generally recognized as having great potential value in the data acquisition process of the reserve system, but many of the reserve sites have limited capacity to maximize their use (Schuyler et al., 2002).

NERRS MISSION

NERRS is a multifaceted, integrated program whose mission is "to promote stewardship of the nation's estuaries through science and education using a system of protected areas" (NERRS, 2002c, p. 2). Table 1.2 describes the functional elements of the program. Four major components are recognized:

1. Federal, state, and community partnership of protected areas
2. Informed management and stewardship of the nation's estuarine and coastal habitats
3. Scientific research and monitoring
4. Public education

TABLE 1.2
Functional Elements of the NERRS Program

Representative Protected Areas
Establish, manage, and maintain a national network of protected areas representing the diverse biogeographic and typological estuarine ecosystems of the United States.

Partnership
Mobilize federal, state, and community resources to mutually define and achieve coastal protection and management goals and objectives.

Informed Management and Stewardship
Operate the NERRS as a national program contributing to informed, integrated management of the nation's coastal ecosystem.

Scientific Understanding through Research
Design and implement a comprehensive program of scientific research to address coastal management issues and their fundamental underlying processes.

Education
Design and implement a comprehensive program of education and interpretation based on solid scientific principles to strengthen the understanding, appreciation, and stewardship of estuaries, coastal habitats, and associated watersheds.

Source: National Estuarine Research Reserve System. 1994. National Estuarine Research Reserve System Education: A Field Perspective. National Estuarine Research Reserve System, National Oceanic and Atmospheric Administration, Silver Spring, MD.

NERRS achieves coastal protection and management goals by mobilizing federal, state, and community resources to support work at the designated sites. The program is strengthened by education and outreach initiatives that improve understanding, appreciation, and stewardship of estuaries and watersheds, thereby promoting informed management of coastal resources (NERRS, 1994a).

Many of the reserve sites are generally pristine and undisturbed areas that can serve as reference locations for comparison with systems impacted by anthropogenic activities. NERRS strives to enhance the management of estuarine resources by acting as a coordinated network for the detection and measurement of local, regional, and national trends in estuarine conditions. Increasing and competing demands for coastal resources require an integrated program such as NERRS to improve coastal zone management efforts (NOAA, 1999). Research and education programs of NERRS can guide estuarine and watershed management for sustained support of coastal resources.

As stated in the Code of Federal Regulations 15 CFR Section 921.1 (b), the goals of the NERR system are to:

1. Ensure a stable environment for research through long-term protection of the reserve sites
2. Address coastal management issues identified as significant through coordinated estuarine research within the system

3. Enhance public awareness and understanding of estuarine areas and provide suitable opportunities for public education and interpretation
4. Promote federal, state, public, and private use of one or more reserves within the system when such entities conduct estuarine research
5. Conduct and coordinate estuarine research within the system, gathering and making available information necessary for improved understanding and management of estuarine areas

To achieve these goals, NOAA makes available five categories of federal awards for NERRS programs:

1. Predesignation
2. Acquisition and development
3. Operations and management
4. Research and monitoring
5. Education and interpretation

Participating states provide matching funds for the federal awards. The Code of Federal Regulations (15 CFR Part 921, Appendix F) contains more details on NOAA funding.

A state can nominate an estuarine system for NERR status under Section 315 of the CZMA, if the site meets the following stipulations:

1. The area is representative of its biogeographical region, is suitable for long-term research, and contributes to the biogeographical and typological balance of the system.
2. The laws of the coastal state provide long-term protection for the proposed reserve's resources to ensure a stable environment for research.
3. Designation of the site as a reserve will serve to enhance public awareness and understanding of estuarine areas and provide suitable opportunities for public education and interpretation.
4. The coastal state has complied with the requirements of any regulations issued by the Secretary of Commerce.

Prior to receiving federal designation, a prospective NERR site can receive predesignation awards for site selection. Acquisition and development awards may be obtained for acquiring interest in land and water areas, performing minor construction, preparing plans and specifications, developing the final management plan, and hiring staff for the reserve. A supplemental acquisition and development award can be granted after a reserve receives federal designation. The reserve site can use this award to obtain additional property, construct research and education facilities, and conduct restoration projects approved by the program. NERRS conducted a land acquisition inventory and strategy in 2002. To manage the reserve and operate programs, annual awards are available for operation and management, education, and monitoring. NOAA conducts performance evaluations of a reserve at least once every three years as required by Sections 312 and 315 of the CZMA. This evaluation is to ensure that the operation and management of the reserve are in compliance with NERRS regulations and consistent with the mission and goals of the national program.

Several criteria are considered when assessing prospective sites for NERRS status. An important initial consideration is whether the site fills a void in biogeographical representation in NERRS. It is also useful to document that the prospective site fills a void in ecosystem representation, comprises a significant part of a coastal ecosystem, or encompasses an entire ecological unit. A review of the quality of the estuarine environment should be conducted, examining primary ecological characteristics such as the biological productivity, diversity of flora and fauna, and various ecological values and functions. The effect of human activities on the estuary and its adjoining transitional habitats and adjacent uplands should likewise be a point of focus. The long-term management of coastal resources depends on understanding the complex and profound anthropogenic influences operating in the coastal zone. Another vital factor is determining whether the prospective site is desirable as a research and monitoring site and valuable as a natural field laboratory. Will the site effectively address coastal resource problems that have local, regional, or even national significance? Equally important is whether the site will provide opportunities for educational and interpretive programs that enhance understanding of the estuarine system and its resources. Finally, management considerations must be pursued to ascertain whether the site constitutes a stable environment for research and education activities that will enhance the NERRS program.

NERRS PROGRAM COMPONENTS

MONITORING AND RESEARCH: SYSTEM-WIDE MONITORING PROGRAM

According to the Regulations and Strategic Plan of NERRS, the development of reserve site programs involves the incorporation of three major elements:

1. Environmental characterization
2. The site profile
3. A systematic long-term monitoring program

Environmental characterization is necessary to inventory conditions at the reserve site and to formulate a comprehensive site description. The site profile is a detailed environmental report providing a synthesis of existing data and information on the reserve. Environmental monitoring entails the periodic collection of selected data using many of the same parameters and, ideally, the same sampling techniques system-wide. Monitoring consists of three phased components: water quality (abiotic factors), habitat and species diversity, and land use and land cover analysis (JCNERR, 1998).

In 1992, NERRS proposed the development of a coordinated, ecosystem-based monitoring network to track the health and functionality of representative estuarine ecosystems and coastal watersheds in the U.S. The mission of this System-wide Monitoring Program (SWMP), as stated previously, is to develop quantitative measurements of short-term variability and long-term changes in the water quality, biotic diversity, and land use/land cover characteristics of estuarine reserves for purposes of contributing to effective coastal zone management. In addition, SWMP serves as an excellent platform to evaluate the response of reserve sites to episodic events such as tropical storms and droughts (Wenner et al., 2001; Sanger et al., 2002). In so doing, SWMP yields a long-term database of great value in establishing national estuarine trends and measuring changes in environmental conditions and ecological processes (NERRS, 2002a).

SWMP represents one of the most comprehensive and effective programs for monitoring the water quality conditions of the nation's estuaries. It is unique because it generates standardized information on national estuarine environmental trends while allowing the flexibility to assess coastal management issues of local or regional concern (NOAA, 1999). The standardization of sampling protocols facilitates data comparisons among reserve sites. As the SWMP databases increase through time, they become more valuable for identifying changes in estuarine conditions, interpreting and predicting responses to change, and delineating anthropogenic stressors that must be addressed by coastal decision makers. In addition, long-term SWMP databases will afford greater understanding of how estuaries function and change over time, thereby forming a foundation for devising solutions to coastal management problems.

SWMP Development

Operation grants from NOAA serve as the main funding base for SWMP at the reserve sites. NOAA provided the initial funding for SWMP in 1994, and at that time NERRS proposed a phased monitoring approach targeting the following three key components:

Abiotic factors
    a. Physical–chemical parameters (water temperature, salinity, dissolved oxygen, pH, turbidity, and depth)
    b. Water quality parameters (nutrients and contaminants)
    c. Atmospheric parameters (air temperature, wind speed and direction, barometric pressure, relative humidity, precipitation, and photosynthetically active radiation)
Biological monitoring (population characteristics, biodiversity, and habitat)
Watershed and land use classifications (changes in spatial coverage, GIS applications)

Gradual changes in SWMP were implemented during each succeeding year to address the aforementioned elements as discussed below (Edmiston et al., 2002; NERRS, 2002a). For example, in 1995–1996, each reserve established two field sites to monitor abiotic parameters deemed to be scientifically valid indicators of water quality (temperature, salinity, dissolved oxygen, pH, turbidity, and water depth) using Yellow Springs Instrument Company (YSI™) 6-series data loggers to record the data every 30 minutes (NERRS, 2002a). Turbidity probes were added to the data loggers in 1996. The CDMO, created in 1994, began to coordinate and store

SWMP data from across the sites and facilitate data analysis and dissemination. To achieve more uniformity on a national basis, NERRS developed system-wide protocols and quality assurance/quality control methods of data assessment in 1996–1997.

Two additional data loggers were procured by each reserve during 1996–1997. In addition, the reserve sites began to use a Campbell Scientific Weather Station to monitor atmospheric conditions, and a weather data management program was developed. In 1997–1998, the emphasis shifted to data utilization, with revisions made to quality assurance/quality control methods to ensure data accuracy. Some reserve sites also expanded their monitoring efforts to include other components (i.e., habitat change, nutrients, or sediment contaminants).

While monitoring efforts continued unabated in SWMP during 1998–1999, emphasis was placed on improved data management. Analysis of data on a system-wide basis was stressed, and an effort was made to post monitoring data on the Web for use by the scientific and management communities. SWMP also began to utilize the Protected Area Geographic Information System (PAGIS) to promote the genesis of a GIS/habitat information database (NERRS, 2002a). Most reserve sites began to set up a Geographic Information System for data processing at this time.

The South Carolina Marine Resources Research Institute conducted a comprehensive synthesis of NERR SWMP water quality data in 1999–2000. This project culminated in the release of a technical report in April 2001 entitled *A Synthesis of Water Quality Data from the National Estuarine Research Reserve's System-wide Monitoring Program*, which detailed water quality conditions in the NERR system as a whole (Wenner et al., 2001). A second SWMP synthesis report in December 2002 entitled, *A Synthesis of Water Quality Data: National Estuarine Research Reserve System-wide Monitoring Program (1995–2000)*, presented additional analysis of water quality data in the reserves from a system-wide perspective (Sanger et al., 2002). Other SWMP developments in 1999–2000 included the initiation of pilot studies employing fluorescence probes and vented-level probes as part of the monitoring program. To provide effective oversight of SWMP programs as well as continuity and standard protocols, an oversight committee was formed during this time period (Edmiston et al., 2002).

Aside from the publication of the first NERRS water quality data synthesis report, several other SWMP projects were completed in 2000–2001. Among them were pilot studies to assess the efficacy and utility of fluorescence probes and vented-level probes for SWMP and the approval of a SWMP plan for program expansion. Systematic monitoring of water quality and weather data continued, with improvement observed in the databases reported by many reserve sites. An effort was made to upgrade older data loggers in the system. In addition, the CDMO received a more complete array of quality assurance/quality control data from the reserve sites (NERRS, 2002a).

Revised monitoring protocols required the deployment of two additional data loggers in 2001–2002, bringing the total number of mandatory SWMP sites to four, one of which was designated as a long-term reference (control) site (NERRS, 2002a). Nutrient monitoring commenced in 2002, with duplicate water samples obtained monthly at all four mandated data logger deployment sites in each reserve. Nutrient monitoring also included the collection of diel water samples (taken at 2- to 3-h intervals over a 24-h, 48-min time period) at one SWMP site every month using an ISCO auto-sampler. Among the Tier I nutrient parameters measured are nitrate, nitrite, ammonium, orthophosphate, and chlorophyll a. A real-time data delivery and management system was approved for implementation in NERRS during 2002 (Ross, 2002). This initiative is designed to facilitate data availability for the scientific community. Finally, discussions for SWMP biomonitoring were undertaken in March and October 2002 and in February 2003 with plans to develop sampling protocols during 2002–2003 (Edmiston et al., 2002).

SWMP Components

Data collected by SWMP have already proven to be useful in estuarine habitat restoration projects. They have also been of value in the investigation of the temporal recovery of estuaries from storm impacts. In addition, SWMP measurements have been utilized as background data for numerous independent research studies.

Phase I

Phase I of SWMP, conducted from 1995 to 2000, focused on monitoring of estuarine water quality and coastal meteorological conditions by measuring a suite of water quality and atmospheric parameters. Currently, the network of 25 reserves measures six water quality parameters (water temperature, salinity, pH, dissolved oxygen, turbidity, and depth) over a range of temporal and spatial scales using YSI 6-series data loggers deployed year-round at four SWMP locations. The NERRS program is in the process of upgrading data loggers from the YSI 6000 model to the YSI 6600 model. The data loggers operate unattended in the field but must be periodically reprogrammed. Thus, the instruments must be switched out with newly programmed data loggers at the end of each deployment period (about every 14 days). The field data are subsequently uploaded to a personal computer or laptop, where they can be processed and analyzed to generate statistics and plots. Telemetry systems also enable interfacing of the data loggers with real-time data collection platforms (NERRS, 2002a). The goal of NERRS is to apply consistent, quality-controlled methodology across the reserve sites to generate a long-term, coordinated monitoring database that will characterize the water quality conditions of the NERRS sites and will help evaluate the health and functionality of the nation's estuaries.

The 25 reserve sites also collect the following meteorological data on a near continuous basis as previously noted: air temperature, wind speed and direction, barometric pressure, relative humidity, precipitation, and photosynthetically active radiation. For most sites, these data are obtained using the Campbell Scientific Weather Station. The meteorological data, which have been shown to strongly influence the aforementioned physical–chemical variables in estuarine waters, are archived at the reserve sites. The water quality and meteorological data must be collected, documented, edited, and submitted, along with metadata, to the CDMO on an annual schedule. A SWMP mandate requires that all reserve sites must submit at least 85% of the data that could be collected to the CDMO. Together with the

SWMP water quality database, the meteorological database consists of some of the most comprehensive physical–chemical measurements ever collected in estuaries. These findings may aid investigators in predicting how estuaries respond to changes in climate and various anthropogenic perturbations (Wenner et al., 2001). They may also enable investigators to determine whether conditions in estuaries are improving, worsening, or remaining unchanged.

NERR-SWMP fills an important void for an integrated national program that evaluates the status of marine environmental resources and the trends in estuarine water quality over protracted periods. Therefore, it differs from most existing nationwide monitoring programs, which concentrate on relatively short-term measurements of environmental variables taken annually or biannually over periods of 3–4 days (NERRS, 2002a).

The ultimate goal of environmental monitoring is the protection of living resources, human health, and the environment (National Research Council, 1990). Three broad categories of environmental monitoring exist:

1. Compliance monitoring
2. Trends monitoring
3. Model validation and verification

NERRS is a most effective trends-monitoring program because it has developed the infrastructure and public support needed to successfully conduct long-term monitoring projects over large geographic regions. NERR-SWMP, for example, is represented in the contiguous U.S. by almost every degree of geographic latitude between 26°N and 43°N (~1600 km north to south) (Wenner et al., 2001). As noted by the National Research Council (1990, p. 19), "Monitoring is most beneficial when it results in more effective management decisions — decisions that protect or rehabilitate the marine environment, its living resources, and uses or resources that society considers important." These monitoring benefits are part of the mission and goals of the NERRS program.

Wenner et al. (2001) have completed a detailed analysis of water quality data collected at 22 reserves (44 sampling sites) in the NERR system during the 1996–1998 time period of SWMP Phase I. After employing various statistical tests (t-tests, Analysis of Variances [ANOVAs], Kruskal-Wallis test, Pearson's correlation analysis, and harmonic regression analysis) on an array of water quality parameters (temperature, salinity, dissolved oxygen, pH, turbidity, and depth), Wenner et al. (2001) reported the following findings:

1. Depth, salinity, and dissolved oxygen (mg/l or % saturation) were generally dominated by 12.42-h cycles at sites that experienced moderate daily tidal amplitudes (2–4 m). Twenty-four hour cycles (i.e., day-night cycle, wind, and man-made perturbations) dominated sites where salinity was characterized as very low (tidal freshwater environment) or very high (marine environment) and where tidal effects were minimal.
2. Twenty-four hour cycles accounted for 30–50% of the water temperature variance at most sites, and 12.42-h cycles accounted for <20% of the

water temperature variance at most sites. In the summer, water temperature fluctuated by as much as 10°C over 24 h at some sites.
3. Hypoxia (dissolved oxygen [DO] < 28% saturation) was strongly influenced by latitude and climate. Half of the sites where hypoxia was observed (on average) for more than 20% of the first 48 h of data logger post-deployment were located in the Gulf of Mexico and Caribbean; however, 92% of hypoxia events persisted less than 8 h. Hypoxia was most frequently observed during summer.
4. The temporal and spatial distribution of supersaturation (DO > 120% saturation) events was not as clear as the temporal and spatial distribution of hypoxia events. Supersaturation events primarily occurred in cooler months, particularly winter; however, supersaturation events were observed in all seasons. Although supersaturation events sometimes co-occurred with hypoxic events during the same day, supersaturation was negatively correlated with hypoxia.
5. At 92% of the sampling sites evaluated, aquatic respiration exceeded aquatic production. Water temperature was significantly correlated with aquatic metabolic rates at most sites; however, salinity was only significantly correlated with aquatic metabolic rates at half of the sites evaluated. Although metabolic rates were not noticeably different among geographic regions, they were strongly influenced by habitat type.
6. Cluster analyses grouped the NERR sites based on geographic region and latitude, demonstrating the significance of biogeography and climate as principal controls of water chemistry in the NERR estuaries (Sanger et al., 2002).

Sanger et al. (2002) analyzed water quality data collected in NERR SWMP over the 6-year (Phase I) period from 1995 to 2000. They focused on the assessment of seasonal and inter-annual variability of parameters (water temperature, salinity, pH, dissolved oxygen, turbidity, and water depth). The following major findings were presented:

1. With respect to season, the application of three-way ANOVA models showed that the highest water temperature, salinity, and number of hypoxia events typically occurred among reserve sites in summer or fall, with lowest values for these parameters generally observed in the winter or spring. Supersaturation peaked in winter. These water quality patterns, particularly for salinity, were significantly influenced by seasonal precipitation and evapo-transpiration.
2. The daily solar radiation cycle greatly influenced water temperature and dissolved oxygen. Tides also impacted dissolved oxygen levels for reserve sites near inlets. Low tide events influenced water temperature in shallow estuaries.
3. The tide, wind patterns, and solar radiation cycles were the principal factors affecting water depth.

4. The passage of tropical storms was linked to changes in temperature and salinity.
5. Multivariate analysis revealed that sites with high salinity were associated with a lower percentage of agricultural land in the watershed and lower turbidity, and sites with low salinity were associated with a higher percentage of agricultural land in the watershed and higher turbidity.
6. While maximum autotrophic conditions generally occurred in the winter or spring, maximum heterotrophic conditions typically developed in the summer or fall. Minimum heterotrophic conditions were noted in the winter or spring.

High-Priority SWMP Initiatives

Several elements of the NERRS program are considered to be high-priority initiatives (i.e., they are to be implemented as soon as funds become available). Aside from ongoing personnel needs and the technical challenges associated with analysis of the extensive water quality database of SWMP, other high-priority initiatives in the program are nutrient monitoring, chlorophyll fluorescence, biomonitoring, land use and habitat change, watershed land use mapping, and benthic habitat mapping. Medium-priority initiatives (i.e., those that require further evaluation prior to implementation) include photopigment analysis, chemical contaminant monitoring, and global sea level rise. Two other major projects of significance are habitat restoration and invasive species.

Water Quality Monitoring A high-priority initiative of SWMP in recent years has been an expansion of water quality monitoring. This has been accomplished, as noted above, by upgrading data loggers, as well as by increasing the number of data loggers and the number of monitoring sites in each reserve. Spatial expansion has been pursued because the reserves initially monitored abiotic conditions at only two sites, which could not provide adequate coverage of the diverse habitats comprising these systems. It is advantageous to conduct water quality monitoring at multiple sites in a reserve to more fully characterize conditions in estuarine waters and their watersheds. Water quality data derived from multiple monitoring sites also yield more information for coastal decision makers to address resource management problems. In addition, these data enable investigators to examine more closely the natural perturbations and anthropogenic disturbances in these valuable coastal ecosystems. According to NERRS (2002a), specific problems that may be targeted by a more extensive water quality database are:

1. Impacts of nonpoint source pollutants and nutrient loadings on water quality
2. Changes in water quality associated with altered freshwater inflow due to human intervention
3. Comparison of water quality between a freshwater-dominated source and a marine-dominated source
4. Comparison of water quality in a restored marsh with that in a natural marsh
5. Comparison of effects of different land uses on adjacent wetland water quality

6. Effects of saltwater encroachment on tidal freshwater wetlands
7. Correlation of water quality over broad spatial scales with occurrence, density, and distribution of biological resources

Estuaries are physically controlled systems subject to a high level of natural variability at all spatial and temporal scales (e.g., storms, floods, upwelling events, and tidal currents), as well as a wide array of anthropogenic disturbances closely coupled to rapid population growth and uncontrolled or poorly controlled development in coastal watersheds. Estuarine reserves are not immune to these stressors. Potentially significant anthropogenic impacts on these systems are habitat loss and alteration, shoreline development, nonpoint source pollution, chemical contamination, eutrophication, freshwater diversions, overfishing, and nonindigenous species invasions (Kennish, 1997, 2000). Some of these impacts are expected to become more severe during the next 25 years as the coastal population continues to expand (Kennish, 2002). A focus of NERR-SWMP is to improve understanding of estuarine variability associated with both natural processes and anthropogenic activities through measurements of short-term variability and long-term changes in water quality, land use/land cover in watersheds, biotic community structure, and aquatic habitat.

Nutrient Monitoring In February 2002, the NERRS program instituted nutrient monitoring system-wide. Excessive input of nutrients and organic matter to estuaries from various land-based sources (e.g., land runoff, riverine inflow, and wastewater treatment plant discharges), as well as from atmospheric deposition, has been linked to overenrichment or eutrophication problems (Nixon, 1995; Valiela, 1995; Valiela et al., 1997; Smith et al., 1999). Eutrophication may be the most significant water quality problem in U.S. coastal waters today; it has caused the impairment of water quality in estuaries nationwide, resulting in imbalances in their trophic structure and other adverse effects such as periodic toxic or nuisance algal blooms, shading effects, hypoxia, fish kills, and reduced biodiversity (Livingston, 1996, 2000, 2003; Kennish, 2002). As previously noted, Bricker et al. (1999) documented 44 highly eutrophic estuaries and another 40 moderately eutrophic systems in the conterminous U.S., with most of these water bodies found along the Atlantic and Gulf coasts. They also projected that eutrophic conditions would worsen in 86 impacted estuaries by the year 2020.

SWMP nutrient monitoring targets two groups of water quality parameters: Tier I and Tier II components. For Tier I parameters (i.e., ammonium, nitrate, nitrite, orthophosphate, and chlorophyll a), monthly grab sampling (consisting of two replicates) is required at four SWMP data logger sites, and monthly diel sampling (using an ISCO sampler) is mandatory at one SWMP data logger site. For Tier II parameters (i.e., silica, particulate nitrogen, particulate phosphorus, dissolved total nitrogen, dissolved total phosphorus, particulate carbon, dissolved carbon, total suspended solids, and phaeopigments), sampling is optional (NERRS, 2002e).

The protocol for monthly grab sampling entails the collection of water samples on the same day at four SWMP data logger stations; samples are taken at or as near as possible to slack low-tide conditions. Duplicate grab samples are required, with both samples reflective of the water mass sampled by the data logger. Therefore, eight samples are collected every month at each reserve. Depending on hydrographic conditions at the sampling site (i.e., well mixed vs. stratified water column), a surface grab or horizontal and vertical samplers may be used. All samples are filtered as soon as possible after collection (NERRS, 2002e).

Diel nutrient sampling involves the deployment of ISCO auto-samplers at one of the four principal long-term data logger stations in each reserve. Samples are collected at 2- to 3-h intervals over a lunar day (24 h, 48 min) time period. Samples reflect the water mass sampled by the data logger. For example, surface samples are collected within the photic zone or at a fixed depth of 0.5 m from the surface. Bottom samples are collected at a fixed depth from the bottom, generally 0.5 m (NERRS, 2002e).

Storm events should not influence either diel or grab sampling. A dry period 72 h prior to sampling is desirable. All parameters must be analyzed in the laboratory using standard and approved methods.

Chlorophyll Fluorescence Experimental testing has been undertaken at a number of NERR sites using a YSI 6025 chlorophyll probe to monitor chlorophyll levels. Chlorophyll determinations are important because the long-term trends in phytoplankton biomass are an excellent indicator of eutrophication in estuaries (NERRS, 2002a). Although results of deployment of the YSI 6025 chlorophyll probe in the NERR pilot study have been variable, the application of chlorophyll fluorescence will continue to be examined in the future.

NERRS also proposes to conduct photopigment analysis to measure the relative abundance of characteristic algal groups, although this is a medium-priority initiative. Data derived from photopigment analysis will be useful in examining long-term trends and interannual variability in microalgal community composition and biomass, as well as yielding data on the relative abundance of algal groups. Phytoplankton community responses to nutrient availability, therefore, may be clearly evident from photopigment analysis. For example, phytoplankton communities in nutrient-enriched, eutrophic estuaries are typified by reduced species diversity and accelerated productivity. NERRS will employ high-performance liquid chromatography (HPLC) for photopigment-based characterization of phytoplankton at reserve sites (NERRS, 2002a).

Biomonitoring NERRS is currently developing a biological monitoring program. The major goal of this long-term monitoring program parallels that of the water quality initiative of SWMP: to delineate the patterns of short-term variability and long-term changes at the reserve sites. Quantitative biological measures will be used to assess biotic community composition, species abundance, and species distributions. The data will yield baseline biotic conditions, characterize biotic diversity, and help to detect invasive species. The data will also be useful in investigating indicators of estuarine condition and function and enable insightful comparisons to be made among the reserve sites. Such information will be of value to NERRS for developing an estuarine and marine classification system. Databases generated in the program will be not only of academic interest but also of practical utility for coastal resource management and decision making (NERRS, 2002a). The biological monitoring program will consist of the following focus areas:

1. Plankton (specifically larvae)
2. Nekton (fish, decapod crustaceans, and other swimming animals)
3. Benthos (benthic invertebrates and benthic algae)
4. Submerged aquatic vegetation (seagrasses) and emergent vegetation (marsh plants)

NERRS workgroups are establishing monitoring protocols for each of the aforementioned focus areas. Components of these protocols will consist of (NERRS, 2002a):

1. Mandatory requirements for sampling to be carried out by all participating reserves on specific program elements
2. A menu of additional options that can be implemented by the reserves as circumstances and funding allow

Because this approach requires all reserve sites to follow mandatory monitoring protocols on specific studies, statistically valid comparisons can be made of the databases across sites, forging a regional assessment of reserves as well as a national synthesis of trends. In addition to the mandatory requirements for sampling at all reserve sites, a menu of optional, site-specific monitoring protocols will be formulated to address local rather than national biological issues of concern. This flexible sampling schedule will result in maximum benefits for effective resource management and estuarine conservation at the reserve sites (NERRS, 2002a).

Watershed Land Use Mapping A major objective of future NERRS efforts is to document land use at the reserve sites and to determine how the ecological condition and function of valued habitats at these sites change in space and time (NERRS, 2002a). An important part of this effort is the examination of links between watershed land use and coastal habitat quality as well as the tracking of changes in the extent and distribution of wetland and upland habitat types (Nieder et al., 2002). Habitat destruction and fragmentation associated with watershed development cause significant biotic impacts, including loss of feeding and reproductive grounds, elimination of species, and reduction in species diversity. Development leads to an increase of impervious land cover, which promotes stormwater runoff, an increase in erosion, accelerated sediment loading to estuaries, greater frequency of flooding, and diminished infiltration and ground water recharge (Arnold and Gibbons, 1996).

Remote sensing applications are useful in discerning large-scale changes in land cover. Of particular utility is high-resolution digital imagery, which can generate detailed databases from multiple time periods that enable accurate analysis of land use/land cover changes. GIS can be used to document areas where changes have occurred due to either natural or anthropogenic factors. This is a high priority application of the NERRS program. GIS is valuable for long-term monitoring and conservation of resources because change detection studies can identify even subtle shifts of resource distribution patterns over time that yield significant information for coastal managers and other decision makers concerned with assessing the status and trends of coastal watersheds. It is also useful for wetland and terrestrial scientists interested in comparing biotic communities, habitats, and ecosystem processes in developed watersheds to those at reference locations. The application of system-wide standardized analytical procedures to document habitat change is a particularly powerful approach for NERRS since it will enable the status and trends of habitat change to be evaluated consistently across local, regional, and national scales. This effort will facilitate the inventory of reserve habitats in need of restoration and create a baseline land use/land cover dataset in the reserve system of use in identifying those coastal land use practices most effective in ameliorating habitat degradation (Nieder et al., 2002).

The application of remote sensing is a valuable approach for watershed land use mapping. High-resolution airborne imaging is especially attractive to NERRS because it offers the resolution capabilities necessary to accurately characterize the patterns of land use change in watershed areas. For example, Digital Orthophoto Quarter Quads (DOQQ), a U.S. Geological Survey airborne imagery program yielding 1-m resolution on a 5-year cycle, enables relatively rapid comparisons to be made of regional differences in watershed land use patterns. These data can then be used to assess watershed habitat quality (NERRS, 2002a).

Several watershed land use classes (e.g., residential, commercial, and agricultural lands) will be investigated as part of the watershed land use mapping effort. These classes can be mapped and categorized based on their relative contribution to specific altering activities of such factors as:

1. Freshwater inflow
2. Nutrients
3. Heavy metals
4. Pesticides
5. Polycyclic aromatic hydrocarbons (PAHs)
6. Soil erosion

Because the Coastal Change Analysis Program (C-CAP) of NOAA is developing a nationally standardized database on land cover and habitat change in U.S. coastal regions, it behooves NERRS to collaborate with C-CAP on watershed land use mapping. As stated by NERRS (2002a, p. 27), "By comparing regional differences in watershed land use patterns and their influence on estuarine habitat quality, the activity patterns most detrimental or beneficial to estuarine habitat quality will be identified. Furthermore, differences, if any, in this relationship at each reserve will allow an examination of regional differences in the sensitivities of estuarine habitats. These two products will provide the basis for regional recommendations regarding coastal land use planning, including habitat restoration, to benefit estuarine habitat quality."

Benthic Habitat Mapping It is necessary to characterize benthic habitats in the NERRS program to effectively link SWMP water quality monitoring with assessment of aquatic habitat change. Benthic intertidal and subtidal habitat mapping is essential to define the type and extent of aquatic habitats in each reserve system. It is valuable not only for delineating the reserve sites but also for ascertaining future habitat change caused by human activities in adjoining lowland areas and adjacent uplands as well as in the riverine and estuarine water bodies themselves. This information enables investigators to formulate remedial strategies to mitigate anthropogenic impacts.

Conventional benthic habitat mapping has typically entailed the use of grab or coring devices. However, this method of *in situ* data collection is labor intensive, often temporally and spatially limited in scope, and generally not cost effective when investigating extensive areas. Remote sensing applications, together with ground-truthing efforts, may be a more practical approach when examining aquatic habitat change over broad spatial scales.

When used in conjunction with grab or core sampling, sediment profile imagery (SPI) yields more comprehensive data on soft-bottom benthic habitats, including determinations of sediment texture and other characteristics (Rhoads and Germano, 1986; Nieder et al., 2002). SPI consists of a camera and video array that photographs the sediment–water interface, generating detailed images of the upper sediment layers of the estuarine floor. A sediment-profile camera may produce up to 200 images of the estuarine bottom, covering an area of several square kilometers, in one survey day. Photographs taken by a sediment-profile camera can be used to determine sediment type, sediment texture, depositional and erosional regimes, bedforms, depth of the redox potential discontinuity, epifauna, infauna, and bottom habitat type (e.g., bare bottom, seagrasses, and oyster bars). Successional mosaics derived from sediment-profile imaging are useful in tracking changes in benthic community structure. Reconnaissance maps of successional series produced by sediment-profile imaging have great potential value for managing estuarine resources, such as commercially and recreationally important shellfish beds. Photographic images from SPI are also useful in elucidating natural and anthropogenic disturbance gradients along the estuarine floor (Rhoads and Germano, 1986). SPI has been successfully applied to benthic studies in the Hudson River NERR and Apalachicola NERR, as well as in many other estuarine systems in the U.S.

The use of autonomous and remotely operated vehicles offers several advantages over traditional grab or coring devices in benthic habitat mapping. For example, when fitted with acoustic imaging devices (i.e., side-scan systems) or electromagnetics, remotely operated and autonomous vehicles can rapidly survey extensive stretches of the estuarine floor. The resulting acoustic images produce the high resolution necessary to effectively characterize estuarine bottom types. According to Robin Bell (Lamont–Doherty Geological Observatory, personal communication, 2002), remotely operated vehicles have proven to be reliable for studying areas typically inaccessible to traditional mapping technologies, and autonomous vehicles are powerful for obtaining coincident data sets. They may not only support benthic mapping requirements for site characterization but may also track anthropogenic disturbance of the estuarine floor associated with habitat loss and alteration (e.g., dredging, propeller scarring of motorized boats, sediment contamination, and oil spills).

REMUS (Remote Environmental Monitoring UnitS) are relatively low cost autonomous vehicles with considerable capability for mapping bottom habitats of shallow water systems. These preprogrammed systems are navigated via a transponder network. Equipped with lithium batteries, REMUS can cover a distance of more than 75 km at a speed of ~5 km/h. Experimental deployment of a REMUS vehicle in the Jacques Cousteau NERR has shown that side-scan sonar imaging of the Mullica River–Great Bay estuary is effective at depths greater than 1 m.

The Remote Observation Vehicle Earth Resources (ROVER), a remotely operated vehicle of the Lamont–Doherty Geological Observatory, provides low-impact access to very shallow (0.2–10 m) water environments. When fitted with a Mala Ground Penetrating Radar (GPR) sensor, ROVER can yield greater resolution of shallow stratigraphy than shown by standard acoustic frequency modulated subbottom profiles. This is particularly true where bottom sediments have a high gas content (Robin Bell, Lamont–Doherty Geological Observatory, personal communication, 2002).

Low-cost autonomous and remotely operated vehicles, such as REMUS and ROVER, have great potential application for benthic mapping of NERR sites systemwide. They represent two of the most practical means of mapping benthic habitats across broad spatial scales. This information will guide coastal managers who must make informed decisions regarding the long-term viability of habitats and resources in these systems.

For benthic mapping of spatially limited areas, such as SWMP sites, NERRS proposes the use of benthic grabs or cores as the most cost-effective method (NERRS, 2002a). The protocol is to deploy these instruments to determine sediment texture (composition), percent organics, benthic infauna, depth contours, and the distribution of benthic habitat type. The determination of sediment nutrient concentrations is also deemed to be important.

Benthic Community Surveys In concert with benthic habitat mapping, surveys of benthic communities are also important components of future NERR-SWMP efforts. Water and sediment quality data collected at NERRS sites are useful in evaluating changes in benthic species composition, abundance, and diversity. These data may also be related to the effects of watershed development and other human activities at the NERRS sites.

Medium-Priority SWMP Initiatives

Global Sea Level Rise Among the array of human impacts on estuaries, global sea level rise coupled to emissions of radiatively active gases (e.g., carbon dioxide, methane, nitrous oxides) remains a long-term concern. Rising sea level can have a profound effect on coastal wetland habitats, which often cannot accrete rapidly enough to maintain their position (Kennish, 2001b). Therefore, they may recede quickly landward in response to rising sea level, with the loss of fringing wetlands habitat as a resulting impact. The U.S. has lost more than 50% of its original salt marsh habitat, in part because of rising sea level (Kennish, 2001b). This loss has had dramatic impacts on wildlife and fisheries in some regions. Land subsidence must also be considered when assessing the effect of relative sea level rise. The long-term stability of coastal wetland systems is dependent on the relative rates of sediment accretion and land submergence (Mitsch and Gosselink, 1993).

Aside from inland migration of wetlands, an increase in shoreline erosion and coastal flooding together with a decrease in estuarine beach habitat are expected with continued sea level rise. This will be most evident in estuaries located between

20°N and 20°S latitude, where a greater frequency of storms and storm surges is anticipated. More extreme weather conditions may develop with escalating global warming, such as severe droughts punctuated by periods of excessive precipitation (Kennish, 2002). These radical shifts in atmospheric conditions would likely result in marked changes in freshwater inflow to estuaries over short time intervals.

Rising sea level could threaten the infrastructure of shore communities, rendering them more susceptible to flooding events and property damage. Greater upstream penetration of seawater would also increase the probability of saltwater intrusion into groundwater supplies. Current land cover and land use modifications in coastal watersheds could contribute significantly to future infrastructure degradation by facilitating sea level rise via the removal of natural vegetative cover, which now serves as a buffer against flooding and other associated impacts.

Important changes in the physical, chemical, and biological characteristics of estuarine water bodies would also result from global sea level rise. Examples are altered tidal prisms and salinities, which could greatly affect circulation and biotic communities in these systems (Kennish, 2000). Estuarine configurations and dimensions would likewise be altered.

Global mean sea surface temperature has increased 0.6 ± 0.2°C during the past century, with much of this increase ascribed to global warming due largely to the release of greenhouse gases by human activities (IPCC, 2001). Mean sea level has risen 10–25 cm during this time (Kennish, 2002). For the two warmest decades on record, the 1980s and 1990s, the rate of change of mean sea level rise has amounted to ~1.8 ± 0.3 mm/year (IPCC, 2001; Kennish, 2002). Since global climate records began 140 years ago, the highest annual global mean surface temperatures reported by the World Meteorological Organization occurred in 1998 (14.58°C) and 2001 (14.42°C).

Because of the serious potential impacts of rising sea level on adjacent wetland and upland habitats, the NERRS program is considering the implementation of a coastal monitoring effort to assess sediment elevation change relative to sea level rise. The strategy of this effort is to establish a sedimentation–erosion table (SET) plot within a wetland habitat (salt marsh or mangrove) of each NERR site. Observations made at each site will enable a regional assessment of habitat change (NERR, 2002a).

Chemical Contaminant Monitoring NERRS lists chemical contaminant monitoring as a medium priority concern. Halogenated hydrocarbons, PAHs, and heavy metals are the three major classes of chemical contaminants that pose the greatest threat to estuarine organisms and habitats (Kennish, 1992, 1997, 2002). Because these contaminants tend to readily sorb to sediments and other particulate matter, they generally accumulate on the estuarine floor. Hence, NERRS proposes to have surficial bottom sediments sampled for contaminant loadings at two to four sites in each reserve. The sediment samples will be analyzed using appropriate analytical methodology to obtain the precise composition and concentrations of chemical contaminants in the samples. Sampling will be conducted once every 5 years to provide status and trends of the contaminant loadings in each reserve (NERRS, 2002a).

Special High-Priority Initiatives

Habitat Restoration

A major initiative of the NERRS program is the restoration of habitat (NERRS, 2002f). Most of the reserve sites have either initiated small- to medium-scale (0.2–100 ha) restoration projects or are planning such projects, focusing on the re-establishment of salt marshes and seagrass beds where ecological functions and natural resource values have declined. The previous decade has seen the application of innovative restoration and enhancement technologies in these wetland habitats (Simenstad and Thom, 1992; Thayer, 1992; Fonseca et al., 1998). However, the complexities associated with restoring the hydrologic and biotic regimes of impacted wetland habitats can be formidable, and thus sound scientific methods must be used. To test the success of such habitat restoration projects, field sites must be carefully monitored for years. To date, there has been relatively poor predictability of restoration outcomes. Restoration ecology is a rather young science, and few long-term databases have been developed on specific restoration sites (Zedler, 2001).

The proposed goal of the NERRS restoration strategy is "to provide the scientific basis and technical expertise to restore, enhance, and maintain estuarine ecosystems by developing and transferring effective approaches to identify, prioritize, restore, and monitor degraded or lost coastal habitat. Success will require a partnered approach, education and community involvement, regional coordination, and additional resources" (NERRS, 2002f, p. 9). NERRS (2002f) has proposed the following priorities for restoration science:

1. Designing restoration projects
2. Developing effective approaches to restoration
3. Monitoring restoration response
4. Assessing restoration success
5. Serving as local reference or control sites
6. Translating/transferring restoration information
7. Supporting policy and regulatory decisions by providing scientific and technological advice
8. Building constituency for support of restoration science
9. Coordinating regional science

Because of its unique research and monitoring capabilities, NERRS is well suited to undertake habitat restoration initiatives. The reserve system has the site platforms, technical proficiency, and long-term monitoring strategies needed to effectively conduct restoration projects. NERRS is also closely affiliated with the Cooperative Institute for Coastal and Estuarine Environmental Technology (CICEET), a NOAA-funded center that has successfully supported innovative and cost-effective restoration technologies at reserve sites. Most reserves have disturbed habitat where restoration activities can be targeted. To date, more than 40,000 ha of altered habitat within reserve boundaries have been physically restored. Additional habitat restoration efforts are underway or being planned.

Funding is a major limiting factor to large-scale habitat restoration. However, with the passage of the Estuaries and Clean Waters Act of 2000, Congress authorized release of $275 million over 5 years to restore 400,000 ha of estuarine habitat by 2010. Title I of the Act requires the preparation of a National Strategy for Coastal Habitat Restoration.

Most habitat restoration projects in NERRS have concentrated on salt marsh, seagrass, riparian, and native habitat restoration. In addition, hydrologic restoration and controlled burning in fire-dependent habitats have been pursued. Among notable projects are salt marsh restoration in the Gulf of Maine at the Wells NERR and at the Tijuana River NERR in Southern California, eelgrass restoration at the Narragansett Bay NERR, and hydrologic restoration at the South Slough NERR to re-establish tidal circulation. In total, about 75% of the estuarine reserves have undertaken some type of habitat restoration work (NERRS, 2002f). Restoration and mitigation projects in NERRS are expected to increase substantially during the next decade to address ongoing habitat loss and degradation problems in estuarine and watershed habitats. NERRS has a well-organized and integrated research, education, and stewardship approach that will facilitate restoration projects.

Invasive Species

Among the growing concerns of the NERRS program are the effects of nonindigenous species invasions on biotic communities and habitats at reserve sites. This is so because more than 400 invasive species have been documented along U.S. coasts (Steve Rumrill, South Slough NERR, personal communication, 2003). Introduced exotic nuisance species can fundamentally alter the trophic organization of estuaries by outcompeting native forms and, in severe cases, causing their local extinction. Diseased invasives can inoculate detrimental pathogens into estuarine systems. Heavily invaded estuarine communities may exhibit reduced species diversity and shifts in recreational and commercial fisheries. Invasive species often lack natural controls in their adapted estuarine habitats and thus can increase dramatically in abundance to attain overwhelming dominance. The invasion of the common reed (*Phragmites australis*) often displaces or dramatically reduces native marsh plants (e.g., *Spartina alterniflora*) over extensive areas along the East Coast. The common reed has been particularly detrimental to brackish marsh habitat (Weinstein et al., 2000). The function of these habitats may likewise change, with some impacted sites displaying increased runoff and erosion, modified nutrient cycles, accelerated inputs of chemical contaminants, and altered biotic community structure.

Systems most heavily affected by invasive species are those with major shipping ports or centers for commercial fishing and aquaculture (Cohen and Carlton, 1998). A number of locations, such as San Francisco Bay and the Great Lakes, have been especially hard hit by invasive species. In addition, numerous estuaries along the Atlantic, Pacific, and Gulf coasts of the U.S., including NERRS sites, now harbor potentially threatening nonindigenous species (Wasson et al., 2002).

More than 200 nonnative species have been introduced into the San Francisco Bay/Delta region over the years, making it one of the most heavily invaded estuaries in the world (Cohen and Carlton, 1998; Kennish, 2000). Many of the dominant species in the bay are introduced forms, including nearly all macroinvertebrates along the inner shallows. Almost half of the fishes in the delta are exotic species (Herbold and Moyle, 1989). A few invasive species have caused a significant disruption of the endemic flora and fauna. For example, the smooth cordgrass (*Spartina alterniflora*) has successfully invaded salt marsh habitat in the bay area and appears to be supplanting the native cordgrass (*S. foliosa*). The most conspicuous faunal invasive in the estuary is the Asian clam (*Potamocorbula amurensis*), which has reached densities as high as 30,000 individuals/m^2. This species has undergone population irruptions over broad areas, disrupting planktonic and benthic communities, as well as finfish assemblages. It is responsible for an estimated $1 billion in costs each year, and it remains a threat to the bay's ecological health (Orsi and Mecum, 1986; Cohen and Carlton, 1998; Kennish, 2000).

Some invasive faunal species have been quite explosive in estuarine environments and have altered the structure of biotic communities in a number of systems. Examples of potentially serious nuisance invasive species are the zebra mussel (*Dreissena polymorpha*), Chinese mitten crab (*Eriocheir sinensis*), and European green crab (*Carcinus maenas*). Because of the great damage that some invasives can inflict on native estuarine populations, there is growing interest in NERRS to periodically conduct surveys of the species composition, abundance, and geographical distribution of introduced species at selected locations to delineate the spread and impacts of nonindigenous forms.

Wasson et al. (2002) note that nonnative species invasions contribute significantly to the global extinctions of susceptible organisms. They therefore advocate the formation of a nationally coordinated invasive species monitoring program in estuarine habitats and propose a framework for such a program. Because the NERRS network of estuarine sites establishes broad spatial coverage of the Atlantic, Gulf, and Pacific coasts, it serves as an ideal platform for detecting and tracking the range expansions of exotic species in this country. According to Wasson et al. (2002), NERRS has identified more than 85 problematic nonnative species at its constituent sites. They stress that certain habitats (i.e., marshes, submerged aquatic vegetation, and shellfish beds), as well as several taxa (i.e., macroalgae, mammals, and fish), should be the target of invasion monitoring. The implementation of a national invasive monitoring program is an important goal of NERRS.

Education and Outreach

Environmental education and stewardship are also major elements of the NERRS program. These two components represent a vital link between research and coastal management. In essence, education and outreach specialists translate findings of the scientific community into a form readily understood by resource managers, coastal decision makers, resource users, and the general public. Interns and volunteers typically support these education and outreach efforts. The education component of the reserve system essentially bridges the gap between scientific research and public understanding. Interpretive efforts are important when obtaining public support for the application of research findings to practical problems in estuarine waters and nearby watersheds. Most reserves have visitor centers and interpretive trails (NERRS, 1994a).

The NERRS Education, Outreach, and Interpretation Program is designed to forge greater understanding of estuarine ecosystems and nearby coastal watersheds. It also promotes environmental awareness of the adverse effects of human activities on these sensitive water bodies. In addition, the program encourages individual responsibility and stewardship of estuarine habitats, fostering conservation and protection of coastal resources. Through education and outreach, environmental managers and decision makers can learn to make more informed decisions on environmental problems and the utilization of resources along the coast. These decisions will be grounded in sound scientific research and monitoring. The program cultivates appreciation for estuaries and their exceptional biotic communities, habitats, recreational and commercial value, and aesthetics (JCNERR, 1998).

Prior to the formulation of the NERRS Strategic Education Plan in 1994, each reserve's education program operated relatively independently with its own goals and objectives. The Strategic Education Plan connected all reserve education programs through common long-range goals. It was developed at a meeting of representatives from NERRS, staff from the National Marine Sanctuaries Program at Old Woman Creek in Huron, Ohio, and federal representatives from NOAA (Sanctuaries and Reserves Division) (NERRS, 1994a). A major focus of this meeting was the formulation of a unifying strategy among reserve sites to link education, research, and resource management. The Strategic Education Plan lists the following guiding principles for designing and implementing an education program in the reserve system (NERRS, 1994b):

1. Provide on-site and off-site educational experiences
2. Educate holistically about estuaries, including ecological, cultural, historical, sociological, aesthetic, and economic aspects
3. Promote a sense of stewardship and individual responsibility
4. Address coastal issues from local, state, regional, national, and global perspectives
5. Approach estuarine education through a perspective that includes watersheds, bioregions, and biogeographic regions
6. Increase understanding and appreciation of the National Estuarine Research Reserve System

Basic objectives of the NERRS Education, Outreach, and Interpretation Program are as follows (NERRS, 1994a, b):

1. Implement an education and interpretive program that provides state-of-the-art information for NERRS reserves
2. Impart knowledge of broader coastal issues and values of estuarine/marine (protected) areas to educators and students, coastal resource decision makers, environmental professionals, resource users, and the general public
3. Convey the local, regional, and national importance of estuaries
4. Encourage understanding of estuaries for the general public

5. Promote networking among local, state, and regional decision makers to more effectively manage coastal resources of the reserve region
6. Provide local, state, and regional decision makers with access to information and resources of the NERRS research and monitoring program
7. Foster strong interrelationships between research and education initiatives in the reserve system
8. Offer educational services at the reserves for educators, students, and the visiting public that facilitate awareness of the estuary and promote sustainable use of estuarine resources
9. Cultivate mentoring and networking among educators participating in reserve educational programs
10. Establish venues for direct field experiences, such as teacher enrichment and internships for students and educators, where participants work alongside scientists
11. Target educational programming to groups underrepresented in science and technology
12. Participate in local, state, and regional programs such as National Estuaries Day

An array of educational and outreach programs has been implemented at NERRS sites and targeted to students, teachers, and local officials. These consist of in-class presentations, seminars, tours, demonstrations, interpretive displays, exhibits, printed materials (e.g., brochures, newsletters, and special publications), videos, audio tapes, and workshops. Student site visits are an important element of the NERRS educational program as are summer teacher training opportunities and residential teaching institutes. Local organizations and schools (K–12) are encouraged to participate in reserve functions. Educator enrichment programs focus on a range of topics, such as environmental monitoring and training, remote sensing technology, and GIS applications. Educators and their students utilize interactive learning modules, conduct electronic field trips, and engage in informal lectures with scientists in the field. Video/data links between reserve sites and educational institutions connected to the Internet facilitate this educational experience (JCNERR, 1998).

A number of creative educational programs have also been instituted in the reserve system. For example, at the Jacques Cousteau NERR in New Jersey, workshops are offered each year for K–12 teachers, focusing on hands-on training in estuarine and marine science. The teachers engage in basic skills training, problem solving, and critical thinking about coastal environmental issues. They can then disseminate this training and educational experience to their students. Some teachers visiting the Jacques Cousteau NERR site participate directly in field research, working with Rutgers University scientists and reserve staff to design classroom and field-based marine activities that incorporate their experience into Web-based interactive exercises. The Marine Activities, Resources, and Education (MARE) Summer Institute for K–8 educators is a core educational program at the Jacques Cousteau NERR site. It offers a supplementary marine science curriculum that encourages cooperative learning and hands-on classroom activities, with educators attending a 6-day intensive summer institute designed to foster greater understanding of estuaries and associated wetlands. Thousands of students ultimately benefit each year from this summer teacher training.

One of the most important initiatives of NERRS is its Coastal Training Program (CTP). The goal of CTP is "better-informed decision making to improve coastal stewardship at local and regional levels" (NERRS, 2003, p. 1). A focus of CTP is to enable coastal decision makers to improve their scientific understanding of NERRS monitoring and research efforts. This includes greater resolution of NERRS priority issues such as nutrient monitoring, biomonitoring, biodiversity, habitat mapping, habitat restoration, invasive species, and estuarine processes. To this end, coastal decision makers interact with CTP representatives to access SWMP data and other NERRS material to address resource problems. It is hoped that by becoming more knowledgeable about the science-based programs of NERRS, coastal decision makers will be able to deal more effectively with the myriad issues associated with coastal resource management. CTP is also part of an effort by NERRS to develop site-based coastal training institutes.

SUMMARY AND CONCLUSIONS

The NERRS program consists of 25 active reserve sites that encompass more than a million hectares of estuarine, wetland, and upland habitats in the major biogeographical regions of the U.S. The purpose of this program is to improve the health of the nation's coastal habitats by developing and providing information that promotes sound resource management (NERRS, 1995). Established by passage of the
Coastal Zone Management Act of 1972 and its reauthorization in 1985, NERRS also helps to address current and potential degradation of coastal resources resulting from increasing and competing human demands (JCNERR, 1998; NERRS, 2002a). This network of protected areas represents a federal, state, and community partnership in which environmental monitoring and research, as well as a comprehensive program of education and outreach, strengthen understanding, appreciation, and stewardship of estuaries, coastal habitats, and associated watersheds (NERRS, 1994a, b, 1995, 2002a).

The NERRS program has two principal elements:

1. Monitoring and research
2. Education and outreach

Monitoring and research are part of a System-wide Monitoring Program (SWMP) developed to identify and track short-term variability and long-term changes in the integrity and biodiversity of reserve estuaries and their coastal watersheds. Education and outreach concentrate on improving public awareness of estuarine-related issues and coastal resource problems and providing interpretive information based on sound scientific research to decision makers. SWMP is one of the most comprehensive and effective programs for monitoring and coordinating water quality conditions of the nation's estuaries, and it provides the platform for making systematic, long-term observations of vital ecosystem parameters. SWMP databases are most useful in delineating how estuaries function and change over time (NERRS, 2002a). Aside from water quality monitoring, major SWMP initiatives include investigating biotic communities, assessing habitat change in estuaries, and examining land use change in nearby watersheds. Other priority projects of the NERRS program are nutrient monitoring, biomonitoring, watershed land use mapping, benthic habitat mapping, invasive species studies, and habitat restoration. NERRS represents a national network of sites for the integration and coordination of environmental monitoring and research; as such, it serves a vital role in improving public awareness and understanding of estuarine environments and yielding valuable data for more effective management of coastal resources.

REFERENCES

Alongi, D.M. 1998. *Coastal Ecosystem Processes*. CRC Press, Boca Raton, FL.

Arnold, C.L. and C.J. Gibbons. 1996. Impervious surface coverage. *Journal of the American Planning Association* 62: 243–258.

Bricker, S.B., C.G. Clement, D.E. Pirhalla, S.P. Orlando, and D.R.G. Farrow. 1999. National Estuarine Eutrophication Assessment: A Summary of Conditions, Historical Trends, and Future Outlook. Technical Report, National Oceanic and Atmospheric Administration, National Ocean Service, Special Projects Office and the National Centers for Coastal Ocean Service, Silver Spring, MD.

Cohen, A.N. and J.T. Carlton. 1998. Accelerating invasion rate in a highly invaded estuary. *Science* 279: 555–562.

Cohen, J.E., C. Small, A. Mellinger, J. Gallup, and J. Sachs. 1997. Estimates of coastal populations. *Science* 278: 1211–1215.

Edmiston, L., R. Scarborough, S. Ross, and M. Crawford. 2002. Information Package for New Research Coordinators within the National Estuarine Research Reserve. Technical Report, National Estuarine Research Reserve System, Silver Spring, MD.

Fonseca, M.S., W.J. Kenworthy, and G.W. Thayer. 1998. Guidelines for the Conservation and Restoration of Seagrasses in the United States and Adjacent Waters. NOAA Coastal Ocean Program — Decision Analysis Series No. 12, NOAA Coastal Ocean Program, Silver Spring, MD.

Goldberg, E.D. 1994. Coastal Zone Space — Prelude to Conflict? UNESCO Technical Report, UNESCO, Paris.

Greene, D. and D.D. Trueblood. 1999. NERRS System-wide Monitoring Program: Considerations for Program Development and Implementation. NOAA Technical Memorandum NOS/OCRM/ERD-99-01, National Oceanic and Atmospheric Administration, Silver Spring, MD.

Herbold, B. and P.B. Moyle. 1989. The Ecology of the Sacramento–San Joaquin Delta: A Community Profile. Biological Report No. 85(7.22), U.S. Fish and Wildlife Service, Washington, D.C.

IPCC. 2001. *Climate Change 2001: Impacts, Adaptation, and Vulnerability*. J.J. McCarthy, O.F. Canziani, N.A. Leary, D.J. Dokken, and K.S. White (Eds.). Cambridge University Press, Cambridge.

Jacques Cousteau National Estuarine Research Reserve. 1998. Jacques Cousteau National Estuarine Research Reserve at Mullica River–Great Bay, New Jersey: Final Management Plan. Technical Report, Jacques Cousteau National Research Reserve, Rutgers University, New Brunswick, NJ.

Kennish, M.J. 1992. *Ecology of Estuaries: Anthropogenic Effects*. CRC Press, Boca Raton, FL.

Kennish, M.J. (Ed.). 1997. *Practical Handbook of Estuarine and Marine Pollution.* CRC Press, Boca Raton, FL.

Kennish, M.J. (Ed.). 2000. *Estuary Restoration and Maintenance: The National Estuary Program.* CRC Press, Boca Raton, FL.

Kennish, M.J. (Ed.). 2001a. *Practical Handbook of Marine Science,* 3rd ed. CRC Press, Boca Raton, FL.

Kennish, M.J. 2001b. Coastal salt marsh systems in the U.S.: a review of anthropogenic impacts. *Journal of Coastal Research* 17: 731–748.

Kennish, M.J. 2002. Environmental threats and environmental future of estuaries. *Environmental Conservation* 29: 78–107.

Livingston, R.J. 1996. Eutrophication in estuaries and coastal systems: relationships of physical alterations, salinity stratification, and hypoxia. In: Vernberg, F.J., W.B. Vernberg, and T. Siewicke (Eds.). *Sustainable Development in the Southeastern Coastal Zone.* University of South Carolina Press, Columbia, SC, pp. 285–318.

Livingston, R.J. 2000. *Eutrophication Processes in Coastal Systems: Origin and Succession of Plankton Blooms and Secondary Production in Gulf Coast Estuaries.* CRC Press, Boca Raton, FL.

Livingston, R.J. 2003. *Trophic Organization in Coastal Systems.* CRC Press, Boca Raton, FL.

Mitsch, W.J. and J.G. Gosselink. 1993. *Wetlands,* 2nd ed. Van Nostrand-Reinhold, New York.

National Estuarine Research Reserve System. 1994a. National Estuarine Research Reserve System Education: A Field Perspective. National Estuarine Research Reserve System, National Oceanic and Atmospheric Administration, Silver Spring, MD.

National Estuarine Research Reserve System. 1994b. The National Estuarine Research Reserve System Strategic Plan for Education. National Estuarine Research Reserve System, National Oceanic and Atmospheric Administration, Silver Spring, MD.

National Estuarine Research Reserve System. 1995. A Strategic Plan for the National Estuarine Research Reserve System: A State and Federal Partnership. National Estuarine Research Reserve System, National Oceanic and Atmospheric Administration, Silver Spring, MD.

National Estuarine Research Reserve System. 2002a. The National Estuarine Research Reserve's System-wide Monitoring Program (SWMP): A Scientific Framework and Plan for Detection of Short-term Variability and Long-term Change in Estuaries and Coastal Habitats of the United States. Technical Report, National Estuarine Research Reserve System, National Oceanic and Atmospheric Administration, Silver Spring, MD.

National Estuarine Research Reserve System. 2002b. System-wide Monitoring Program: Deployment Plans for Phase 1 (Abiotic Factors). Technical Report, National Estuarine Research Reserve System, National Oceanic and Atmospheric Administration, Silver Spring, MD.

National Estuarine Research Reserve System. 2002c. National Estuarine Research Reserve System Strategic Plan 2003–2008. Technical Report, National Estuarine Research Reserve System, National Oceanic and Atmospheric Administration, Silver Spring, MD.

National Estuarine Research Reserve System. 2002d. Action Plan. Technical Report, National Estuarine Research Reserve System, National Oceanic and Atmospheric Administration, Silver Spring, MD.

National Estuarine Research Reserve System. 2002e. NERR SWMP Nutrient Sampling Protocol. Technical Document, National Estuarine Research Reserve System, National Oceanic and Atmospheric Administration, Silver Spring, MD.

National Estuarine Research Reserve System. 2002f. Restoration Science Strategy: A Framework. Technical Report, National Estuarine Research Reserve System, National Oceanic and Atmospheric Administration, Silver Spring, MD.

National Estuarine Research Reserve System. 2003. NERRS Coastal Training Program. Technical Document, National Estuarine Research Reserve System, National Oceanic and Atmospheric Administration, Silver Spring, MD.

National Oceanic and Atmospheric Administration. 1998. Population: Distribution, Density, and Growth. NOAA's State of the Coast Report, National Oceanic and Atmospheric Administration, Silver Spring, MD.

National Oceanic and Atmospheric Administration. 1999. National Estuarine Research Reserve System: A National Asset for Improving Management of Estuaries. Summary Document, National Oceanic and Atmospheric Administration, Silver Spring, MD.

National Research Council. 1990. *Managing Troubled Waters: The Role of Marine Environmental Monitoring.* National Academy Press, Washington, D.C.

Nieder, C., D. Porter, S. Rumrill, K. Wasson, E. Wenner, B. Stevenson, and M. Treml. 2002. Land Use and Land Cover Analysis in the National Estuarine Research Reserve System. Technical Report, National Estuarine Reserve Program, National Oceanic and Atmospheric Administration, Silver Spring, MD.

Nixon, S.W. 1995. Coastal marine eutrophication: a definition, social causes, and future concerns. *Ophelia* 41: 199–219.

Orsi, J.J. and W.L. Mecum. 1986. Zooplankton distribution and abundance in the Sacramento–San Joaquin Delta in relation to certain environmental factors. *Estuaries* 9: 326–334.

Rhoads, D.C. and J.D. Germano. 1986. Interpreting long-term changes in benthic community structure: a new protocol. *Hydrobiologia* 142: 291–202.

Ross, S.W. 2002. Establishing a Real-time Delivery and Management System for the NERRS Environmental Data. Technical Report, National Estuarine Research Reserve System, Silver Spring, MD.

Sanger, D.M., M.D. Arendt, Y. Chen, E.L. Wenner, A.F. Holland, D. Edwards, and J. Caffrey. 2002. A Synthesis of Water Quality Data: National Estuarine Research Reserve System-wide Monitoring Program (1995–2000). National Estuarine Research Reserve Technical Report Series 2002:3, South Carolina Department of Natural Resources, Marine Resources Division Contribution No. 500, Charleston, SC.

Schuyler, Q., W. Stevenson, H. Recksiek, M. Crawford, and M. Treml. 2002. Addressing Habitat Issues with Remote Sensing in the National Estuarine Research Reserve System. NERRS Needs Assessment Report, National Estuarine Research Reserve System, Silver Spring, MD.

Simenstad, C.A. and R.M. Thom. 1992. Restoring wetland habitats in urbanized Pacific Northwest estuaries. In: Thayer, G. (Ed.). *Restoring the Nation's Marine Environment.* Publication UM-SG-TS-92–06, Maryland Sea Grant College, College Park, MD.

Smith, V.H., G.D. Tilman, and J.C. Nekola. 1999. Eutrophication: impacts of excess nutrient inputs on freshwater, marine, and terrestrial ecosystems. *Environmental Pollution* 100: 179–187.

Thayer, G.W. (Ed.). 1992. *Restoring the Nation's Marine Environment.* Publication UM-SG-TS-92–06, Maryland Sea Grant College, College Park, MD.

Valiela, I. 1995. *Marine Ecological Processes,* 2nd ed. Springer-Verlag, New York.

Valiela, I., I. Kremer, K. Lajtha, M. Geist, B. Seely, J. Brawley, and C.H. Sham. 1997. Nitrogen loading from coastal watersheds to receiving estuaries: new method and application. *Ecological Applications* 7: 358–365.

Wasson, K. 2002. An Approach to Biological Monitoring in the National Estuarine Research Reserves. Technical Document, National Estuarine Research Reserve System, Silver Spring, MD.

Wasson, K., D. Lohrer, M. Crawford, and S. Rumrill. 2002. Non-native Species in our Nation's Estuaries: A Framework for an Invasion Monitoring Program. Technical Report, National Estuarine Research Reserve System, Silver Spring, MD.

Weber, M.L. 1995. Healthy Coasts, Healthy Economy: A National Overview of America's Coasts. Technical Report, Coast Alliance, Washington, D.C.

Weinstein, M.P., K.R. Philipp, and P. Goodwin. 2000. Castastrophes, near-catastrophes, and the bounds of expectation: success criteria for macroscale marsh restoration. In: Weinstein, M.P. and D.A. Kreeger (Eds.). *Concepts and Controversies in Tidal Marsh Ecology.* Kluwer Academic Publishers, Dordrecht, pp. 777–804.

Wenner, E.L., A.F. Holland, M.D. Arendt, D. Edwards, and J. Caffrey. 2001. A Synthesis of Water Quality Data from the National Estuarine Research Reserve System-wide Monitoring Program. Technical Report, Centralized Data Management Office, National Estuarine Research Reserve System, University of South Carolina, Charleston, SC.

Zedler, J. (Ed.). 2001. *Handbook for Restoring Tidal Wetlands.* CRC Press, Boca Raton, FL.

Case Study 1

2 Waquoit Bay National Estuarine Research Reserve

INTRODUCTION

Waquoit Bay is a shallow lagoon-type estuary that lies along a glacial outwash plain on the south shore of Cape Cod (Figure 2.1). The bay covers an area of 600 ha, and it supports rich and diverse biotic communities. Although the bay only averages 1.8 m in depth (maximum depth 3 m), the water column is typically stratified (D'Avanzo and Kremer, 1994). Surface water and groundwater inflows from the watershed mix with waters from Nantucket Sound and Vineyard Sound. Characterized as a multiple inlet estuary, Waquoit Bay is bounded along its southern perimeter by barrier beaches that are breached at two permanent locations (Crawford, 2002). A navigation channel trending north–south bisects the main embayment into eastern and western sections. Proceeding upestuary, the bay is bounded by salt marshes, and it gives way to brackish ponds, freshwater tributaries, freshwater ponds, and upland habitat. Flat, Sage Lot, Hamblin, and Jehu Ponds are brackish ponds, and Bog, Bourne, and Caleb Ponds are freshwater ponds.

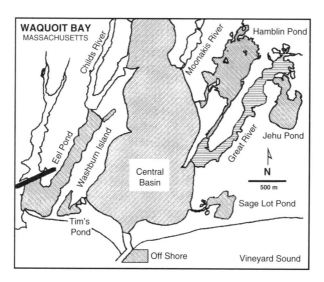

FIGURE 2.1 Map of Waquoit Bay showing sub-basins of the estuary. (From Short, F.T. and D.M. Burdick. 1996. *Estuaries* 19: 730–739.)

Waquoit Bay is the main aquatic component of the Waquoit Bay National Estuarine Research Reserve (Waquoit Bay NERR), which is centered in the towns of Falmouth and Mashpee. Most of the reserve area consists of channels and open waters (~510 ha). Uplands cover ~300 ha, marshes (fresh-, brackish-, and salt-water marshes) >120 ha, and subtidal meadows ~70 ha (Geist and Malpass, 1996).

The reserve encompasses an area of ~14.9 km^2. It includes, in addition to the site headquarters (11.3 ha), public lands within South Cape Beach State Park (175 ha) and Washburn Island (133 ha). The Waquoit Bay NERR was designated in 1988 as the 15th site of the National Estuarine Research Reserve (NERR) system (Geist and Malpass, 1996).

WATERSHED

The Waquoit Bay watershed covers more than 5000 ha. It stretches northward for ~8 km from the head of Waquoit Bay. Cambareri et al. (1992) delineated seven subwatersheds in the Waquoit Bay watershed:

1. Eel Pond
2. Childs River
3. Head of the Bay
4. Quashnet River
5. Hamblin Pond
6. Jehu Pond
7. Sage Lot Pond

Figure 2.2 shows the boundaries of these subwatersheds.

The Waquoit Bay watershed is comprised of a wide array of habitats, notably upland pitch pine/oak forests, pine barrens, wetlands (fresh-, brackish-, and salt-water marshes), riparian habitats, sandplain grasslands, vernal pools, and coastal plain pond shores, as well as barrier beaches and sand dunes. These habitats support numerous plant and animal populations, including some endangered, threatened, and rare species. Concern is growing with regard to future development and associated anthropogenic impacts in the watershed habitats.

UPLAND PITCH PINE/OAK FORESTS

The primary forest community in the Waquoit Bay watershed consists of a complex of pitch pines (*Pinus rigida*) and scrub oak trees (*Quercus ilicifolia*). It has formed on the acidic, well-drained sandy soils of the glacial outwash plain. A mix of sand and gravel, together with pebbles and small boulders, is evident along the surface in barren areas of the watershed (Malpass and Geist, 1996).

In watershed areas north of the Waquoit Bay NERR, a pine barrens community of pitch pine (*Pinus rigida*)/scrub oak (*Quercus ilicifolia*) has become established in response to periodic fires, which generate nutrients from ashes in an otherwise nutrient-deficient habitat. This community, similar to that observed in the watershed areas of the Jacques Cousteau NERR in New Jersey, consists

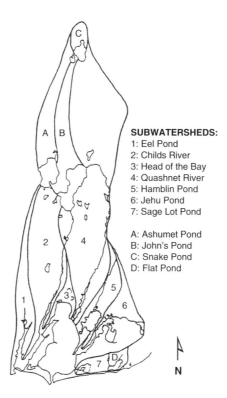

FIGURE 2.2 Map showing Waquoit Bay subwatershed areas. (From Geist, M.A. 1996. In: *The Ecology of the Waquoit Bay National Estuarine Research Reserve*, Geist, M.A. (Ed.). Technical Report, Waquoit Bay National Estuarine Research Reserve, Waquoit, MA, pp. II-1 to II-22.)

of a unique complex of pitch pines and an understory of scrub oak and huckleberry (*Gaylussacia baccata*) growing on relatively flat terrane. Among the predominant low-lying vegetation found under the larger trees are lichens (*Cladonia* spp.), bearberry (*Arctostaphylos uva-ursi*), lowbush blueberry (*Vaccinium angustifolium*), and sweetfern (*Comptonia peregrina*). Frequent fire shapes the pine barrens vegetative complex and appears to enhance the species diversity of the floristic assemblage, demonstrating the selective action of this natural process (McCormick, 1998). The lack of fire favors the development of a climax forest of pitch pine and scrub oak trees.

Sandplain Grasslands

Another floral community type in the uplands maintained by fire, as well as by grazing, is the sandplain grassland complex. Consisting of treeless grasslands, this community occupies several areas of the highly porous sandy deposits of the uplands. However, increasing development poses a long-term threat to this habitat. Species of plants commonly reported in the sandplain grasslands include the little blue-stem

(*Schizachyrium scoparium*), sandplain gerardia (*Agalinis acuta*), bird's foot violet (*Viola pedata*), and New England blazing star (*Liatris scariosa* var. *novae-angliae*) (Malpass and Geist, 1996).

VERNAL POOLS AND COASTAL PLAIN POND SHORES

The Mashpee outwash plain is marked by numerous water-filled depressions (i.e., kettles) formed during the Wisconsinan glacial epoch. Many of these depressions are vernal ponds that fill with freshwater during the winter and spring but often dry out in summer due to excessive heat and evaporation. Although these ponds may be seasonally ephemeral, they provide valuable habitat for numerous anurans and other organisms. Several amphibian species breed here and thus depend on the habitat for successful reproduction. The yellow-spotted salamander (*Ambystoma maculatum*) is one such species. Examples of other anurans that breed in vernal ponds are the American toad (*Bufo americanus*), green frog (*Rana clamitans melanota*), and red-spotted newt (*Notophthalmus viridescens viridescens*).

The shoreline and surrounding areas of the vernal ponds are also important feeding and resting sites for many organisms. Similar habitat values exist in the perimeter areas of coastal plain ponds, such as at Achumet Pond and Caleb Pond. These groundwater-fed ponds are less transitory than the vernal ponds. Rare species habitats typically surround them (Malpass and Geist, 1996).

RIPARIAN HABITATS

Willows (*Salix* spp.), alder (*Alnus rugosa*), and other low-lying vegetation inhabit banks and moist perimeter areas of coastal plain streams in the Waquoit Bay watershed. These plants grade into border forests of pitch pine (*Pinus rigida*) and scrub oak (*Quercus ilicifolia*). Phreatophytic vegetation proliferates in the moist soils of the riparian zone, which is characterized by thick shrub vegetation.

While the coastal plain streams support an array of algal and vascular plant species, numerous invertebrates, various finfish populations (e.g., eastern brook trout, *Salvelinus fontinalis*; white sucker, *Catostomus commersoni*; white perch, *Morone americana*; blueback herring, *Alosa aestivalis*; and alewife, *A. pseudoharengus*), insects (e.g., mosquitos, caddisflies, and mayflies), and other organisms, the surrounding land areas serve as important habitat for anurans (frogs and toads), reptiles (snakes and turtles), small mammals (e.g., rabbits, raccoons, squirrels, and skunks), and birds (waterfowl, song birds, and raptors). These riparian habitats provide protection and rich sources of food for numerous fauna. Many species also nest and reproduce here (Malpass and Geist, 1996).

FRESHWATER WETLANDS

The common cattail (*Typha latifolia*) and common reed (*Phragmites australis*) dominate many freshwater wetland areas in the Waquoit Bay watershed. Other plant species frequently encountered in these habitats are the sheep laurel (*Kalmia angustifolia*), sweet gale (*Myrica gale*), and twig rush (*Cladium marascoides*). *Sphagnum* sp. is likewise a significant constituent. As is the case for riparian habitats in the watershed, freshwater wetlands support a wide variety of reptilian, mammalian, and avian species, which use these habitats for feeding, breeding, reproduction, and loafing activities.

A number of ponds, cranberry bogs, streams, and rivers in the Waquoit Bay NERR are bordered by luxuriant freshwater marshes. For example, freshwater marshes harboring diverse assemblages of plant and animal species occur along the shoreline of Johns Pond north of the bay and parts of South Cape Beach State Park. They continue to the south on the Childs River, which originates at Johns Pond. In addition to these areas, freshwater marshes also abut Ashumet, Bourne, Snake, and Fresh Ponds north of the bay, as well as Grassy, Flashy, and Martha's Ponds. Other freshwater marsh habitat can be found along the perimeter of the Quashnet River and Red Brook. Cranberry bogs and marginal areas of kettle hole ponds likewise support freshwater marshes (Malpass and Geist, 1996).

SALT MARSHES

The Waquoit Bay NERR includes ~120 ha of salt marsh habitat, primarily at the head of Eel Pond and Waquoit Bay, in shoreline areas of Washburn Island, at the mouths of the Childs and Moonskis Rivers, and at the head of the Great River, as well as at Jehu, Sage Lot, and Hamblin Ponds. Smooth cordgrass (*Spartina alterniflora*) dominates the low marsh intertidal zone, and salt marsh hay (*S. patens*) predominates in the high marsh zone. Tidal action is a major controlling factor. Low marsh develops in protected areas subjected to semidiurnal tidal inundation, whereas high marsh forms at greater elevations affected only by extreme high tide (Malpass and Geist, 1996).

Although the low marsh appears to be comprised of monotypic stands of *Spartina alterniflora*, sea lavender (*Limonium nashii*) and glassworts (*Salicornia* spp.) may also occur in this habitat. Aside from *Spartina patens* and *Salicornia* spp., the most common species of plants observed in the high marsh include the spike grass (*Distichlis spicata*), black rush (*Juncus gerardi*), and marsh elder (*Iva frutescens*) (Malpass and Geist, 1996). Howes and Teal (1990) have compiled a comprehensive list of salt marsh species in the Waquoit Bay NERR (Table 2.1). They describe three distinct types of salt marsh wetlands in the reserve complex. The most expansive salt marshes occur at Hamblin Pond and Jehu Pond. At these sites, plant zonations and transition zones are broader than at other locations in the system. Species diversity is also greater here. Salt marsh habitat is likewise more extensive, and species diversity is greater along rivers than in the main body of the bay. Salt marshes surrounding the bay are spatially restricted with narrow plant zonations.

MUDFLATS AND SANDFLATS

Tidal flats are not well developed in the Waquoit Bay estuarine system, mainly because the tidal range only averages ~0.5 m. However, tidal flats are conspicuous in three areas (Malpass and Geist, 1996):

TABLE 2.1
Salt Marsh Plants Occurring in the Waquoit Bay Estuarine System[a]

Common Name	Scientific Name
Salt marsh cordgrass	*Spartina alterniflora*
Salt reed grass	*Spartina cynosuroides*
Salt marsh hay	*Spartina patens*
Spike grass	*Distichlis spicata*
Black rush	*Juncus gerardi*
Glasswort	*Salicornia europa*
Glasswort	*Salicornia bigelovii*
Woody glasswort	*Salicornia virginica*
Sea lavender	*Limonium carolinianum*
Chair-maker's rush	*Scirpus americanus*
Salt marsh bullrush	*Scirpus maritimus*
Robust bulrush	*Scirpus robustus*
Seaside goldenrod	*Solidago sempervirens*
Marsh elder	*Iva frutescens*
Halberd-leaved orach	*Atriplex patulah*
Reed grass	*Phragmites communis*
Dusty miller	*Artemisia stelleriana*
Narrow leaved cattail	*Typha angustifolia*
Salt marsh fleabane	*Pluchea purpurascens*
Poison ivy	*Rhus radicans*
Beach grass	*Ammophila breviligulata*
Beach pea	*Lathhyrus japonicus*
Salt marsh aster	*Aster tenufolius*
Bayberry	*Myrica pensylvanica*
Salt spray rose	*Rosa rugosa*

[a] Species listed in order of emergence.

Source: Howes, B.G. and J.M. Teal. 1990. Waquoit Bay — A Model Estuarine Ecosystem: Distribution of Fresh and Salt Water Wetland Plant Species in the Waquoit Bay National Estuarine Research Reserve. Final Technical Report, National Oceanic and Atmospheric Administration, Washington, D.C.

1. At the eastern shore of Washburn Island
2. At the eastern shore of the head of the bay
3. At the outlet of the Moonakis River

These protean habitats support a wide array of bivalves, gastropods, polychaetes, crustaceans, and other invertebrates. Among the most notable species encountered in these habitats are the gem clam (*Gemma gemma*), soft-shelled clam (*Mya arenaria*), and hard-shelled clam (*Mercenaria mercenaria*). Burrowing amphipods (*Corophium* sp.) build U-shaped tubes in the sediments. The horseshoe crab (*Limulus polyphemus*) and blue crab (*Callinectes sapidus*) also frequent these environments. Polychaetes observed burrowing in tidal flat sediments include such forms as clam worms (*Nereis virens*) and capitellids (e.g., *Heteromastus filiformis*). Abundant infaunal species constitute a rich food supply for birds and other wildlife (Whitlach, 1982).

BEACHES AND DUNES

Two barrier beaches lie at the seaward end of Waquoit Bay, one extending eastward from the southern margin of Washburn Island and the other extending westward from South Cape Beach. Together, they stretch for more than 40 ha, enclosing most of Waquoit Bay and Eel Pond. Two jetties have been constructed on the east and west sides of the main inlet to Waquoit Bay. The two barrier beaches are highly dynamic features, which are constantly modified by the action of wind, waves, tides, and currents. Major storms and heavy winds periodically cause the overwash of sediment into the back beach and lower bay areas, resulting in shoaling of the lower bay areas (Geist, 1996a).

Plants trap windblown sand and promote the development of dunes on the barrier beaches. This process creates important habitat. Beach grass (*Ammophila breviligulata*) is an initial colonizer and dune stabilizer. Beach heather (*Hudsonia tomentosa*), beach pea (*Lathyrus japonicus* var. *glaber*), seaside goldenrod (*Solidago sempervirens*), and dusty miller (*Artemisia stelleriana*) are also important primary dune stabilizers along the barrier beaches (Cullinan and Botelho, 1990). Back dune areas harbor beach plum (*Prunus maritima*), bayberry (*Myrica pensylvanica*), salt spray rose (*Rosa rugosa*), and poison ivy (*Rhus radicans*).

The dunes and associated vegetation form valuable habitat for shorebirds that forage, rest, reproduce, and nest on the barrier beaches. For example, herring gulls (*Larus argentatus*), laughing gulls (*L. atricilla*), and roseate terns (*Sterna dougallii*) forage along the beaches. Other species commonly rest here, including greater black-backed gulls (*L. marinus*), ring-billed gulls (*L. delawarensis*), and various species of terns (e.g., common terns, *S. hirundo*; least terns, *S. antillarum*; and Arctic terns, *S. paradisaea*). Least terns also use barrier beach habitats for breeding. Common terns and piping plovers (*Charadrius melodus*) utilize these habitats for nesting. Other shorebird species frequently observed on the barrier beaches are the semipalmated plover (*C. semipalmatus*), black-bellied plover (*Pluvialis squatarola*), willet (*Catotrophorus semipalmatus*), dunlin (*Calidris alpina*), least sandpiper (*C. minutilla*), semipalmated sandpiper (*C. pusilla*), sanderling (*C. alba*), short-billed dowitcher (*Limnodromus griseus*), ruddy twinstone (*Arenaria interpres*), lesser yellowlegs (*Tringa flavipes*), and greater yellowlegs (*T. melanoleuca*). Migrating shorebirds that stop over on Waquoit Bay beach and dune habitats during the spring and fall generally gain a significant amount of weight by foraging heavily in nearby coastal and estuarine waters. Waterfowl (e.g., buffleheads, *Bucephala albeola*; eiders, *Somateris mollissima*; scoters, *Melanitta* sp.; and mergansers, *Mergus serrator*) often utilize the bay habitats as well, especially during the winter months (Malpass and Geist, 1996).

The ongoing sea level rise associated with eustatic and isostatic changes and its effect on the long-term condition of the barrier beaches, salt marshes, and back bay waters of the system are a growing concern. The relative sea level rise in the Waquoit Bay area amounts to ~3 mm/yr, with land subsidence responsible for about two-thirds of this increase and eustatic sea level rise responsible for the remaining one-third (Giese and Aubrey, 1987). The barrier beaches are responding to the rising sea level by slowly migrating landward; the net movement of sand is from the forebeach to the back beach zone via wave and overwash action. Salt marshes behind the barrier beaches are also slowly migrating landward despite accretion rates in Waquoit Bay ranging from 2.8 to 4.6 mm/yr (Orson and Howes, 1992). Another result of rising sea level, according to Orson and Howes (1992), is the formation of freshwater swamps and bogs (e.g., at South Cape Beach). Greater human development and attempts to stabilize coastal features, however, act in opposition to dynamic natural forces shaping the beach and dune environment and the back-bay areas.

ESTUARY

Floral and faunal communities are rich and diverse in Waquoit Bay and contiguous tidal creeks and channels. Salt ponds (e.g., Sage Lot, Jehu, and Hamblin Ponds) also support numerous organisms. Benthic algae, phytoplankton, zooplankton, benthic invertebrates, finfish, and shellfish are well represented. Several species are of considerable recreational or commercial importance, such as the American eel (*Anguilla rostrata*), winter flounder (*Pseudopleuronectes americanus*), hard clam (*Mercenaria mercenaria*), soft clam (*Mya arenaria*), and bay scallop (*Argopecten irradians*) (Crawford, 1996a).

TIDAL CREEKS AND CHANNELS

Great River and Little River are two tidal creeks in the Waquoit Bay complex. Great Bay connects Waquoit Bay to Jehu Pond, and Little River links bay waters to Hamblin Pond. Tidal creeks also feed Bog Pond and Caleb Pond, as well as Sage Lot Pond.

Malpass and Geist (1996) discussed the benthic flora and fauna as well as the fish assemblages occurring in the tidal creeks and channels. Benthic macroalgae are observed along the bottom of the tidal creeks and channels. While some macroalgal species drift passively over tidal creek floors (e.g., *Ulva lactuca* and *Cladophora vagabunda*), other, attached forms (e.g., *Codium fragile* and *Fucus* spp.) are anchored to the bottom. *C. fragile* often attaches via a holdfast to shell substrate and other hard surfaces that lie on bottom sediments.

Common invertebrates in the tidal creek and channel habitats include barnacles (*Balanus* spp.), sea squirts (*Molgula manhattensis*), blue crabs (*Callinectes sapidus*), lady crabs (*Ovalipes ocellatus*), and mussels (*Geukensia demissa*). Other arthropods that may be encountered in these habitats are *Cymadusa compta*, *Erichsoniella filiformis*, *Hippolyte zostericola*, *Microdeutopus gryllotalpa*, *Neopanope texana*, and *Palaemonetes vulgaris*. Polychaetes (e.g., *Scoloplos fragilis*) and echinoderms (e.g., *Leptosynapta* sp. and *Sclerodactyla briarias*) may also be found in the tidal creeks and channels.

Forage fishes (e.g., mummichogs, *Fundulus heteroclitus*; striped killifish, *Fundulus majalis*; Atlantic silversides, *Menidia menidia*; and sheepshead minnows, *Cyprinodon variegatus*) dominate fish assemblages in the tidal creeks and channels. These species spend most of their lives in these habitats. Other fish species residing in these waters are those forms utilizing the habitat as a nursery area. Examples are the bluefish (*Pomatomus saltatrix*), Atlantic menhaden (*Brevoortia tyrannus*), and tautog (*Tautoga onitis*).

WAQUOIT BAY

Environment

As a shallow coastal system, Waquoit Bay is highly responsive to local meteorological conditions, and it thus exhibits relatively large seasonal changes in water temperature. Over an annual period, water temperature in the bay ranges from near 0°C to >25°C. Salinity, in turn, ranges from <10‰ to ~32‰. Bottom sediments consist of silt and clay in deeper areas of the central bay, while coarser sands and shell predominate elsewhere in the system, particularly in nearshore habitats (Valiela et al., 1990; Ayvazian et al., 1992; Crawford, 2002).

Organisms

Benthic Organisms

Eelgrass (*Zostera marina*) once covered much of the Waquoit Bay bottom, but progressive eutrophication and disease during the past several decades have essentially eliminated the beds in the bay (Crawford, 2002). In contrast, benthic macroalgae (e.g., *Cladophora vagabunda* and *Gracilaria tikvahiae*) have become increasingly more abundant in the bay, carpeting extensive areas of the bottom (D'Avanzo and Kremer, 1994). Valiela et al. (1992) reported that the annual mean biomass of macroalgae in the Childs River exceeds 300 g/m^2. This subestuary of the bay, bordered by the highest housing density in the area, receives elevated nutrient loads, which enhance algal growth. Greater inputs of nutrients also increase phytoplankton production and epiphytic growth in the bay; this accelerated plant growth leads to shading of the benthos, further impacting submerged aquatic vegetation.

Macroalgal mats have become the dominant bottom-dwelling plant forms in the estuary complex. Dense mats of the filamentous green macroalga, *Cladophora vagabunda*, and the filamentous red macroalga, *Gracilaria tikvahiae*, predominate. Both of these algal species form thick floating mats that drift above the bay bottom (Hersh, 1996). The extensive mats have created a relatively new habitat type in the estuary. Other commonly occurring green algae in the system include *Codium fragile*, *Enteromorpha* spp., and *Ulva lactuca*. Aside from *G. tikvahiae*, several additional red macroalgal species (*Agardhiella tenera*, *Chondras crispus*, *Polysiphonia urceolata*, and *Grinnellia americana*) have been reported in the estuary. Brown macroalgae of note are *Petroderma maculiforme*, *Pseudolithoderma* spp., *Fucus* spp., *Laminaria agardhii*, and *Ralfsia* spp.

Table 2.2 is a list of invertebrates identified in the Waquoit Bay complex. Eelgrass once provided a major habitat for many of the species, but its disappear-

TABLE 2.2
Estuarine Invertebrates Identified in the Waquoit Bay Complex

Annelids

Autolytus sp.
Capitella spp.
Cirratulus grandis
Eteone lactea
Hypaniola gray
Mediomastus ambiseta
Nereis arenaceodonta
Nereis grayi
Nereis virens
Parahesione luteola
Podarke obscura
Polydora cornuta
Polydora ligni
Prionospio heterobranchia
Prionospio spp.
Sabella micropthalma
Scolecolepides viridis
Scoloplos fragilis
Stauronereis rudolphi
Tharyx sp.

Mollusks

Anachis sp.
Bittium alternatum
Anadara ovalis
Crepidula fornicata
Elysia chlorotica
Haminoea solitaria
Hydrobia tonenii
Littorina littorea
Polinices duplicatus
Lunatia heros
Mitrella lunata
Busycon canaliculatum
Busycon carica
Eupleura caudata
Urosalpinx cinerea
Nassarius obsoletus
Anomia simplex
Argopecten irradians
Crassostrea virginica
Ensis directus
Gemma gemma

Arthropods

Ampelisca vandorum
Ampelisca agassizi
Ampithoe longimana
Corophium insidiosum
Cymadusa compta
Gammarus mucronatus
Lysianopsis alba
Microdeutopus sp.
Microdeutopus gryllotalpa
Neopanope texana
Leucon americanus
Hippolyte zostericola
Palaemonetes vulgaris
Crangon septemspinosa
Pagurus longicarpus
Libinia dubia
Emerita talpoida
Callinectes sapidus
Carcinus maenas
Ovalipes ocellatus
Uca pugilator
Uca pugnax
Cyathura polita
Edotea triloba
Idotea baltica
Erichsoniella filiformis
Balanus improvisus
Balanus eburneus
Limulus polyphemus
Callipallene brevirostris

Echinoderms

Cucumaria pulcherrima
Leptosynapta sp.
Sclerodactyla briarias
Ophioderma brevispina

Nemerteans

Lineus ruber
Zygeupolia rubens

Platyhelminthes

Euplana polynyma

TABLE 2.2 (CONTINUED)
Estuarine Invertebrates Identified in the Waquoit Bay Complex

Mollusks
Geukensia demissa
Laevicardium mortoni
Mercenaria mercenaria
Mya arenaria
Mytilus edulis
Petricola pholadiformis
Spisula polynyma
Spisula solidissima
Loligo peali

Urochordates
Molgula manhattensis
Botryllus schlosseri
Amarouciun stellatum
Lysianopsis alba
Cyclichna occulata

Sponges
Haliclona loosanoffi

Source: Malpass, W. and M.A. Geist. 1996. In: *The Ecology of the Waquoit Bay National Estuarine Research Reserve,* Geist, M.A. (Ed.). Technical Report, Waquoit Bay National Estuarine Research Reserve, Waquoit, MA, pp. III-1 to III-26.

ance has had a marked impact on some of them. For example, the bay scallop (*Argopecten irradians*) has declined appreciably in abundance concomitant with the loss of eelgrass habitat. As a result, the hard clam (*Mercenaria mercenaria*) and soft clam (*Mya arenaria*) now dominate the bay shellfisheries. These species have exhibited improved growth in areas dominated by macroalgae (Chalfoun et al., 1994). Other invertebrate species relying heavily on eelgrass beds for food sources and protection from predators, however, have also been adversely affected by the disappearance of the plants.

Benthic macroalgae serve as protective habitat for various invertebrate species. Sogard and Able (1991), for example, demonstrated that sea lettuce (*Ulva lactuca*) is an important habitat for decapod crustaceans (i.e., blue crabs, *Callinectes sapidus*; sand shrimp, *Crangon septemspinosa*; and grass shrimp, *Palaemonetes vulgaris*) in areas of shallow New Jersey coastal bays lacking eelgrass. They showed that only one decapod species, *Hippolyte pleuracanthus*, was more abundant at eelgrass sites than at sea lettuce sites in the coastal bays. Both eelgrass and sea lettuce supported higher densities of decapod crustaceans than did adjacent unvegetated substrates.

The benthic invertebrate community of Waquoit Bay consists of a wide array of epifaunal and infaunal populations. Bivalves, gastropods, crustaceans, polychaetes, and echinoderms are well represented (Table 2.2). Among commonly occurring bivalves in the bay are hard clams, soft clams, razor clams (*Ensis directus*), and mussels (*Geukensia demissa* and *Mytilus edulis*). Gastropods of significance include whelks (*Busycon carica* and *B. canaliculatum*), moon snails (*Lunatia heros* and *Polinices duplicatus*), and slipper shells (*Crepidula fornicata* and *C. plana*). Barnacles (*Balanus eburneus* and *B. improvisus*), blue crabs (*Callinectes sapidus*), green crabs (*Carcinus maenus*), mud crabs (*Neopanope texana*), horseshoe crabs (*Limulus polyphemus*), spider crabs (*Libinia dubia*), sand shrimp (*Crangon septemspinosa*), and grass shrimp (*Palaemonetes vulgaris*) are important arthropods. A number of polychaetes attain relatively high abundance in some areas. Notable in this regard are *Capitella* spp., *Nereis* spp., *Polydora* spp., *Prionospio* spp., and *Mediomastus ambiseta* (Malpass and Geist, 1996).

Finfish

Waquoit Bay represents a transition zone for fish assemblages, harboring both warm-temperate and cold-water fish fauna (Ayvasian, 1992). As a result, the fish assemblages observed in Waquoit Bay are highly diverse (Table 2.3). Aside from the resident species that spend most of their lives in estuarine waters (e.g., Atlantic silverside, *Menidia menidia*; mummichog, *Fundulus heteroclitus*; and sheepshead minnow, *Cyprinodon variegatus*), other species found in the bay may be classified as warm-water migrants or cool-water migrants, as well as freshwater and marine strays. Diadromous forms (e.g., American eel, *Anguilla rostrata*; blueback herring, *Alosa aestivalis*; and alewife, *A. pseudoharengus*) are common during spawning migrations to freshwater. The most abundant species appear to be the smaller forage fishes (Malpass and Geist, 1996).

Some of the larger predatory species are those that enter the bay in spring and summer to feed on smaller prey. Examples are the bluefish (*Pomatomus saltatrix*) and striped bass (*Morone saxatilis*). Other marine forms use the estuarine waters as spawning, feeding, and nursery grounds during winter, and they leave in the spring and summer. The winter flounder (*Pseudopleuronectes americanus*) and scup (*Stenotomus chrysops*) provide examples. The bay serves as an important nursery for some marine species that are present almost exclusively as juveniles. Pollack (*Pollachius virens*), Atlantic tomcod (*Microgadus tomcod*), and white hake (*Urophycis tenuis*) fall into this category (Ayvasian et al., 1992). The diversity of fish species in the bay peaks during the warmer months of the year when many species enter the system to feed and spawn. Because Waquoit Bay represents a transition zone where both cold- and warm-water fishes with overlapping biogeographic ranges coexist, species diversity is enhanced (Ayvasian et al., 1992).

Although the absolute abundance of fishes in the bay varies considerably from year to year, the relative abundance is reasonably consistent. The most abundant forms, as mentioned previously, are small forage species, largely resident in the estuary, or young and juveniles of marine species that occur only seasonally but feed and grow rapidly. While some species are more widely distributed in the bay (e.g., bay anchovy, *Anchoa mitchilli*), others such as the threespine stickleback (*Gasterosteus aculeatus*) and fourspine stickleback (*Apeltes quadracus*) appear to be more spatially restricted and habitat specific. The result is a system heavily utilized by a wide variety of fish species (Malpass and Geist, 1996).

TABLE 2.3
Finfish Species Found in the Waquoit Bay Estuarine Complex

Common Name	Scientific Name
Marine Species	
Striped anchovy	*Anchoa hepsetus*
Pollack	*Pollachius virens*
Striped bass	*Morone saxatilis*
Black sea bass	*Centropristis striata*
Scup	*Stenotomus chrysops*
White mullet	*Mugil curema*
American sand lance	*Ammodytes americanus*
Northern searobin	*Prionotus carolinus*
Striped searobin	*Prionotus evolans*
Longhorn sculpin	*Myoxocephalus octodecemspinosus*
Summer flounder	*Paralichthys dentatus*
Windowpane flounder	*Scophthalmus aquosus*
Yellowtail flounder	*Limanda ferruginea*
Estuarine Resident Species	
Mummichog	*Fundulus heteroclitis*
Striped killifish	*Fundulus majalis*
Sheepshead minnow	*Cyprinodon variegatus*
Inland silverside	*Menidia beryllina*
Tidewater silverside	*Menidia peninsulae*
Oyster toadfish	*Opsanus tau*
Rainwater killifish	*Lucania parva*
Threespine stickleback	*Gasterosteus aculeatus*
Fourspine stickleback	*Apeltes quadracus*
Ninespine stickleback	*Pungitius pungitius*
Blackspotted stickleback	*Gasterosteus wheatlandi*
Northern pipefish	*Syngathus fuscus*
Northern kingfish	*Menticirrhus saxatilis*
Naked goby	*Gobiosoma bosci*
Rock gunnel	*Pholis gunnellus*
Grubby	*Myoxocephalus aenaeus*
Hogchoker	*Trinectes maculatus*
Northern puffer	*Sphoeroides maculatus*
Estuarine Nursery Species	
Bay anchovy	*Anchoa mitchilli*
Atlantic silverside	*Menidia menidia*
Atlantic menhaden	*Brevoortia tyrannus*
Atlantic herring	*Clupea harengus*
Atlantic tomcod	*Microgadus tomcod*
Atlantic needlefish	*Strongylura marina*
Bluefish	*Pomatomus saltatrix*

(continued)

TABLE 2.3 (CONTINUED)
Finfish Species Found in the Waquoit Bay Estuarine Complex

Common Name	Scientific Name
Tautog	*Tautoga onitis*
Cunner	*Tautogolabrus adspersus*
Striped mullet	*Mugil cephalus*
Winter flounder	*Pseudopleuronectes americanus*
White hake	*Urophycis tenuis*
Freshwater/Brackish Water Species	
Banded killifish	*Fundulus diaphanus*
Marsh killifish	*Fundulus confluentus*
White perch	*Morone americana*
Golden shiner	*Notemigonus crysoleucas*
Bridled shiner	*Notropis bifrenatus*
Blacknose shiner	*Notropis heterolepis*
White sucker	*Catostomus commersoni*
Tesselated darter	*Etheostoma olmstedi*
Eastern brook trout	*Salvelinus fontinalis*
Brown trout	*Salmo trutta*
Tiger trout	*Salvelinus fontinalis* × *Salmo trutta* (hybrid)
Brown bullhead	*Ameiurus nebulosus*
Pumpkinseed	*Lepomis gibbosus*
Largemouth bass	*Micropterus salmoides*
Diadromous Species	
American eel	*Anguilla rostrata*
Blueback herring	*Alosa aestivalis*
Alewife	*Alosa pseudoharengus*
American shad	*Alosa sapidissima*
Rainbow trout	*Osmerus mordax*
Adventitious Visitors	
Crevalle jack	*Caranx hippos*
Ladyfish	*Elops saurus*
Ballyhoo	*Hemiramphus brasiliensis*
Barrelfish	*Hyperoglyphe perciformis*
Atlantic cod	*Gadus morhua*
Lumpfish	*Cyclopterus lumpus*

Source: Malpass, W. and M.A. Geist. 1996. In: *The Ecology of the Waquoit Bay National Estuarine Research Reserve,* Geist, M.A. (Ed.). Technical Report, Waquoit Bay National Estuarine Research Reserve, Waquoit, MA, pp. III-1 to III-26.

ANTHROPOGENIC IMPACTS

Increasing development and human activities in the Waquoit Bay watershed during the past several decades have contributed significantly to the alteration of environmental conditions in Waquoit Bay (Gault, 1996; Geist, 1996b). Most significant has been excessive nitrogen loading via inputs from septic systems to groundwater that enters the bay. This nitrogen loading, as well as the influx from secondary sources, has been responsible for considerable estuarine eutrophication, manifested by the decline of eelgrass beds, accelerated growth of macroalgae, and episodes of hypoxia and anoxia due to high rates of benthic community respiration. Periods of dissolved oxygen depletion in summer, particularly in the upper reaches of the bay, have caused large episodic "kills" of fish and invertebrates, which are typically short-lived (1–2 days) and limited in extent (D'Avanzo and Kremer, 1994). Short and Burdick (1996) correlated the progressive loss and fragmentation of eelgrass beds to the degree of housing development and associated nitrogen loading in various estuarine subwatersheds. Valiela et al. (1990, 1992) noted that nutrient enrichment has had far-reaching effects on the Waquoit Bay ecosystem, altering the structure and function of biotic communities via bottom-up controls of the estuarine food web.

While eutrophication is the most serious environmental problem currently plaguing the Waquoit Bay estuarine complex, other anthropogenic factors also adversely affect the system. For example, the input of pathogens from malfunctioning septic systems has caused water quality degradation and the closure of shellfish beds, as demonstrated by impacted areas along the Moonakis River. Herbicides and pesticides used for lawn maintenance and agriculture constitute a source of organochlorine compounds that accumulate in the estuary via land runoff. Stormwater runoff transports other chemical contaminants, such as heavy metals, to the estuary as well (Gault, 1996). Motorboat engines are a source of aliphatic and aromatic hydrocarbons that concentrate in bottom sediments. Since more than 1000 boats operate in the bay, their aggregate effect can be significant (Crawford, 1996b, 2002). These particle-reactive contaminants also concentrate in the surface water microlayer, and they can result in both lethal and sublethal impacts on plants and animals exposed to them (Albers, 2002). The action of boat engine propellers roils bottom sediments and disturbs the surface water microlayer, facilitating the remobilization and dispersal of the contaminants (Kennish, 2002). Propeller dredging damages submerged aquatic vegetation, excavates bottom sediments, scars the substrate, and increases sediment resuspension and turbidity in the water column (Crawford, 1996b, 2002; Kenworthy et al., 2002). Maintenance dredging of inlet and channel areas likewise damages the benthic habitat, displaces benthic organisms, and remobilizes chemical constituents.

EUTROPHICATION

Valiela et al. (1997) examined nitrogen inputs to the Waquoit Bay watershed and estuary (Table 2.4). They determined that the principal sources of nitrogen to the watershed and estuary are atmospheric deposition, fertilizer use, and domestic wastewater. In terms of nitrogen loading to the watershed, atmospheric

TABLE 2.4
Nitrogen Loading Estimates from Atmospheric Deposition, Fertilizer, and Wastewater to Waquoit Bay and Losses during Transport through Different Land Covers of the Watershed[a]

Nitrogen Source	Nitrogen Input	Nitrogen Load (%) to Watershed	Nitrogen Input (%) Lost within Watershed	Total Nitrogen to Bay	Nitrogen Load (%) to Estuary
Atmospheric Deposition to:					
Natural vegetation	47,036	42	91	4,447	20
Turf	9,934	9	90	957	4
Cranberry bogs	1,713	1.5	90	165	0.8
Other agricultural land	90	0.08	90	9	0.04
Roofs and driveways	625	0.5	90	60	0.3
Roads	1,685	1.5	75	429	1.9
Ponds	894	0.8	56	394	1.8
Fertilizer Use:					
On lawns	7,019	6	84	1,095	5
On golf courses	5,889	5	84	918	4
On cranberry bogs	3,198	3	54	1,485	6.8
On other agricultural land	816	0.7	84	127	0.6
Wastewater	31,323	28	67	10,241	47
Ponds upgradient[b]	2,574	2	35	1,673	7
Grand total	112,797	100	81	22,000	99

[a] Values in kg/N/yr.
[b] Import from larger ponds and lakes deep enough to intercept the flow through the aquifer.

Source: Valiela, I., G. Collins, J. Kremer, K. Lajtha, M. Geist, B. Seely, J. Brawley, and C.H. Sham. 1997. *Ecological Applications* 7: 358–380

deposition accounts for the largest fraction (55%), followed by domestic wastewater (28%) and fertilizer (15%). However, most (90%) of the nitrogen derived from atmospheric input does not leave the watershed because only 29% of the nitrogen load entering the bay derives from this source. Nitrogen influx from septic systems in the watershed is quantitatively more significant, accounting for 47% of the total nitrogen load to the estuary. About 67% of septic system–derived nitrogen is lost in the watershed. Fertilizer-derived nitrogen comprises 16% of the total nitrogen load to the estuary, with 78% of the nitrogen lost during transport in the watershed. The breakdown of nitrogen loading (by source) to the Waquoit Bay watershed and estuary clearly indicates that much of the nitrogen in the watershed is lost through several processes (i.e., adsorption, uptake, volatilization, and denitrification) during travel in soils, subsoils, salt ponds, and the downgradient aquifer (Table 2.4).

With increasing development in the Waquoit Bay watershed after 1960, the number of septic system installations increased dramatically. The Waquoit Bay watershed does not have any centralized sewage treatment facility. Instead, each house constructed in the watershed has a septic system, which represents a potential source of additional nitrogen for the watershed and estuary. Fertilizer use in the watershed has also increased concomitantly with escalating development. The net effect has been greater nitrogen inputs to Waquoit Bay in recent years (Gault, 1996; Geist, 1996b).

Primary production by phytoplankton and macroalgae has increased substantially with greater nitrogen loading to the estuary (Valiela et al., 1992). Larger macroalgal biomass in the system has been detrimental to eelgrass beds, which rapidly declined between 1987 and 1992 (Short and Burdick, 1996). Thick mats (50–75 cm) of *Cladophora vagabunda* and *Gracilaria tikvahiae* overlie broad areas of the estuarine floor, effectively shading the benthos. Reduction or extinction of photosynthetically active radiation (PAR) by this process has contributed markedly to the demise of eelgrass in Waquoit Bay. Further, hypoxic conditions often develop along the bottom of the macroalgal mats and may threaten survival of benthic invertebrates and finfish (D'Avanzo and Kremer, 1994). Other factors that have also played roles are the shading effects of phytoplankton blooms, most evident in Jehu Pond and Great River, and epiphytic growth in Eel Pond (Short et al., 1992). Motorboat-induced sediment resuspension must be considered as well, although this is not likely to be a primary factor (Crawford, 2002).

The catastrophic effects of eutrophication on seagrass habitat in the Waquoit Bay system are well documented (Costa, 1988; Valiela et al., 1992; Short et al., 1995, 1996; Short and Burdick, 1996). Costa (1988) and Costa et al. (1992) detailed the timeline of eelgrass changes in the bay from 1951 to 1987. While eelgrass covered much of the Waquoit Bay bottom in 1951, it had disappeared in deeper areas by the mid-1960s. The decrease in eelgrass distribution continued through the late 1960s, and by the mid-1970s, most shallower areas of the bay were also devoid of this vascular plant. Short and Burdick (1996) reported a further decline of eelgrass in the bay through the early 1990s, as did Crawford (2002) through the mid-1990s. According to Crawford (2002), eelgrass disappeared from the bay proper about 1995. Today, only small patches of eelgrass remain in the estuarine system — in salt ponds and other protected, spatially restricted sites (C. Weidman, Waquoit Bay NERR, personal communication, 2002).

The loss of valuable eelgrass habitat has resulted in secondary biotic impacts such as the decline of shellfisheries in the bay, most notably bay scallops (*Argopecten irradians*) and blue crabs (*Callinectes sapidus*) (Crawford, 1996). However, other changes in the composition and structure of benthic faunal communities are evident. For example, Valiela et al. (1992) ascertained that, in the lower parts of Waquoit Bay where macroalgae flourish, benthic invertebrates exhibit lower abundance and species richness. Dense macroalgal canopies, therefore, can have profound effects on the viability and health of major biotic components of the Waquoit Bay system.

SUMMARY AND CONCLUSIONS

The Waquoit Bay NERR covers an area of nearly 1000 ha on the south coast of Cape Cod, Massachusetts, centered in the towns of Falmouth and Mashpee. The reserve consists of an array of watershed and estuarine habitats that are biologically productive. The Waquoit Bay watershed is characterized by a wide variety of lowland and upland habitats, including wetlands (fresh-, brackish-, and salt-water marshes), riparian habitats, mudflats and sandflats, barrier beaches and sand dunes, coastal plain pond shores, vernal pools, sandplain grasslands, pine barrens, and upland pitch pine/oak forests. The Waquoit Bay estuary is defined by tidal creeks and channels as well as open waters of the embayment.

Waquoit Bay remains a highly productive system despite ongoing eutrophication problems, which are largely responsible for the disappearance of eelgrass beds in the bay proper as well as altered structure and function of biotic communities. Phytoplankton and macroalgae are the dominant primary producers in the estuary. Both plant groups have been linked to shading impacts on eelgrass beds. Dense macroalgal mats of *Cladophora vagabunda* and *Gracilaria tikvahiae* have also been coupled to hypoxic and anoxic events and periodic "kills" of fish and invertebrate populations. The loss of eelgrass habitat has resulted in the decline of fishery resources in the bay, notably the bay scallop (*Argopecten irradians*).

Despite the demise of eelgrass in the Waquoit Bay estuary during the past several decades, the benthic invertebrate community is well established, as evidenced by the wide array of bivalves, gastropods, crustaceans, polychaetes, echinoderms, and other taxa represented in benthic samples. Both epifauna (e.g., *Balanus* spp. and *Molgula manhattensis*) and infauna (e.g., *Capitella* spp., *Prionospio* spp., and *Mercenaria mercenaria*) are relatively abundant in the system. Many of these organisms serve as a rich food supply for finfish and shorebird populations.

Fish assemblages in the estuary consist of resident species that spend most of their lives there, warm- and cool-water migrants, freshwater and marine strays, and diadromous forms. Forage species (e.g., bay anchovy, *Anchoa mitchilli*; mummichog, *Fundulus heteroclitus*; and sheepshead minnow, *Cyprinodon variegatus*) support many migratory fish (e.g., bluefish, *Pomatomus saltatrix* and striped bass, *Morone saxatilis*) present only seasonally in the bay. The absolute abundance of fish populations in Waquoit Bay is highly variable from year to year and likely reflects the flux of environmental conditions in the region.

Development in the Waquoit Bay watershed is largely responsible for nutrient overenrichment in the estuary, particularly nitrogen derived from septic systems. This nitrogen, together with nitrogen derived from lawn fertilizers and atmospheric deposition, promotes eutrophic conditions in the estuary. Other anthropogenic impacts on Waquoit Bay originate from stormwater runoff of chemical contaminants, such as organochlorine compounds and heavy metals, as well as the influx of hydrocarbons from boat engine emissions. More than 1000 boats operate on the bay, and their collective physical and chemical impacts may be significant. Finally, maintenance dredging modifies the benthic habitat and disrupts benthic communities. These impacts may extend to the plankton and nekton communities as well.

REFERENCES

Albers, P.H. 2002. Sources, fate, and effects of PAHs in shallow water environments: a review with special reference to small watercraft. *Journal of Coastal Research* Special Issue 37, pp. 143–150.

Ayvazian, S.G., L.A. Deegan, and J.T. Finn. 1992. Comparison of habitat use by estuarine fish assemblages in the Acadian and Virginian zoogeographic provinces. *Estuaries* 15: 368–383.

Cambareri, T.C., E.M. Eichner, and C.A. Griffeth. 1992. Submarine Groundwater Discharge and Nitrate Loading to Shallow Coastal Embayments. Proceedings of Focus, Eastern Regional Groundwater Conference, October 13–15, Newton, MA, pp. 1–23.

Chalfoun, A., J. McClelland, and I. Valiela. 1994. The effect of nutrient loading on the growth rate of two species of bivalves, *Mercenaria mercenaria* and *Mya arenaria*, in estuaries of Waquoit Bay, Massachusetts. *Biological Bulletin* 187: 281–283.

Costa, J.E. 1988. Eelgrass in Buzzards Bay: Distribution, Production, and Historical Changes in Abundance. Technical Report No. EPA 503/4-88-002, U.S. Environmental Protection Agency, Boston, MA.

Costa, J.E., B.L. Howes, A.E. Giblin, and I. Valiela. 1992. Monitoring nitrogen and indicators of nitrogen loading to support management action in Buzzards Bay. In: McKenzie, D.H., D.E. Hyatt, and V.J. McDonald (Eds.). *Ecological Indicators*. Elsevier Applied Science, Research Triangle Park, NC, pp. 499–531.

Crawford, R.E. 1996a. Example life histories of commercially important species found in Waquoit Bay. In: Geist, M.A. (Ed.). *The Ecology of the Waquoit Bay National Estuarine Research Reserve*. Technical Report, Waquoit Bay National Estuarine Research Reserve, Waquoit, MA, pp. IV-1 to IV-15.

Crawford, R.E. 1996b. Recreational boating impacts. In: Geist, M.A. (Ed.). *The Ecology of the Waquoit Bay National Estuarine Research Reserve*. Technical Report, Waquoit Bay National Estuarine Research Reserve, Waquoit, MA, pp. V-19 to V-22.

Crawford, R.E. 2002. Secondary wake turbidity from small boat operation in a shallow sandy bay. *Journal of Coastal Research* Special Issue 37, pp. 50–65.

Cullinan, M. and P.J. Botelho. 1990. South Cape Beach State Park, Mashpee, Massachusetts. Final Environmental Impact Report, Massachusetts Department of Environmental Management, Region 1, Boston, MA.

D'Avanzo, C. and J. Kremer. 1994. Diel oxygen dynamics and anoxic events in an eutrophic estuary of Waquoit Bay. *Estuaries* 18: 131–139.

Gault, C. 1996. Management issues in Waquoit Bay. In: Geist, M.A. (Ed.). *The Ecology of the Waquoit Bay National Estuarine Research Reserve*. Technical Report, Waquoit Bay National Estuarine Research Reserve, Waquoit, MA, pp. VI-1 to VI-20.

Geist, M.A. 1996a. The physical environment of Waquoit Bay. In: Geist, M.A. (Ed.). *The Ecology of the Waquoit Bay National Estuarine Research Reserve*. Technical Report, Waquoit Bay National Estuarine Research Reserve, Waquoit, MA, pp. II-1 to II-22.

Geist, M.A. 1996b. Estuarine ecosystem functioning in response to urban development. In: Geist, M.A. (Ed.). *The Ecology of the Waquoit Bay National Estuarine Research Reserve*. Technical Report, Waquoit Bay National Estuarine Research Reserve, Waquoit, MA, pp. V-1 to V-21.

Geist, M.A. and W. Malpass. 1996. An introduction to the reserve. In: Geist, M.A. (Ed.). *The Ecology of the Waquoit Bay National Estuarine Research Reserve*. Technical Report, Waquoit Bay National Estuarine Research Reserve, Waquoit, MA, pp. I-1 to I-14.

Giese, G. and D.G. Aubrey. 1987. Losing coastal upland to relative sea level rise: three scenarios for Massachusetts. *Oceanus* 30: 16–22.

Hersh, D. 1996. Abundance and Distribution of Intertidal and Subtidal Macrophytes in Cape Cod: The Role of Nutrient Supply and Other Controls. Ph.D. thesis, Boston University, Boston, MA.

Howes, B.G. and J.M. Teal. 1990. Waquoit Bay — A Model Estuarine Ecosystem: Distribution of Fresh and Salt Water Wetland Plant Species in the Waquoit Bay National Estuarine Research Reserve. Final Technical Report, National Oceanic and Atmospheric Administration, Washington, D.C.

Kennish, M.J. 2002. Sediment contaminant concentrations in estuarine and coastal marine environments: potential for remobilization by boats and personal watercraft. *Journal of Coastal Research* Special Issue 37, pp. 151–178.

Kenworthy, W.J., M.S. Fonseca, P.E. Whitfield, and K.K. Hammerstrom. 2002. Analysis of seagrass recovery in experimental excavations and propeller-scar disturbances in the Florida Keys National Marine Sanctuary. *Journal of Coastal Research* Special Issue 37, pp. 75–85.

Malpass, W. and M.A. Geist. 1996. Habitats and communities of the Waquoit Bay reserve. In: Geist, M.A. (Ed.). *The Ecology of the Waquoit Bay National Estuarine Research Reserve*. Technical Report, Waquoit Bay National Estuarine Research Reserve, Waquoit, MA, pp. III-1 to III-26.

McCormick, J. 1998. The vegetation of the New Jersey Pine Barrens. In: Forman, R.T.T. (Ed.). *Pine Barrens: Ecosystem and Landscape*. Rutgers University Press, New Brunswick, NJ, pp. 229–243.

Orson, R.A. and B.L. Howes. 1992. Salt marsh development studies at Waquoit Bay, Massachusetts: influence of geomorphology on long-term plant community structure. *Estuarine, Coastal and Shelf Science* 35: 453–471.

Short, F.T. and D.M. Burdick. 1996. Quantifying eelgrass habitat loss in relation to housing development and nitrogen loading in Waquoit Bay, Massachusetts. *Estuaries* 19: 730–739.

Short, F.T., D.M. Burdick, J. Wolf, and G.F. Jones. 1992. Declines of Eelgrass in Estuarine Research Reserves along the East Coast, U.S.A.: Problems of Pollution and Disease and Management of Eelgrass Meadows in East Coast Research Reserves. Technical Report, National Oceanic and Atmospheric Administration, Silver Spring, MD.

Short, F.T, D.M. Burdick, and J.E. Kaldy. 1995. Mesocosm experiments quantify the effects of eutrophication on eelgrass, *Zostera marina* L. *Limnology and Oceanography* 40: 740–749.

Short, F.T., D.M. Burdick, S. Granger, and S.W. Nixon. 1996. Long-term decline in eelgrass, *Zostera marina* L., linked to housing development. In: Kuo, J., R.C. Phillips, D.I. Walker, and H. Kirkman (Eds.). *Seagrass Biology*. Proceedings of an International Workshop, Rottnest Island, Western Australia, 25–29 January 1996, Nedlands, Western Australia, pp. 291–298.

Sogard, S.M. and K.W. Able. 1991. A comparison of eelgrass, sea lettuce macroalgae, and marsh creeks as habitats for epibenthic fishes and decapods. *Estuarine, Coastal and Shelf Science* 33: 501–519.

Valiela, I., J. Costa, K. Foreman, J.M. Teal, B. Howes, and D. Aubrey. 1990. Transport of groundwater-borne nutrients from watersheds and their effects on coastal waters. *Biogeochemistry* 10: 177–197.

Valiela, I., K. Foreman, M. LaMontagne, D. Hersh, J. Costa, P. Peckol, B. DeMeo-Anderson, C. D'Avanzo, M. Babione, C.-H. Sham, J. Brawley, and K. Lajtha. 1992. Couplings of watersheds and coastal waters: sources and consequences of nutrient enrichment in Waquoit Bay, Massachusetts. *Estuaries* 15: 443–457.

Valiela, I., G. Collins, J. Kremer, K. Lajtha, M. Geist, B. Seely, J. Brawley, and C.H. Sham. 1997. Nitrogen loading from coastal watersheds to receiving waters: new method and application. *Ecological Applications* 7: 358–380.

Whitlach, R.B. 1982. The Ecology of New England Tidal Flats: A Community Profile. Technical Report No. FWS/OBS-81/01, U.S. Fish and Wildlife Service, Washington, D.C.

Case Study 2

3 Jacques Cousteau National Estuarine Research Reserve

INTRODUCTION

The Jacques Cousteau National Estuarine Research Reserve (JCNERR) is the 22nd program site of the National Estuarine Research Reserve System (NERRS). It was officially dedicated on October 20, 1997. The reserve, which covers an area of more than 45,000 ha, lies along the south-central New Jersey coastline about 15 km north of Atlantic City (Figure 3.1). The terrestrial and aquatic habitats are highly diverse, ranging from upland pine–oak forests and woodland swamps in the alluviated stream valleys of the New Jersey Pinelands to tidal marshes and open estuarine and coastal waters. Only 553 ha of developed landscape (>1% of the area) occur in the reserve. Forest cover and marsh habitat account for an additional 4616 ha (~10% of the reserve area) and 13,034 ha (>28% of the reserve area), respectively. The most extensive habitat is open water; it spans 27,599 ha (~60% of the reserve area).

Because of sparse development in watershed areas of the reserve as well as the bordering New Jersey Pinelands, the JCNERR exhibits exceptional environmental quality. Nearly all of the land area surrounding open waters of the reserve is in public ownership. It mainly consists of state wildlife management areas, state forests, and federal reserves. The open waters of lower Barnegat Bay, Little Egg Harbor, Great Bay, Mullica River, and the back-bays (i.e., Little Bay, Reeds Bay, and Absecon Bay) as far south as Absecon support rich populations of finfish, shellfish, and wildlife. Similarly, numerous organisms, including some endangered and threatened species, inhabit tidal creeks along fringing *Spartina* marshes, as well as brackish and freshwater marshes to the west. The seaward part of the reserve extends to the barrier islands (dune and beach habitats) and open waters of the adjacent inner continental shelf out to the Long-Term Ecosystem Observatory (LEO-15), a 2.8 km^2 offshore research platform of Rutgers University located about 9 km offshore of Little Egg Inlet, which is designed to continuously sample and sense the local marine environment. The JCNERR is the only reserve system with such seaward boundaries in the Atlantic Ocean (Figure 3.1).

Although biotic communities of the coastal bays in the JCNERR are replete with numerous species of planktonic, nektonic, and benthic organisms, a limited number of taxa often predominate in terms of total abundance. For example, copepods generally dominate the zooplankton community in the Mullica River–Great Bay Estuary, with *Acartia tonsa*, *Eurytemora affinis*, and *Oithona similis* the most abundant species. Nearly 150 species of benthic fauna occur in this system. In

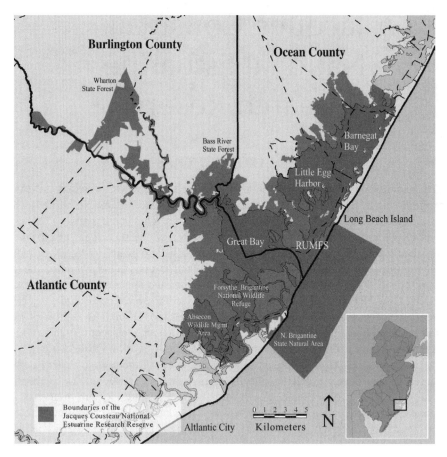

FIGURE 3.1 Map showing the location of the Jacques Cousteau National Estuarine Research Reserve.

addition, more than 60 species of fish inhabit the estuary as well (Durand and Nadeau, 1972; Able et al., 1996; Szedlmayer and Able, 1996; Jivoff and Able, 2001; Kennish, 2001a–c). The U.S. Fish and Wildlife Service (1996) recorded 275 species of macroinvertebrates, 91 species of fish, and 350 species of algae in inland habitats of the Mullica River and its tributaries. Watershed areas of the JCNERR support many species of shorebirds, wading birds, waterfowl, raptors, and songbirds. Amphibians, reptiles, and land mammals also utilize wetlands, riparian buffer, and upland habitats of the JCNERR and contiguous pinelands (Zampella et al., 2001).

Rutgers University (Institute of Marine and Coastal Sciences) oversees research and monitoring in the JCNERR. Other partners in the reserve include Richard Stockton College of New Jersey, the New Jersey Department of Environmental Protection (Division of Fish, Game, and Wildlife at Nacote Creek), the U.S. Fish and Wildlife Service, Tuckerton Seaport, and the Pinelands Commission. These partners are interacting to assess water quality and habitat conditions in the coastal bays and neighboring watershed areas of the JCNERR.

ENVIRONMENTAL SETTING

The JCNERR site lies in the gently sloping Atlantic Coastal Plain and is characterized by low and relatively flat terrain. The Mullica River Basin, which covers an area of 1474 km^2, borders most of the JCNERR coastal bays along their western perimeter, and the barrier island complex forms the eastern boundary for these water bodies. Several major tributaries of the Mullica River drain surrounding land areas of the pinelands. These are the Hammonton Creek, Nescochague Creek, Sleeper Branch, Atsion (Upper Mullica) River, Batsto River, Wading River, Oswego River, Bass River, and Lower Mullica River. The Batsto River, Atsion (Upper Mullica) River, Sleeper Branch, and Nescochague Creek join near the town of Batsto to form the main stem of the Mullica River. Mean monthly streamflow of the Mullica River ranges from ~1.7 to 4.2×10^8 l/d (Rhodehamel, 1998). Base flow accounts for most of this flow, which discharges along the northwest side of Great Bay.

Several smaller volume streams that flow through the lower Barnegat Bay watershed to the north discharge into Little Egg Harbor. These include Tuckerton Creek, Westecunk Creek, Cedar Run, and Mill Creek. Parker Run, Dinner Point Creek, Ezras Creek, and Thompson Creek also occur in the lower Barnegat Bay watershed and terminate near the upland–salt marsh boundary. Absecon Creek, located approximately 12 km south of Great Bay, drains into the shallow waters of Absecon Bay.

The Mullica River and lower Barnegat Bay watershed areas consist largely of sandy, siliceous, and droughty soils with low concentrations of nutrients. The porous substrate enables rainfall to percolate rapidly down to the shallow water table, thereby limiting surface water runoff. Along estuarine shorelines and surrounding wetlands, however, organic-rich soils and thick layers of peat contrast markedly with the upland soils.

Temperate climatic conditions dominate New Jersey coastal areas. At the JCNERR, air temperatures average 0 to 2.2°C in winter and 22 to 24°C in summer. Northwesterly winds predominate from December through March. Winds progressively shift directions in the spring; from late spring through summer, southerly winds prevail. Sea breezes usually reduce air temperatures at the JCNERR during the summer months. Wind speeds are generally less than 15 km/h at the reserve site. Precipitation is well distributed year-round, amounting to a total of ~100 to 125 cm/yr. Northeasters, extratropical storms, and hurricanes occasionally deliver large amounts of precipitation (10 cm or more) in relatively short periods of time. These storms can cause significant flooding and erosion problems (Forman, 1998).

Several distinct tidal water bodies with unique physical and hydrologic characteristics occur in the JCNERR (i.e., Lower Barnegat Bay, Little Egg Harbor, Great Bay, Little Bay, Reeds Bay, and Absecon Bay). They form a backbarrier lagoon system separated from the Atlantic Ocean by a Holocene barrier island complex that is breached at Little Egg Inlet, Brigantine Inlet, and Absecon Inlet. The Mullica River–Great Bay Estuary is a drowned river valley that communicates directly with the Atlantic Ocean through Little Egg Inlet. Lower Barnegat Bay, Little Egg Harbor, Little Bay, Reeds Bay, and Absecon Bay are shallow coastal back-bays behind stabilized barrier island units. Little Bay, Reeds Bay, and Absecon Bay comprise the smallest lagoon-type estuaries in the JCNERR.

The shallow microtidal estuaries of the JCNERR are polyhaline embayments with mean depths of less than 2 m. Because they are extremely shallow, the estuaries are highly responsive to air temperatures. Over an annual cycle, water temperatures in the coastal bays range from ~2 to 30°C. Salinity, in turn, ranges from ~10 to 32‰.

MULLICA RIVER–GREAT BAY ESTUARY

Tidal influence extends a considerable distance up streams and rivers in the Mullica River watershed. For example, in Pine Barrens streams the salt water–freshwater interface typically occurs 8 to 16 km upstream of the head of the bay. While tidal effects are evident over the lower 40 km of the Mullica River, the upper limit of salt water inundation is at Lower Bank located ~25 km upstream of the head of Great Bay. Hence, Lower Bank marks the upper end of the Mullica River–Great Bay Estuary, and a well-defined salinity gradient is observed from near 0‰ upriver of Lower Bank to >30‰ at Little Egg Inlet. Along the Mullica River, the type of marsh vegetation encountered reflects the gradual increase in salinity levels downestuary. Freshwater tidal marshes along tributary streams and the headwaters of the Mullica River give way to brackish marshes downriver and extensive (*Spartina*) salt marshes near the river mouth and along the perimeter of Great Bay.

Water circulation in Great Bay follows a counterclockwise pattern. Tidal currents (>2 m/sec) enter at Little Egg Inlet and flow along the northern part of the bay. Water discharging from the Mullica River flows along the southern part of the bay (Durand, 1988). A counterclockwise gyre occurs in the central region. Periodic episodes of coastal upwelling inject cold, high-density seawater into the bay from the continental shelf. The Institute of Marine and Coastal Sciences of Rutgers University recorded 12 coastal upwelling events in 2000 at the LEO-15 site in the JCNERR.

Sediments in the eastern bay, which originate mainly from marine sources, consist of large amounts of well-sorted fine sand. Sediments transported into the bay through Little Egg Inlet tend to accumulate in sand bars (tidal deltas) landward of the inlet. In the western part of the bay, the amounts of silt and clay increase appreciably. These finer sediments largely derive from discharges of the Mullica River and shoreline marshes. Sediments entering the bay from marine and land sources also accumulate in sandflats and mudflats, which cover more than 1300 ha in the system (U.S. Fish and Wildlife Service, 1996). In addition, sediments derived from land-based sources promote accretion of salt marsh habitat bordering the estuary.

Chant (2001) showed that coastal pumping, remotely forced by coastal sea level, is the predominant factor controlling subtidal motion in coastal bays of the JCNERR. For example, he attributed 70% of subtidal motion in Little Egg Harbor to this process. Little Egg Harbor is a shallow (1 to 7 m), irregularly shaped tidal basin with tidal currents less than 1 m/sec. Weak salinity and thermal stratification characterize this system.

WATER QUALITY

The Mullica River–Great Bay Estuary has been the target of a number of water quality studies (Durand and Nadeau, 1972; Zimmer, 1981; Durand, 1988, 1998;

Zampella, 1994; Dow and Zampella, 2000; Kennish and O'Donnell, 2002). Zampella (1994) and Dow and Zampella (2000) correlated decreasing water quality in the Mullica River watershed with increasing development. They reported a gradient of increasing pH, specific conductance, and nutrients (i.e., total nitrite and nitrate as nitrogen, total ammonia as nitrogen, and total phosphorus) along a watershed disturbance gradient of increasing development, agricultural land-use intensity, and wastewater flow in the Mullica River drainage basin. Areas of degraded water quality have been shown to alter the structure and function of affected biotic communities (Zampella and Laidig, 1997; Zampella and Bunnell, 1998). Hunchak-Kariouk et al. (2001) and Lathrop and Conway (2001) have likewise documented degraded water quality in areas of high development in the Barnegat Bay watershed.

Nutrient concentrations are relatively low in streams discharging to the coastal bays of the JCNERR. Nitrate is the primary limiting nutrient to plant growth in the coastal bays. In the Mullica River, nitrogen levels are as follows: ammonium (0 to <10 µgat N/l), nitrate (0 to >70 µgat N/l), nitrite (0 to <2 µgat N/l), and total organic nitrogen (0 to >60 µgat N/l). Phosphate concentrations, in turn, range from 0 to <5 µgat P/l (Durand and Nadeau, 1972; Zimmer, 1981; Durand, 1988, 1998; Zampella, 1994).

Water quality in the estuary has been investigated most intensely since initiation of the JCNERR System-wide Monitoring Program (SWMP) in August 1996. Rutgers University scientists deployed Yellow Springs Instrument Company (YSI™) Model 6000 UPG data loggers at the following locations in the JCNERR during the summer and fall of 1996:

1. Buoy 126 in Great Bay (August)
2. Buoy 139 in Great Bay (August)
3. Chestnut Neck in the Mullica River (September)
4. Lower Bank in the Mullica River (October)

They subsequently deployed three additional data loggers at Little Sheepshead Creek (April 1997), Nacote Creek (May 1997), and Tuckerton Creek (November 1998). These instruments record six water quality parameters (water temperature, salinity, dissolved oxygen [mg/l and % saturation], pH, turbidity, and depth) semicontinuously (i.e., every 30 min). While the instruments operate unattended in the field, they must be periodically reprogrammed and calibrated. At these times, approximately every 2 weeks, data stored in internal memory are uploaded to a personal computer and later analyzed. Except during icing periods in winter, the data loggers are deployed year-round at each monitoring site.

The most continuous and complete water quality database developed from data logger deployment exists for Buoy 126, Chestnut Neck, and Lower Bank. Buoy 139 was discontinued as a monitoring site in July 1999; however, it was reinstituted as a monitoring site in June 2002. The Buoy 126, Chestnut Neck, and Lower Bank SWMP monitoring sites are important because they lie along the salinity gradient of the Mullica River–Great Bay Estuary (Figure 3.2).

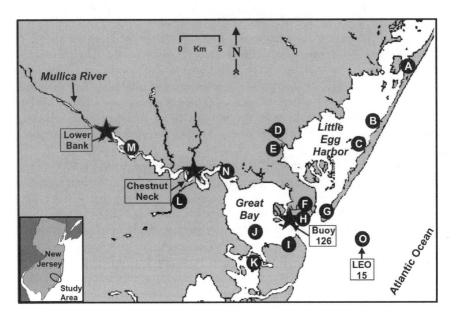

FIGURE 3.2 Map showing temporary and permanent water quality monitoring sites in the Jacques Cousteau National Estuarine Research Reserve.

Figure 3.3 through Figure 3.9 show measurements of physical–chemical parameters by the data loggers at the three aforementioned SWMP sites during the 1999–2000 study period. Temperatures at this time ranged from −1.7 to 27.9°C at Buoy 126, −1.3 to 29.4°C at Chestnut Neck, and 0.7 to 31.5°C at Lower Bank. A conspicuous seasonal temperature cycle characteristic of mid-latitude estuarine systems is evident (Figure 3.3). Polyhaline conditions predominate at Buoy 126, mesohaline conditions at Chestnut Neck, and oligohaline conditions at Lower Bank (Figure 3.4). Mean salinities at Buoy 126, Chestnut Neck, and Lower Bank for the study period amounted to 29.5‰, 15.1‰, and 2.6‰, respectively. Salinity differences at the three sites were statistically significant ($P < 0.05$). Seasonal dissolved oxygen values at the three SWMP sites generally ranged from 6 to 12 mg/l, with highest values observed in the winter and lowest values in the summer (Figure 3.5). All three sites are well oxygenated, with mean % saturation values of 75 to 120% (Figure 3.6). Hypoxia has never been observed in the Mullica River–Great Bay Estuary. The pH levels progressively increase from upriver areas to the open waters of Great Bay. For example, during the study period the pH measurements increased from 6.2 at Lower Bank and 7.2 at Chestnut Neck to 8.0 at Buoy 126 (Figure 3.7). The low pH values at the river stations are due to the high concentrations of tannins and humic acids originating in the Mullica River watershed. Differences in pH levels are statistically significant ($P < 0.05$) at the three monitoring sites. Mean turbidity levels ranged from ~5 to 32 Nephelometry Turbidity Units (NTU) during 1999–2000 (Figure 3.8). Highest values occurred in the bay at Buoy 126; values at the river sites were substantially lower. Turbidity was generally greatest during the spring and winter seasons. Mean water depths

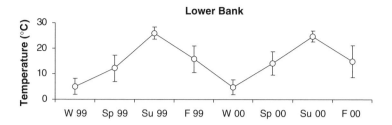
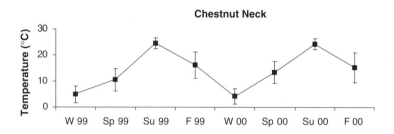
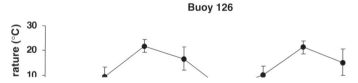

FIGURE 3.3 Mean seasonal water temperature and standard deviation values at three SWMP sites in the Jacques Cousteau National Estuarine Research Reserve during the 1999 and 2000 sampling period. (From Kennish, M.J. and S. O'Donnell. 2002. *Bulletin of the New Jersey Academy of Science* 47: 1–13.)

at Buoy 126 exceeded 2 m during both 1999 and 2000, but water depths were less than 2 m at Chestnut Neck and Lower Bank (Figure 3.9).

The Mullica River–Great Bay Estuary has excellent water quality. This is principally attributed to the limited development and low anthropogenic impacts in the Mullica River watershed. As a result, the Mullica River–Great Bay Estuary serves as an important reference location to assess more heavily impacted coastal bays in New Jersey and elsewhere.

WATERSHED BIOTIC COMMUNITIES

PLANT COMMUNITIES

Salt Marshes

Spartina salt marshes form the dominant habitat surrounding the shorelines of the coastal bays in the JCNERR. These marshes also extend some distance inland along

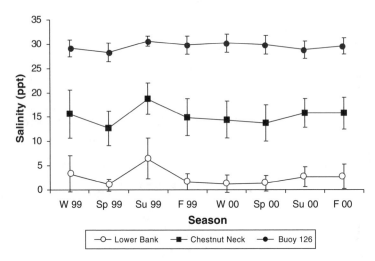

FIGURE 3.4 Mean seasonal salinity and standard deviation values at three SWMP sites in the Jacques Cousteau National Estuarine Research Reserve during the 1999 and 2000 sampling period. (From Kennish, M.J. and S. O'Donnell. 2002. *Bulletin of the New Jersey Academy of Science* 47: 1–13.)

stream and river banks, where they are gradually replaced by brackish marshes in lower salinity areas. For example, salt marshes extend ~25 km up the Mullica River to Lower Bank. In the Mullica River–Great Bay Estuary alone, salt marsh vegetation covers nearly 9000 ha. The most extensive salt marshes in the JCNERR system occur in the Great Bay Boulevard Wildlife Management Area, the Brigantine portion of the Forsythe National Wildlife Refuge, the Barnegat portion of the Forsythe National Wildlife Refuge, and the Holgate Unit of the Forsythe National Wildlife Refuge.

Salt marsh vegetation in the JCNERR exhibits a zoned pattern with smooth cordgrass (*Spartina alterniflora*) forming nearly monotypic stands in low marsh areas. Here, tall-form *S. alterniflora* predominates along tidal creek banks, and short-form *S. alterniflora* concentrates in other low marsh areas (Smith and Able, 1994). Three species (i.e., salt-meadow cordgrass, *S. patens*; spike grass, *Distichlis spicata*; and black grass, *Juncus gerardii*) are the most abundant plants in the high marsh areas. Several other species (i.e., marsh fleabane, *Pluchea purpurascens*; orach, *Atriplex patula*; perennial glasswort, *Salicornia virginica*; saltwort grass, *S. bigelovii*; and samphir, *S. europea*) proliferate in salt pannes. Along the marsh–upland border, five plant species are characteristic (i.e., salt-meadow cordgrass, *Spartina patens*; marsh elder, *Iva frutescens*; seaside goldenrod, *Solidago sempervirens*; salt marsh pink, *Sabatia stellaris*; and common reed, *Phragmites australis*). The invasive common reed is a growing concern because it appears to be replacing native species in some areas (Able and Hagen, 2000).

Brackish Tidal Marshes

Several plant species dominate the brackish tidal marshes of the JCNERR, including the big cordgrass (*Spartina cynosuroides*), Olney three-square bulrush (*Scirpus*

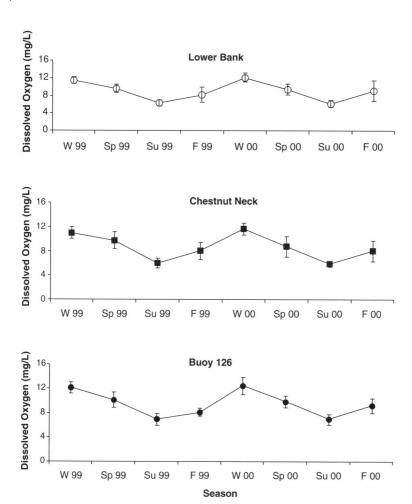

FIGURE 3.5 Mean seasonal dissolved oxygen and standard deviation values at three SWMP sites in the Jacques Cousteau National Estuarine Research Reserve during the 1999 and 2000 sampling period. (From Kennish, M.J. and S. O'Donnell. 2002. *Bulletin of the New Jersey Academy of Science* 47: 1–13.)

americanus), narrow-leaved cattail (*Typha angustifolia*), and common reed (*Phragmites australis*). Among the submerged aquatic plants encountered in these marshes are widgeon grass (*Ruppia maritima*), slender pondweed (*Potamogeton pusillus*), redhead grass (*P. perfoliatus*), horned pondweed (*Zanniuchellia palustris*), and water celery (*Vallisneria americana*). A number of other species appear as freshwater tidal reaches are approached; these are the Nuttall's pondweed (*P. epihydrus*), bulrush (*Scirpus* spp.), American mannagrass (*Glyceria grandis*), and arrowheads (*Sagittaria engelmanniana, S. latifolia*, and *S. spatulata*). Brackish tidal marshes are best developed along the Mullica River, Bass River, Wading River, Landing Creek, and Nacote Creek (JCNERR, 1999).

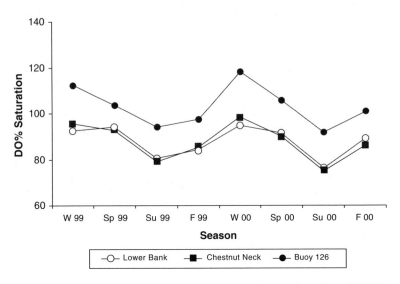

FIGURE 3.6 Mean seasonal dissolved oxygen (% saturation) levels at three SWMP sites in the Jacques Cousteau National Estuarine Research Reserve during the 1999 and 2000 sampling period. (From Kennish, M.J. and S. O'Donnell. 2002. *Bulletin of the New Jersey Academy of Science* 47: 1–13.)

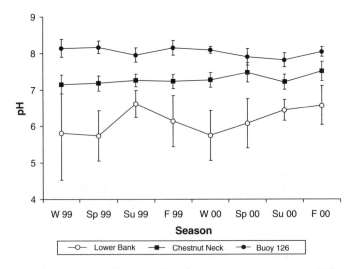

FIGURE 3.7 Mean seasonal pH and standard deviation values at three SWMP sites in the Jacques Cousteau National Estuarine Research Reserve during the 1999 and 2000 sampling period. (From Kennish, M.J. and S. O'Donnell. 2002. *Bulletin of the New Jersey Academy of Science* 47: 1–13.)

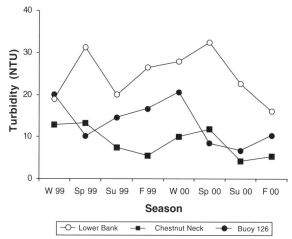

FIGURE 3.8 Mean seasonal turbidity levels at three SWMP sites in the Jacques Cousteau National Estuarine Research Reserve during the 1999 and 2000 sampling period. (From Kennish, M.J. and S. O'Donnell. 2002. *Bulletin of the New Jersey Academy of Science* 47: 1–13.)

Able and Hagen (2000) recently reported on the invasion of *Phragmites australis* in brackish water marsh habitat (i.e., Hog Islands) along the upper reaches of the Mullica River. This highly invasive species has spread rapidly. Between 1971 and 1991, its vegetative coverage increased from 3.2 to 83.1% (Windom and Lathrop, 1999). The spread of *P. australis* is significant because its presence influences the composition of marsh fauna. For example, Able and Hagen (2000) showed that the occurrence of the common reed affected fish and decapod use of the marsh surface at Hog Islands. Although *P. australis* had little or no effect on larger fish and decapods, it adversely affected larval and small fish, notably the mummichog, *Fundulus heteroclitus*. Abundance of recently hatched *F. heteroclitus* was significantly less in the *Phragmites*-dominated marsh than in the *Spartina*-dominated marsh. In addition, overall use of the *Phragmites*-dominated marsh by small fishes was consistently less than that of the *Spartina*-dominated marsh. With regard to decapods, *Rhithropanopeus harrisii* was most abundant in the *P. australis* marsh, whereas *Callinectes sapidus* and *Palaemonetes* spp. were most abundant in the *Spartina* marsh.

Freshwater Marshes

Proceeding upriver in the Mullica River, Wading River, and other tributary systems, an array of plant species forms luxuriant freshwater tidal marsh communities. These species grow in three distinct zones:

1. Low-tide zone
2. Mid-tide zone
3. Upper tidal zone

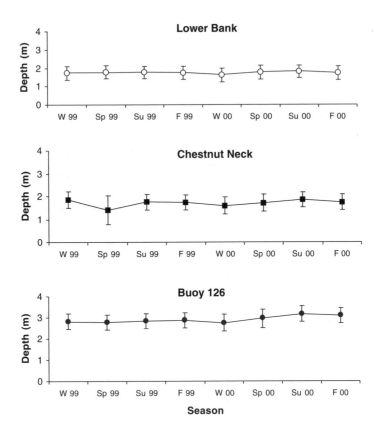

FIGURE 3.9 Mean seasonal water depth at three SWMP sites in the Jacques Cousteau National Estuarine Research Reserve during the 1999 and 2000 sampling period. (From Kennish, M.J. and S. O'Donnell. 2002. *Bulletin of the New Jersey Academy of Science* 47: 1–13.)

The low tidal marsh, which is only exposed at low tide, consists primarily of bluntscale bulrush (*Scirpus smithii* var. *smithii*), Parker's pipewort (*Eriocaulon parkeri*), riverbank guillwort (*Isoetes riparia*), and arrowheads (Hudson arrowhead, *Sagittaria subulata*; grass-leaved arrowhead, *S. graminea*; and stiff arrowhead, *S. rigida*). Wild rice (*Zinzania aquatica*), water hemp (*Amaranthus cannabinus*), three-square bulrush (*Scirpus pungens*), spatterdock (*Nurphur advena*), pickerel weed (*Ponderia cordata*), dotted smartweed (*Polygonum punctatum*), and arrow arum (*Peltandra virginica*) are the principal species comprising marshes in the mid-tide zone. A diverse assemblage of marsh plants occupies the upper tidal zone, although cattails (*Typha angustifolia* and *T. glauca*) predominate. Among the commonly observed species in the upper tidal zone are the common reed (*Phragmites australis*), purple loosestrife (*Lythrum salicaria*), knob-styled dogwood (*Cornus amomum*), button bush (*Cephalanthus occidentalis*), sensitive fern (*Onaclea sensibilis*), smooth burmarigold (*Bidens laevis*), swamp rose (*Rosa palustris*), swamp rose mallow (*Hibiscus moscheutos*), sweet flag (*Acorus calamus*), orange jewelweed (*Impatiens capensis*), and arrowheads (*Sagittaria* spp.) (JCNERR, 1999).

Lowland Plant Communities

Lowland vegetation in the pinelands region consists of six main types of plant communities:

1. Atlantic white cedar swamp forests
2. Broadleaf swamp forests
3. Pitch pine lowland forests
4. Pine transition forests
5. Herbaceous wetland communities
6. Shrubby wetland communities

McCormick (1998) has examined these communities in detail. Atlantic white cedar swamp forests, together with the broadleaf swamp forests, comprise the most extensive plant communities of the lowland area. Atlantic white cedar (*Chamaecyparis thyoides*), trident red maple (*Acer rubrum*), black gum (*Nyssa sylvatica*), and sweetbay magnolia (*Magnolia virginiana*) form most of the canopy in the cedar swamp forests. Various shrubs constitute the understory, notably sweet pepperbush (*Clethra alnifolia*), swamp azalea (*Rhododendron viscosum*), fetterbush (*Leucothoe racemosa*), bayberry (*Myrica pensylvanica*), and dangleberry (*Gaylussacia frondosa*). Species dominating the herbaceous ground cover in the cedar swamp forests include the partridge berry (*Mitchella repens*), sundew (*Drosera capillaris*), pitcher plant (*Sarracenia purpurea*), curly-grass fern (*Schizaea pusilla*), swamp pink (*Helonias bullata*), and *Sphagnum* moss.

The most abundant tree in the broadleaf forest community is the trident red maple (*Acer rubrum*). However, two other species, Atlantic white cedar (*Chamaecyparis thyoides*) and pitch pine (*Pinus rigida*), are also locally important components of the canopy. Several other species found in this community, albeit in lower abundances, are the sweetbay magnolia (*Magnolia virginiana*), gray birch (*Betula populifolia*), sassafras (*Sassafras albidum*), and black gum (*Nyssa sylvatica*). Swamp azalea (*Rhododendron viscosum*), leatherleaf (*Chamaedaphne calyculata*), sheep laurel (*Kalmia angustifolia*), fetterbush (*Leucothoe racemosa*), dangleberry (*Gaylussacia frondosa*), and black huckleberry (*G. baccata*) predominate in the shrub layer. Ground cover here consists mainly of mosses and lichens.

The canopy in the pitch pine lowland forests consists primarily (90%) of pitch pine (*Pinus rigida*). Of secondary importance are gray birch (*Betula populifolia*), trident red maple (*Acer rubrum*), and black gum (*Nyssa sylvatica*). The principal species in the understory are sheep laurel (*Kalmia angustifolia*), leatherleaf (*Chamaedaphne calyculata*), black huckleberry (*Gaylussacia baccata*), and dangleberry (*G. frondosa*). *Sphagnum* moss, bracken fern (*Pteridium aquilinum*), turkeybeard (*Xerophyllum asphodeloides*), and wintergreen (*Gaultheria procumbens*) are the main ground cover species.

Pine transition communities occur between the Atlantic white cedar forests or broadleaf swamp forests and the upland forests. Pitch pine (*Pinus rigida*) dominates these transition communities. Secondary canopy species are the trident red maple (*Acer rubrum*), gray birch (*Betula populifolia*), and black gum (*Nyssa sylvatica*).

Bayberry (*Myrica pensylvanica*), sheep laurel (*Kalmia angustifolia*), winterberry (*Ilex verticillata*), dangleberry (*Gaylussacia frondosa*), black huckleberry (*G. baccata*), and grouseberry (*G. dumosa*) are the dominant species of the shrub layer. Ground cover is generally sparse in the pine transition forests, covering only ~2% of the area. Principal herbs and shrubs forming the ground cover in this community are *Sphagnum* moss, turkey-beard (*Xerophyllum asphodeloides*), bracken fern (*Pteridium aquilinum*), cinnamon fern (*Osmunda cinnamonea*), and wintergreen (*Gaultheria procumbens*).

Perimeter areas of ponds and streams in the Mullica River watershed support rich herbaceous wetland communities. Several submerged and floating leaf plants, such as bladderworts (*Utricularia* spp.), white water lilies (*Nymphaea odorata*), and bullhead lilies (*Nuphar variegatum*), are important members of these communities. Emergent plants (e.g., rushes, *Juncus* spp.; sedges, *Carex* spp.; chain ferns, *Woodwardia* spp.; and pipeworts, *Eriocaulon* spp.) concentrate along the shore.

Aside from the aforementioned communities, vernal pond (coastal plain intermittent pond) plant communities also exist in the lowland areas. Panic and muhly grasses (*Panicum capillare, P. mattamuskeettense, P. verrucosum,* and *Muhlenbergia torreyana*) and sedges (*Carex* sp., *Cladium mariscoides, Eleocharis microcarpa,* and *Scleria reticularis*) dominate these communities. Other species that may be found are rose tickseed (*Coreopsis rosea*), drowned beaked-rush (*Rhynchospora inundata*), short-beaked bald-rush (*R. nitens*), Long's bulrush (*Scirpus longii*), knotted spikerush (*Eleocharis equisetoides*), Wright's panic grass (*Panicum wrightianum*), awned meadow beauty (*Rhexia aristosa*), floating heart (*Nymphoides cordata*), dwarf white bladderwort (*Utricularia olivacea*), Boykin's lobelia (*Lobelia boykinii*), and slender water-milfoil (*Myriophyllum tenellum*).

Along the margins of some ponds, shrubby wetland communities are also delineated. Sheep laurel (*Kalmia augustifolia*), leatherleaf (*Chamaedaphne calyculata*), highbush blueberry (*Vaccinium corymbosum*), staggerbush (*Lyonia mariana*), and *Sphagnum* moss are the primary constituents of these communities. Shrubby wetland communities have likewise been observed in the channels of intermittent streams in the Mullica River watershed. The flood plains of some streams in the watershed provide habitat for wet meadow communities (savannas) dominated by sedges and grasses. Button sedge (*Carex bullata*), coast sedge (*Carex exilis*), lowland broomsedge (*Andropogon virginicus* var. *virginicus*), golden crest (*Lophiola aurea*), and Torrey's dropseed (*Muhlenbergia torreyana*) typically dominate these communities. Table 3.1 provides a list of plants found along streams in the Mullica River Basin.

Cedar swamps, as well as sphagnum and cranberry bogs, support an array of herbaceous plants and other vegetation. In these habitats, the Atlantic white cedar is usually the dominant tree. Highbush blueberry (*Vaccinium corymbosum*), dangleberry (*Gaylussacia frondosa*), fetterbush (*Leucothoe racemosa*), and swamp azalea (*Rhododendron viscosum*) are the commonly encountered herbaceous plant species. The ground cover consists mainly of *Sphagnum* mosses with lesser amounts of bladderworts (*Ultricularia* spp.), sundews (*Drosera* spp.), and pitcher plants (*Sarracenia purpurea*).

TABLE 3.1
Taxonomic List of Plants Identified along Stream Vegetation Sites in the Mullica River Basin

Common Name	Scientific Name
Herbaceous Plants	
Ticklegrass	*Agrostis hyemalis*
Ticklegrass	*Agrostis hyemalis* var. *scabra*
Upland bent grass	*Agrostis perennans*
Upland bent grass	*Agrostis perennans* var. *elata*
Small water plantain	*Alisma subcordatum*
Garlic	*Allium* sp.
Pursh's millet grass	*Amphicarpum purshii*
Bushy beard grass	*Andropogon virginicus* var. *abbreviatus*
Broomsedge	*Andropogon virginicus* var. *virginicus*
Groundnut	*Apios americana*
Wild sarsaparilla	*Aralia nudicaulis*
Arethusa	*Arethusa bulbosa*
Swamp milkweed	*Asclepias incarnata*
Bushy aster	*Aster dumosus*
Bog aster	*Aster nemoralis*
New York aster	*Aster novi-belgii*
Heath aster	*Aster pilosus* var. *pringlei*
Small white aster	*Aster racemosus*
Twining bartonia	*Bartonia paniculata*
Yellow bartonia	*Bartonia virginica*
Purple-stemmed beggar ticks	*Bidens connata*
Northern tickseed-sunflower	*Bidens coronata*
Small beggar ticks	*Bidens discoidea*
Beggar ticks	*Bidens frondosa*
False nettle	*Boehmeria cylindrica*
Blue-joint grass	*Calamagrostis canadensis*
Nuttall's reed grass	*Calamagrostis cinnoides*
Larger water starwort	*Callitriche heterophylla*
Pennsylvania bitter cress	*Cardamine pensylvanica*
Greenish-white sedge	*Carex albolutescens*
Atlantic sedge	*Carex atlantica*
Howe's sedge	*Carex atlantica* var. *capillacea*
Button sedge	*Carex bullata*
Silvery sedge	*Carex canescens*
Collins' sedge	*Carex collinsii*
Fringed sedge	*Carex crinita*
Coast sedge	*Carex exilis*
Long sedge	*Carex folliculata*
Bladder sedge	*Carex intumescens*
Livid sedge	*Carex livida*

(continued)

TABLE 3.1 (CONTINUED)
Taxonomic List of Plants Identified along Stream Vegetation Sites in the Mullica River Basin

Common Name	Scientific Name
Long's sedge	*Carex longii*
Sallow sedge	*Carex lurida*
Pennsylvania sedge	*Carex pensylvanica*
Pointed broom sedge	*Carex scoparia*
Awl-fruited sedge	*Carex stipata*
Walter's sedge	*Carex striata*
Tussock sedge	*Carex stricta*
Blunt broom sedge	*Carex tribuloides*
Three-fruited sedge	*Carex trisperma*
Dark green sedge	*Carex venusta*
Prickly hornwort	*Ceratophyllum echinatum*
Slender spike grass	*Chasmanthium laxum*
Wood reed	*Cinna arundinacea*
Twig rush	*Cladium mariscoides*
Dodder	*Cuscuta* sp.
Toothed cyperus	*Cyperus dentatus*
Red-rooted cyperus	*Cyperus erythrorhizos*
Coarse cyperus	*Cyperus odoratus*
Pine Barrens cyperus	*Cyperus retrorsus*
Straw-colored cyperus	*Cyperus strigosus*
Silky wild oat grass	*Danthonia sericea* var. *epilis*
Swamp loosestrife	*Decodon verticillatus*
Common wild yam	*Dioscorea villosa*
Thread-leaved sundew	*Drosera filiformis*
Spatulate-leaved sundew	*Drosera intermedia*
Round-leaved sundew	*Drosera rotundifolia*
Spinulose wood fern	*Dryopsteris carthusiana*
Dulichium	*Dulichium arundinaceum*
American barnyard grass	*Echinochloa muricata*
Needle spike rush	*Eleocharis acicularis*
Green spike rush	*Eleocharis flavescens* var. *olivacea*
Small-fruited spike rush	*Eleocharis microcarpa*
Blunt spike rush	*Eleocharis ovata*
Robbin's spike rush	*Eleocharis robbinsii*
Slender spike rush	*Eleocharis tenuis*
Tubercled spike grass	*Eleocharis tuberculosa*
Nuttall's water weed	*Elodea nuttallii*
Purple-leaved willow herb	*Epilobium coloratum*
Pilewort	*Erechtites hieracifolia*
Plume grass	*Erianthus giganteus*
Seven-angled pipewort	*Eriocaulon aquaticum*
Flattened pipewort	*Eriocaulon compressum*

TABLE 3.1 (CONTINUED)
Taxonomic List of Plants Identified along Stream Vegetation Sites in the Mullica River Basin

Common Name	Scientific Name
Ten-angled pipewort	*Eriocaulon decangulare*
Tawny cotton grass	*Eriophorum virginicum*
Eastern joe-pye weed	*Eupatorium dubium*
Boneset	*Eupatorium perfoliatum*
Rough boneset	*Eupatorium pilosum*
Pine Barrens boneset	*Eupatorium resinosum*
Late-flowering boneset	*Eupatorium serotinum*
Ipecac spurge	*Euphorbia ipecacuanhae*
Slender-leaved goldenrod	*Euthamia tenuifolia*
Stiff marsh bedstraw	*Galium tinctorium*
Gill-over-the-ground	*Glechoma hederacea*
Rattlesnake grass	*Glyceria canadensis*
Blunt mannagrass	*Glyceria obtusa*
Fowl mannagrass	*Glyceria striata*
Northern mannagrass	*Glyceria laxa*
Green wood orchid	*Habenaria clavellata*
Ragged fringed orchid	*Habenaria lacera*
Swamp rose mallow	*Hibiscus moscheutos*
Canada Saint John's wort	*Hypericum canadense*
Coppery Saint John's wort	*Hypericum denticulatum*
Dwarf Saint John's wort	*Hypericum mutilum*
Saint Andrew's cross	*Hypericum stragulum*
Spotted touch-me-not	*Impatiens capensis*
Slender blue flag	*Iris prismatica*
Larger blue flag	*Iris versicolor*
Spiny-spored quillwort	*Isoetes echinospora*
Sharp-fruited rush	*Juncus acuminatus*
Two-flowered rush	*Juncus biflorus*
New Jersey rush	*Juncus caesariensis*
Canada rush	*Juncus canadensis*
Common rush	*Juncus effusus*
Bayonet rush	*Juncus militaris*
Brown-fruited rush	*Juncus pelocarpus*
Redroot	*Lachnanthes caroliniana*
Rice cut grass	*Leersia oryzoides*
Duckweed	*Lemna* sp.
Turk's-cap lily	*Lilium superbum*
Short-stalked false pimpernel	*Lindernia dubia*
Canby's lobelia	*Lobelia canbyi*
Cardinal flower	*Lobelia cardinalis*
Nuttall's lobelia	*Lobelia nuttalli*

(*continued*)

TABLE 3.1 (CONTINUED)
Taxonomic List of Plants Identified along Stream Vegetation Sites in the Mullica River Basin

Common Name	Scientific Name
Golden crest	*Lophiola aurea*
Seedbox	*Ludwigia alternifolia*
Water purslane	*Ludwigia palustris*
Foxtail-clubmoss	*Lycopodium alopecuroides*
Southern bog clubmoss	*Lycopodium appressum*
Tree clubmoss	*Lycopodium obscurum*
Northern bugleweed	*Lycopus uniflorus*
Virginia bugleweed	*Lycopus virginicus*
Swamp loosestrife	*Lysimachia terrestris*
Purple loosestrife	*Lythrum salicaria*
Eulalia	*Microstegium vimineum*
Climbing hempweed	*Mikania scandens*
Square-stemmed monkey flower	*Mimulus ringens*
Partridge berry	*Mitchella repens*
Indian pipe	*Monotropa uniflora*
Torrey's dropseed	*Muhlenbergia torreyana*
Late-flowering dropseed	*Muhlenbergia uniflora*
Bullhead lily	*Nuphar variegata*
White water lily	*Nymphaea odorata*
Sensitive fern	*Onoclea sensibilis*
Golden club	*Orontium aquaticum*
Cinnamon fern	*Osmunda cinnamomea*
Royal fern	*Osmunda regalis*
Upright yellow wood sorrel	*Oxalis stricta*
Cowbane	*Oxypolis rigidior*
Deertongue grass	*Panicum clandestinum*
Forked panic grass	*Panicum dichotomum*
Small-leaved panic grass	*Panicum ensifolium*
Panic grass	*Panicum lanuginosum*
Long-leaved panic grass	*Panicum longifolium*
Long-leaved panic grass	*Panicum rigidulum*
Sheathed panic grass	*Panicum scabriusculum*
Eaton's panic grass	*Panicum spretum*
Warty panic grass	*Panicum verrucosum*
Switchgrass	*Panicum virgatum*
Arrow arum	*Peltandra virginica*
Reed canary grass	*Phalaris arundinacea*
Reed	*Phragmites australis*
Pokeweed	*Phytolacca americana*
Clearweed	*Pilea pumila*
Fowl bluegrass	*Poa palustris*
Kentucky bluegrass	*Poa pratensis*

TABLE 3.1 (CONTINUED)
Taxonomic List of Plants Identified along Stream Vegetation Sites in the Mullica River Basin

Common Name	Scientific Name
Rose pogonia	*Pogonia ophioglossoides*
Short-leaved milkwort	*Polygala brevifolia*
Cross-leaved milkwort	*Polygala cruciata*
Halberd-leaved tearthumb	*Polygonum arifolium*
Cespitose knotweed	*Polygonum crespitosum*
Mild water pepper	*Polygonum hydropiperoides*
Dotted smartweed	*Polygonum punctatum*
Arrow-leaved tearthumb	*Polygonum sagittatum*
Pickerel weed	*Ponderia cordata*
Algal-like pondweed	*Potamogeton confervoides*
Half-like pondweed	*Potamogeton diversifolius*
Nuttall's pondweed	*Potamogeton epihydrus*
Oakes' pondweed	*Potamogeton oakesianus*
Small pondweed	*Potamogeton pusillus*
Cut-leaved mermaid weed	*Proserpinaca pectinata*
Bracken	*Pteridium aquilinum*
Maryland meadow beauty	*Rhexia mariana*
Virginia meadow beauty	*Rhexia virginica*
White-beaked-rush	*Rhynchospora alba*
Small-headed beaked-rush	*Rhynchospora capitellata*
Loose-headed beaked-rush	*Rhynchospora chalarocephala*
Marsh yellow cress	*Rorippa palustris*
Lance-leaved sabatia	*Sabatia difformis*
Engelmann's arrowhead	*Sagittaria engelmanniana*
Pitcher plant	*Sarracenia purpurea*
Little bluestem	*Schizachyrium scoparium*
Curly-grass fern	*Schizaea pusilla*
Wool-grass	*Scirpus cyperinus*
Three-square bulrush	*Scirpus pungens*
Water club-rush	*Scirpus subterminalis*
Reticulated nut-rush	*Scleria reticularis*
Sclerolepis	*Sclerolepis uniflora*
Mad-dog skullcap	*Scutellaria lateriflora*
Carrion flower	*Smilax herbacea*
Halberd-leaved greenbrier	*Smilax pseudochina*
Black nightshade	*Solanum nigrum*
Canada goldenrod	*Solidago canadensis*
Rough-stemmed goldenrod	*Solidago rugosa*
Slender-bur-reed	*Sparganium americanum*
Nodding ladies'-tresses	*Spiranthes cernua*
Common stitchwort	*Stellaria graminea*

(continued)

TABLE 3.1 (CONTINUED)
Taxonomic List of Plants Identified along Stream Vegetation Sites in the Mullica River Basin

Common Name	Scientific Name
Common chickweed	*Stellaria media*
Dandelion	*Taraxacum officinale*
Marsh fern	*Thelypteris palustris*
Bog fern	*Thelypteris simulata*
Marsh Saint John's wort	*Triadenum virginicum*
Starflower	*Trientalis borealis*
Broad-leaved cattail	*Typha latifolia*
Stinging nettle	*Urtica dioica*
Horned bladderwort	*Utricularia cornuta*
Fibrous bladderwort	*Utricularia fibrosa*
Hidden-fruited bladderwort	*Utricularia geminiscapa*
Floating bladderwort	*Utricularia inflata*
Purple bladderwort	*Utricularia purpurea*
Zig-zag bladderwort	*Utricularia subulata*
Greater bladderwort	*Utricularia vulgaris*
Blue vervain	*Verbena hastata*
New York ironweed	*Vernonia noveboracensis*
Lance-leaved violet	*Viola lanceolata*
Primrose-leaved violet	*Viola primulifolia*
Woolly blue violet	*Viola sororia*
Netted chain fern	*Woodwardia areolata*
Virginia chain fern	*Woodwardia virginica*
Turkey-beard	*Xerophyllum asphodeloides*
Yellow-eyed grass	*Xyris difformis*
Small's yellow-eyed grass	*Xyris smalliana*
Wild rice	*Zizania aquatica*

Woody Plants

Common Name	Scientific Name
Trident red maple	*Acer rubrum*
Ailanthus	*Ailanthus altissima*
Smooth alder	*Alnus serrulata*
Oblongleaf juneberry	*Amelanchier canadensis*
Coastal juneberry	*Amelanchier obovalis*
Red chokeberry	*Aronia arbutifolia*
Japanese barberry	*Berberis thunbergii*
Black birch	*Betula lenta*
Gray birch	*Betula populifolia*
Common catalpa	*Catalpa bignonioides*
Buttonbush	*Cephalanthus occidentalis*
Atlantic white cedar	*Chamaecyparis thyoides*
Leatherleaf	*Chamaedaphne calyculata*
Yam-leaved clematis	*Clematis terniflora*
Sweet pepperbush	*Clethra alnifolia*

TABLE 3.1 (CONTINUED)
Taxonomic List of Plants Identified along Stream Vegetation Sites in the Mullica River Basin

Common Name	Scientific Name
Persimmon	*Diospyros virginiana*
Fetterbush	*Eubotrys racemosa*
Wintergreen	*Gaultheria procumbens*
Black huckleberry	*Gaylussacia baccata*
Dwarf huckleberry	*Gaylussacia dumosa*
Dangleberry	*Gaylussacia frondosa*
Golden heather	*Hudsonia ericoides*
Bushy Saint John's wort	*Hypericum densiflorum*
Inkberry	*Ilex glabra*
Smooth winterberry	*Ilex laevigata*
American holly	*Ilex opaca*
Winterberry	*Ilex verticillata*
Virginia willow	*Itea virginica*
Red cedar	*Juniperus virginiana*
Sheep laurel	*Kalmia angustifolia*
Mountain laurel	*Kalmia latifolia*
Sand myrtle	*Leiophyllum buxifolium*
Sweet gum	*Liquidambar styraciflua*
Japanese honeysuckle	*Lonicera japonica*
Maleberry	*Lyonia ligustrina*
Staggerbush	*Lyonia mariana*
Sweet bay	*Magnolia virginiana*
Bayberry	*Myrica pensylvanica*
Black gum	*Nyssa sylvatica*
Virginia creeper	*Parthenocissus quinquefolia*
Shortleaf pine	*Pinus echinata*
Pitch pine	*Pinus rigida*
White pine	*Pinus strobus*
Sycamore	*Platanus occidentalus*
Black cherry	*Prunus serotina*
White oak	*Quercus alba*
Scrub oak	*Quercus ilicifolia*
Black-jack oak	*Quercus marilandica*
Black oak	*Quercus velutina*
Post oak	*Quercus stellata*
Swamp azalea	*Rhododendron viscosum*
Swamp rose	*Rosa palustris*
Swamp dewberry	*Rubus hispidus*
Blackberry	*Rubus* sp.
Black willow	*Salix nigra*
Common elder	*Sambuscus canadensis*

(continued)

TABLE 3.1 (CONTINUED)
Taxonomic List of Plants Identified along Stream Vegetation Sites in the Mullica River Basin

Common Name	Scientific Name
Sassafras	*Sassafras albidum*
Glaucous greenbrier	*Smilax glauca*
Laurel-leaved greenbrier	*Smilax laurifolia*
Common greenbrier	*Smilax rotundifolia*
Red-berried greenbrier	*Smilax walteri*
Narrow-leaved meadowsweet	*Spiraea alba* var. *latifolia*
Steeplebush	*Spiraea tomentosa*
Basswood	*Tilia americana*
Poison ivy	*Toxicodendron radicans*
Poison sumac	*Toxicodendron vernix*
American elm	*Ulmus americana*
Highbush blueberry	*Vaccinium corymbosum*
Large cranberry	*Vaccinum macrocarpon*
Early low blueberry	*Vaccinum pallidum*
Southern arrowwood	*Viburnum dentatum*
Naked withe-rod	*Viburnum nudum* var. *nudum*
Fox grape	*Vitis labrusca*

Source: Zampella, R.A., J.F. Bunnell, K.J. Laidig, and C.L. Dow. 2001. The Mullica River Basin. Technical Report, New Jersey Pinelands Commission, New Lisbon, NJ.

Algae are well represented in streams, lakes, ponds, and bogs of the Mullica River watershed. Green algae (Chlorophyta), yellow-green algae (Chlorophyta), and euglenoids (Euglenophyta) are quite diverse, with 350 taxa being registered in the Pine Barrens (Moul and Buell, 1998). Diatoms often predominate in these habitats.

Upland Plant Communities

Pine–oak forests characterize upland habitats of the Mullica River watershed (McCormick, 1998; JCNERR, 1999). Pitch pine (*Pinus rigida*) and several species of oak (i.e., white oak, *Quercus alba*; black oak, *Q. velutina*; scarlet oak, *Q. coccinea*; and chestnut oak, *Q. prinus*) form the predominant upland forest canopy. In some areas, pitch pine is the overwhelmingly dominant species, accounting for more than 50% of the cover. Nearly pure stands of pitch pine occur locally, as do nearly pure stands of oak trees. The understory in these upland forests typically consists of mountain laurel (*Kalmia latifolia*), sweet fern (*Comptonia peregrina*), inkberry *(Ilex glabra)*, huckleberries (*Gaylussacia* spp.), blueberries (*Vaccinium* spp.), and scrub oak (*Q. ilicifolia*).

The pine–oak canopy is well developed in the Bass River State Forest, Penn State Forest, and Wharton State Forest. Pitch pine (*Pinus rigida*) is most abundant, covering ~50 to 80% of the uplands vegetation in these forests. Shortleaf pine (*P. echinata*), also present in the upland forests, is of secondary importance.

Among the species of oak found in the upland communities, the southern red oak (*Quercus falcata*) is the predominant form south of the Mullica River and the black oak (*Q. velutina*) the predominant form to the north. Other species of oak trees identified in these forests include the white oak (*Q. alba*), scarlet oak (*Q. coccinea*), scrub oak (*Q. ilicifolia*), post oak (*Q. stellata*), chestnut oak (*Q. prinus*), and blackjack oak (*Q. marilandica*).

McCormick (1998) described two types of shrub understory in the upland forests:

1. Heath-type vegetation dominated by black huckleberry (*Gaylussacia baccata*) and lowbush blueberry (*Vaccinium vacillans*)
2. Scrub-oak type vegetation

Compared to scrub-oak understory, which grows about 1 to 5 m high, heath-type understory generally grows from 30 to 60 cm high. Among the various species of ground cover documented in the upland forests are *Sphagnum* moss, bearberry (*Arctostaphylos uva-ursi*), bracken fern (*Pteridium aquilinum*), wintergreen (*Gaultheria procumbens*), goatsrue (*Tephrosia virginiana*), and cowwheat (*Melampyrum lineare*).

Dwarf pitch or pygmy pine (*Pinus rigida*) less than ~3 m high and low-growing scrub oaks (*Quercus marilandica* and *Q. ilicifolia*) inhabit areas of the Pine Barrens subject to frequent fires. Dwarf pitch pine communities in the Pine Barrens cover nearly 5000 ha (Good et al., 1998). Species of shrubs and herbs found in these communities include sand myrtle (*Leiophyllum buxifolium*), sweet fern (*Comtonia peregrina*), sheep laurel (*Kalmia angustifolia*), and mountain laurel (*K. latifolia*). Species of ground cover, in turn, consist of wintergreen (*Gaultheria procumbens*), broom crowberry (*Cormea conradii*), trailing arbutus (*Epigaea repens*), and bearberry (*Arctostaphylos uva-ursi*).

Barrier Island Plant Communities

The barrier island complex of the JCNERR consists of both developed and undeveloped areas. Where the barrier island complex is undeveloped, such as along the protected Holgate Unit of the Forsythe National Wildlife Refuge and the North Brigantine State Natural Area, extensive scrub/shrub and woodland communities occur. However, plant communities along developed portions of the barrier island complex have been radically altered or destroyed. The decimated plant communities in developed regions contrast markedly with the plant communities in undisturbed habitat of the undeveloped lands.

Several distinct habitats characterize the barrier island complex; along the ocean side, sand beaches as well as primary and secondary dune systems are characteristic, and along the backbarrier areas, salt marshes and tidal flats predominate. American beach grass (*Ammophila breviligulata*) dominates the primary dune plant community in undisturbed habitats. Sea rocket (*Cakile edentula*), seaside goldenrod (*Solidago sempervirens*), Japanese sedge (*Carex kobomugi*), and beach pea (*Lathyrus maritimus*) may also be present here. The secondary dune plant community typically consists of beach heather (*Hudsonia tomentosa*), beach plum (*Prunus maritima*), pineweed (*Hypericum gentianoides*), salt spray rose (*Rosa rugosa*), and bayberry (*Myrica pensylvanica*) (U.S. Fish and Wildlife Service, 1996).

In the woodland community behind the secondary dune plant community, the red cedar (*Juniperus virginiana*) is the dominant species. Other species comprising the canopy, but in lower abundance, are the black cherry (*Prunus serotina*), sassafras (*Sassafras albidum*), willow oak (*Quercus phellos*), southern red oak (*Q. falcata*), American holly (*Ilex opaca*), and serviceberry (*Amelanchier canadensis*). The understory consists of hackberry (*Celtis occidentalis*), bayberry (*Myrica pensylvanica*), blueberries (*Vaccinium* spp.), multiflora rose (*Rosa multiflora*), and sweet pepperbush (*Clethra alnifolia*). Pitch pine (*Pinus rigida*), Atlantic white cedar (*Chamaecyparis thyoides*), American holly (*I. opaca*), and several species of oak can also be found in some open woodlands. The shrub layer here consists mainly of highbush blueberry (*Vaccinum corymbosum*) and sheep laurel (*Kalmia angustifolia*) (U.S. Fish and Wildlife Service, 1996).

Animal Communities

Amphibians and Reptiles

More than 50 species of herpetofauna have been recorded in the New Jersey Pine Barrens, including 19 snakes, 14 frogs and toads, 11 salamanders, 10 turtles, and 3 lizards (Table 3.2). Thirteen anuran species inhabit the Mullica River Basin (Table 3.3). The acid-water habitats of the New Jersey Pinelands support anuran assemblages uniquely different from those found elsewhere in the state (Conant, 1998; Zampella et al., 2001). Only two anuran species (Pine Barrens treefrog, *Hyla andersonii* and carpenter frog, *Rana virgatipes*) are confined to the Pine Barrens. Five other anuran species (eastern spadefoot, *Scaphiopus holbrooki*; Fowler's toad, *Bufo woodhousii fowleri*; northern spring peeper, *Pseudacris crucifer crucifer*; southern leopard frog, *R. utricularia*; and green frog, *R. clamitans melanota*), although native to the region, are more widely distributed. They have been reported throughout southern New Jersey. Seven other anuran species (i.e., bullfrog, *R. catesbeiana*; pickerel frog, *R. palustris*; wood frog, *R. sylvatica*; northern cricket frog, *Acris crepitans crepitans*; gray treefrogs, *Hyla versicolor* and *H. chrysoscelis*; and New Jersey chorus frog, *Pseudacris triseriata kalmi*) only occur in Pinelands habitat disturbed by anthropogenic activity (Zampella et al., 2001). These latter seven species, therefore, may be valuable as indicators of watershed disturbance.

Several salamander species inhabit the New Jersey Pinelands, but only three of them (four-toed salamander, *Hemidactylium scutatum*; northern red salamander, *Pseudotriton ruber*; and red-backed salamander, *Plethodon cinereus*) are relatively abundant (Conant, 1998). A fourth species (marbled salamander, *Ambystoma opacum*), although not abundant, has been observed in various areas of the Pinelands. Two other species (northern dusky salamander, *Desmognathus fuscus* and northern two-lined salamander, *Eurycea bislineata*) are rare. The eastern tiger salamander (*Ambystoma tigrinum tigrinum*) remains on the endangered species list.

Three species of lizards have been documented in the Pine Barrens: the five-lined skink (*Eumeces fasciatus*), ground skink (*Scincella lateralis*), and northern fence lizard (*Sceloporus undulatus hyacinthinus*). Only the northern fence lizard is

TABLE 3.2
Occurrence and Status of Amphibians and Reptiles in the New Jersey Pine Barrens

Species	Status in Pine Barrens
Salamanders	
Spotted salamander	Uncertain
Marbled salamander	REL, locally common
Eastern tiger salamander	BOR, endangered
Red-spotted newt	REL, few records
Northern dusky salamander	BOR, rare
Red-backed salamander	Abundant
Slimy salamander	Uncertain
Four-toed salamander	REL, numerous records
Eastern mud salamander	Uncertain
Northern red salamander	Abundant
Northern two-lined salamander	BOR, rare
Toads and Frogs	
Eastern spadefoot	Locally common
Fowler's toad	Abundant
Northern cricket frog	BOR, scattered records
Pine Barrens treefrog	PBO, declining
Cope's gray treefrog	PER, not present
Northern spring peeper	Abundant
Barking treefrog	INT, possibly extirpated
Gray treefrog	BOR, scattered records
New Jersey chorus frog	BOR, numerous records
Bullfrog	BOR, scattered records
Green frog	Abundant
Pickerel frog	BOR, few records
Wood frog	BOR, few records
Southern leopard frog	Abundant
Carpenter frog	PBO, common
Turtles	
Common snapping turtle	Common
Stinkpot	Abundant
Eastern mud turtle	Numerous records
Spotted turtle	Abundant records, but declining
Wood turtle	BOR, few records, threatened
Bog turtle	BOR, endangered
Eastern box turtle	Numerous records, but declining
Northern diamondback terrapin	PER, not present
Map turtle	PER, not present
Eastern painted turtle	Abundant
Red-bellied turtle	Common
Eastern spiny softshell	INT, at western edge only

(*continued*)

TABLE 3.2 (CONTINUED)
Occurrence and Status of Amphibians and Reptiles in the New Jersey Pine Barrens

Species	Status in Pine Barrens
Lizards	
Northern fence lizard	Abundant
Ground skink	PBO, uncommon
Five-lined skink	REL, few records
Snakes	
Queen snake	PER, not present
Northern water snake	Abundant
Northern brown snake	Scattered records
Northern red-bellied snake	PBO, numerous records
Eastern ribbon snake	Numerous records, but uncommon
Eastern garter snake	Numerous records
Eastern earth snake	Uncertain
Eastern hognose snake	Locally common, but declining
Northern-southern ringneck snake	Scattered
Eastern worm snake	REL, common
Northern black racer	Locally common, but declining
Rough green snake	PBO, common
Corn snake	PBO, scattered records
Black rat snake	Locally common
Northern pine snake	PBO, locally common
Eastern king snake	PBO, locally common
Eastern milk snake–scarlet king snake (intergrading population)	Numerous records
Northern scarlet snake	PBO, scattered records
Timber–canebrake rattlesnake (intergrading population)	PBO, threatened

Note: PBO, Pine Barrens only; BOR, border entrant; REL, relict in Pine Barrens; PER, peripheral to Pine Barrens; INT, introduced.

Source: Conant, R. 1998. In: Forman, R.T.T. (Ed.). *Pine Barrens: Ecosystem and Landscape.* Rutgers University Press, New Brunswick, NJ, pp. 467–488.

relatively abundant. It inhabits upland pine–oak forest areas. In contrast, the five-lined skink prefers hardwood swamps and other wet woodlands. The ground skink, in turn, occupies open sandy wooded habitats.

A number of snakes reside in the JCNERR and upland Pine Barrens. The eastern king snake (*Lampropeltis getula getula*), eastern ribbon snake (*Thamnophis sauritus*), and northern water snake (*Nerodia sipedon*) are mainly found in wetland habitats. The timber rattlesnake (*Crotalus horridus horridus*), an endangered species, occurs in both wetland and upland habitats. More species prefer upland forest habitat;

TABLE 3.3
Taxonomic List of Anuran Species Found in the Mullica River Basin

Common Name	Scientific Name
Northern cricket frog	*Acris crepitans crepitans*
Pine Barrens treefrog	*Hyla andersonii*
Cope's gray treefrog	*Hyla chrysoscelis*
Gray treefrog	*Hyla versicolor*
Fowler's toad	*Bufo woodhousii fowleri*
Northern spring peeper	*Pseudacris crucifer crucifer*
New Jersey chorus frog	*Pseudacris triseriata kalmi*
Bullfrog	*Rana catesbeiana*
Green frog	*Rana clamitans melanota*
Pickerel frog	*Rana palustris*
Southern leopard frog	*Rana utricularia*
Wood frog	*Rana sylvatica*
Carpenter frog	*Rana virgatipes*

Source: Modified from Zampella, R.A., J.F. Bunnell, K.J. Laidig, and C.L. Dow. 2001. The Mullica River Basin. Technical Report, New Jersey Pinelands Commission, New Lisbon, NJ.

included here are the corn snake (*Elaphe guttata guttata*), northern black racer (*Coluber constrictor constrictor*), eastern worm snake (*Carphophis amoenus amoenus*), eastern hognose snake (*Heterodon platyrhinos*), northern pine snake *(Pituophis melanoleucus melanoleucus)*, northern scarlet snake (*Cemophora coccinea*), rough green snake (*Opheodrys aestivus*), and eastern garter snake (*Thamnophis sirtalis*) (Conant, 1998).

Several species of turtles exist in watershed areas of the JCNERR, with most concentrating near freshwater and brackish water habitats. Those common or abundant in the Pinelands are the snapping turtle (*Chelydra serpentina*), red-bellied turtle (*Chrysemys rubriventris*), eastern painted turtle (*Chrysemys picta*), stinkpot (*Sternotherus odoratus*), spotted turtle (*Clemmys guttata*), eastern box turtle (*Terrapene carolina*), and eastern mud turtle (*Kinosternon subrubrum subrubrum*). The bog turtle (*C. muhlenbergii*), an endangered species, and the wood turtle (*C. insculpta*), a threatened species, also can be found in the Pinelands (Conant, 1998). The northern diamondback terrapin (*Malaclemys terrapin terrapin*) is a year-round resident, nesting on sandy uplands adjacent to tidal creeks and salt marshes (U.S. Fish and Wildlife Service, 1996). Hence, it is frequently observed in *Spartina* marsh habitat of the JCNERR (JCNERR, 1999).

Mammals

More than 30 land-dwelling mammals inhabit the Pine Barrens in proximity to the JCNERR. Based on their size, these mammals have been divided into small, intermediate, and large species groups (Wolgast, 1998). The small mammals are defined as those with an adult body length (excluding tail) of less than 26 cm. Twenty-two species are listed here including the meadow jumping mouse (*Zapus hudsonius*), house mouse (*Mus musculus*), white-footed mouse (*Peromyscus leucopus*), meadow vole (*Microtus pennsylvanicus*), pine vole (*M. pinetorum*), red-backed vole (*Clethrionomys gapperi*), Norway rat (*Rattus norvegicus*), rice rat (*Oryzomys palustris*), southern bog lemming (*Synaptomys cooperi*), long-tailed weasel (*Mustela frenata*), red squirrel (*Tamiasciurus hudsonicus*), gray squirrel (*Sciurus carolinensis*), southern flying squirrel (*Glaucomys volans*), eastern chipmunk (*Tamias striatus*), eastern pipistrelle (*Pipistrellus subflavus*), big brown bat (*Eptesicus fuscus*), little brown myotis (*Myotis lucifugus*), eastern mole (*Scalopus aquaticus*), star-nosed mole (*Condylura cristata*), least shrew (*Cryptotis parva*), short-tailed shrew (*Blarina brevicauda*), and masked shrew (*Sorex cinereus*).

The mammals of intermediate size range from 26 to 76 cm in length (excluding tail). Eleven species belong to this group. These are the raccoon (*Procyon lotor*), mink (*Mustela vison*), muskrat (*Ondatra zibethicus*), beaver *(Castor canadensis)*, river otter (*Lutra canadensis*), gray fox (*Urocyon cinereoargenteus*), red fox (*Vulpes vulpes*), woodchuck (*Marmota monax*), opossum (*Didelphis marsupialis*), eastern cottontail (*Sylvilagus floridanus*), and striped skunk (*Mephitis mephitis*).

The large mammal category includes those forms with an adult body length greater than 1 m. Only two species comprise this group: the white-tailed deer (*Odocoileus virginianus*) and humans (*Homo sapiens*). Humans, of course, have the greatest capacity to impact watershed environments.

Most of the aforementioned species have distinct habitat preferences. For example, more than a dozen species prefer the upland forests, notably the eastern chipmunk (*Tamias striatus*), gray squirrel (*Sciurus carolinensis*), red squirrel (*Tamiasciurus hudsonicus*), southern flying squirrel (*Glaucomys volans*), pine vole (*Microtus pinetorum*), white-footed mouse (*Peromyscus leucopus*), gray fox (*Urocyon cinereoargenteus*), red fox (*Vulpes vulpes*), raccoon (*Procyon lotor*), long-tailed weasel (*Mustela frenata*), striped skunk (*Mephites mephites*), opossum (*Didelphis marsupialis*), and white-tailed deer (*Odocoileus virginianus*). Only a few species inhabit grasslands and shrublands in the watershed, specifically the meadow jumping mouse (*Zapus hudsonius*), meadow vole (*Microtus pennsylvanicus*), woodchuck (*Marmota monax*), and eastern cottontail (*Sylvilagus floridanus*). The red-backed vole (*Clethrionomys gapperi*) inhabits bogs and wetland forests. The muskrat (*Ondatra zibethicus*) occupies both brackish and freshwater marshes. While the beaver (*Castor canadensis*) resides in freshwater tributary systems, the river otter (*Lutra canadensis*) has a broader distribution; it is observed in Pine Barren streams and tidal marshes, as well as bay islands. The mink (*Mustela vison*), southern bog lemming (*Synaptomys cooperi*), and least shrew (*Cryptotis parva*) prefer wetland habitats (Wolgast, 1998).

Birds

The JCNERR lies within the Atlantic Flyway, and consequently numerous species of migrating birds utilize the coastal habitats there. The reserve is replete with a wide diversity of seabirds, shorebirds, waterfowl, songbirds, and raptors. Protected habitat of the Holgate Unit of the Forsythe National Wildlife Refuge, Brigantine and Barnegat portions of the Forsythe National Wildlife Refuge, and the Great Bay Boulevard Wildlife Management Area provide many hectares of unaltered habitat for foraging, staging, and nesting birds.

During the period from July through December 1995, aerial surveys revealed more than 900,000 seabirds migrating along the coast in Cape May County. Loons (*Gavia* spp.), northern gannets (*Sula bassanus*), sooty shearwaters (*Puffinus griseus*), cormorants (*Phalacrocorax* spp.), and Wilson's storm petrels (*Oceanites oceanicus*) were representative seabird populations, among others, recorded in the surveys (U.S. Fish and Wildlife Service, 1996). They have also been observed migrating along the coast in Ocean and Atlantic Counties.

Recent avifaunal surveys identified nearly 300 species of birds in the Forsythe National Wildlife Refuge alone (U.S. Fish and Wildlife Service, 1996). Some of the shorebird species censused in the refuge are the American oystercatcher (*Haematopus palliatus*), semipalmated plover (*Charadrius semipalmatus*), lesser golden plover (*Pluvialis dominica*), black-bellied plover (*Pluvialis squatarola*), willet (*Catoptrophorus semipalmatus*), ruddy turnstone (*Arenaria interpres*), Hudsonian godwit (*Lemosa haemastica*), marbled godwit (*Lemosa fedoa*), greater yellowlegs (*Tringa melanoleuca*), short-billed dowitcher (*Limnodromus griseus*), dunlin (*Calidris alpina*), sanderling (*C. alba*), red knot (*C. canutus*), least sandpiper (*C. minutilla*), western sandpiper (*C. mauri*), white-rumped sandpiper (*C. fuscicollis*), and semipalmated sandpiper (*C. pusilla*). Many of these species actively breed in the refuge.

The barrier island complex and the back-bay lagoonal system provide important habitats for shorebirds not only during the critical spring and fall migration periods but also during other times of the year. Many shorebirds feed along tidal flats and roost in coastal marshes. A number of species (e.g., willet, *Catoptrophorus semipalmatus*; American oystercatcher, *Haematopus palliatus*; black-bellied plover, *Pluvialis squatarola*; and piping plover, *Charadrius melodus*) inhabit dunes and beaches, nesting in undisturbed areas. The Holgate Unit of the Forsythe National Wildlife Refuge is highly favored habitat of the piping plover. While the piping plover and black-bellied plover prefer beach and dune habitats, the willet and American oystercatcher have broader habitat preferences and thus commonly occur in salt marsh habitats as well. Shorebirds that nest on exposed stretches of beaches and dunes, such as the piping plover, are especially susceptible to predators (e.g., raccoons, *Procyon lotor*; foxes, *Vulpes vulpes*; crows, *Corvus* spp.; and gulls, *Larus* spp.).

JCNERR habitats support an array of colonial nesting waterbirds, such as gulls, terns, and waders. For example, the Great Bay Boulevard Wildlife Management Area, the Holgate Unit of the Forsythe National Wildlife Refuge, and parts of the Brigantine National Wildlife Refuge, as well as bay islands, are important locations for nesting colonies of gulls (herring gull, *Larus argentatus*; laughing gull, *L. atricilla*; and great black-backed gull, *L. marinus*) and terns (Forster's tern, *Sterna forsteri*; common tern, *S. hirunda*; and gull-billed tern, *S. nilotica*). Two important beach nesting birds in the JCNERR are the least tern (*S. antillarum*) and black skimmer (*Rynchops niger*). Nesting waders documented in the reserve include egrets (great egret, *Casmerodius albus*; snowy egret, *Egretta thula*; and cattle egret, *Bubulcus ibis*), herons (little blue heron, *E. caerulea*; tri-colored heron, *E. tricolor*; yellow-crowned night heron, *Nycticorax violaceus*; black-crowned night heron, *N. nycticorax*; and green-backed heron, *Butorides striatus*), and the glossy ibis (*Plegadis falcinellus*). These birds commonly nest among shrubs and trees in the lower watershed. The common tern, great egret, and snowy egret are considered to be potentially valuable indicator species of changing habitat conditions in the region (Joanna Burger, Rutgers University, personal communication, 2002).

In the Brigantine Bay and marsh complex, both terns (Forster's tern, *Sterna forsteri*; common tern, *S. hirunda*; and gull-billed tern, *S. nilotica*) and gulls (herring gull, *Larus argentatus*; laughing gull, *L. atricilla*; and great black-backed gull, *L. marinus*) are abundant. Long-legged waders, egrets and herons, are also relatively abundant here. In the Barnegat Bay–Little Egg Harbor estuarine system to the north, aerial colonial waterbird surveys in 1995 registered 435 waders in 14 heronies (U.S. Fish and Wildlife Service, 1996). Species reported in declining order of abundance included the snowy egret (*Egretta thula*), great egret (*Casmerodius albus*), glossy ibis (*Plegadis falcinellus*), black-crowned night heron (*Nycticorax nycticorax*), little blue heron (*E. caerulea*), tri-colored heron (*E. tricolor*), and yellow-crowned night heron (*Nycticorax violaceus*). Based on aerial waterbird surveys, Burger et al. (2001) noted a significant increase in the number of colonies of great egrets, black-crowned night herons, and glossy ibises during the period from 1977 to 1995. In addition, the number of colonies of common terns and black skimmers decreased substantially over the 1976–1999 period. Table 3.4 lists the colonial waterbirds recorded by Burger et al. (2001) in the estuarine system.

Waterfowl in the JCNERR attain peak numbers during the winter season. In the Mullica River–Great Bay estuarine system, mid-winter aerial waterfowl counts have recently revealed an average of more than 12,000 birds. Species observed in descending order of abundance in the system are the American black duck (*Anas rubripes*), American brant (*Branta bernicla*), greater and lesser scaup (*Aythra marila* and *A. affinis*), mallard (*Anas platyrhynchos*), and bufflehead (*Bucephala albeola*). Other species present, albeit in lesser numbers, are the canvasback (*Aythra valisneria*), American wigeon (*Anas americana*), northern pintail (*A. acuta*), gadwall (*A. strepera*), green-winged teal (*A. crecca*), common merganser (*Mergus merganser*), red-breasted merganser (*M. serrator*), hooded merganser (*Lophodytes cucullatus*), tundra swan (*Cygnus columbianus*), oldsquaw (*Clangula hyemalis*), common goldeneye (*Bucephala clangula*), and Canada goose (*Branta canadensis*) (U.S. Fish and Wildlife Service, 1996).

Waterfowl counts in the Brigantine and marsh complex to the south and the Barnegat Bay system to the north are greater than those in the Mullica River–Great Bay Estuary, far exceeding averages of more than 70,000 birds and 50,000 birds, respectively. In the Brigantine and marsh complex, the most abundant species in descending order are the brant, Amerian black duck, snow goose (*Chen caerulescens*), greater and lesser scaup, Canada goose, bufflehead, scoters (*Melanitta* spp.), and mallard. Similarly, in the Barnegat Bay system, the most abundant waterfowl

TABLE 3.4
Colonial Waterbirds of the Barnegat Bay–Little Egg Harbor Estuary

Common Name	Scientific Name
Gulls, Terns, Skimmers	
Common tern	*Sterna hirundo*
Least tern	*Sterna antillarum*
Forster's tern	*Sterna forsteri*
Roseate tern	*Sterna dougallii*
Caspian tern	*Sterna caspia*
Gull-billed tern	*Sterna nilotica*
Laughing gull	*Larus atricilla*
Herring gull	*Larus argentus*
Great black-backed gull	*Larus marinus*
Black skimmer	*Rynchops niger*
Long-Legged Wading Birds	
Great egret	*Casmerodius albus*
Snowy egret	*Egretta thula*
Cattle egret	*Bubulcus ibis*
Great blue heron	*Ardea herodias*
Green-backed heron	*Butorides striatus*
Little blue heron	*Egretta caerulea*
Tri-colored heron	*Egretta tricolor*
Yellow-crowned night-heron	*Nycticorax violaceus*
Black-crowned night-heron	*Nycticorax nycticorax*
Glossy ibis	*Plegadis falcinellus*

Source: Burger, J., C.D. Jenkins, Jr., F. Lesser, and M. Gochfeld. 2001. *Journal of Coastal Research* Special Issue 32, pp. 197–211.

in descending order are the greater and lesser scaup, brant, American black duck, bufflehead, canvasback, mallard, and Canada goose. According to Nichols and Castelli (1997), waterfowl hunting is significant in the coastal bays.

Raptors are well represented in the JCNERR. The primary raptors in the reserve include the osprey (*Pandion haliaetus*), peregrine falcon (*Falco peregrinus*), and northern harrier (*Circus cyaneus*). While the osprey migrates to South America, Central America, and the Southeast U.S., the peregrine falcon and northern harrier nest year-round in the region. Other raptors of note inhabiting the reserve are the sharp-shinned hawk (*Accipiter striatus*), broad-winged hawk (*Buteo platypterus*), short-eared owl (*Asio flammeus*), great horned owl (*Bubo virginianus*), eastern screech owl (*Otus asio*), American kestrel (*Falco sparverius*), and merlin (*Falco columbarius*). The bald eagle (*Haliaeetus leucocephalus*), a state endangered species (Table 3.5), nests along the Mullica River; it roosts and feeds along tidal reaches of the river (U.S. Fish and Wildlife Service, 1996). The aforementioned

TABLE 3.5
Selected List of Endangered and Threatened Plant and Animal Species of the Mullica River Watershed

Scientific Name	Common Name	Federal Status	State Status
Plants			
Amaranthus pumilus	Seabeach amaranth	Threatened	Endangered
Schwalbea americana	Chaffseed	Endangered	Endangered
Rhynchospora knieskernii	Knieskern's beaked-rush	Threatened	Endangered
Aeschynomene virginica	Sensitive joint-vetch	Threatened	Endangered
Helonias bullata	Swamp pink	Threatened	Endangered
Juncus caesariensis	New Jersey rush		Endangered
Lobelia boykinii	Boykin's lobelia		Endangered
Cyperus polystachyos	Coast flat sedge		Endangered
Carex cumulata	Clustered sedge		Endangered
Rhexia aristosa	Awned meadow-beauty		Endangered
Kuhnia eupatorioides	False boneset		Endangered
Eupatorium resinosum	Pine Barren boneset		Endangered
Lemna perpusilla	Minute duckweed		Endangered
Sagittaria teres	Slender arrowhead		Endangered
Sagittaria australis	Southern arrowhead		Endangered
Eleocharis tortilis	Twisted spikerush		Endangered
Cirsium virginianum	Virginia thistle		Endangered
Scleria verticillata	Whorled nut-rush		Endangered
Chenopodium rubrum	Red goosefoot		Endangered
Cardamine longii	Long's bittercress		Endangered
Scirpus longii	Long's woolgrass		Endangered
Corema conradii	Broom crowberry		Endangered
Tofieldia racemosa	False asphodel		Endangered
Aster radula	Low rough aster		Endangered
Glaux maritima	Sea-milkwort		Endangered
Linum intercursum	Sandplain flax		Endangered
Verbena simplex	Narrow-leaf vervain		Endangered
Animals			
Nicrophorus americanus	American burying beetle	Endangered	Endangered
Callophrys irus	Frosted elfin		Threatened
Cicindela dorsalis dorsalis	Northeastern beach tiger beetle	Threatened	Endangered
Boloria selene myrina	Silver-bordered fritillary		Threatened
Atrytone arogos arogos	Arogos skipper		Endangered
Hyla chrysoscelis	Cope's gray treefrog		Endangered
Pseudotriton montanus montanus	Eastern mud salamander		Threatened
Ambysoma tigrinum tigrinum	Eastern tiger salamander		Endangered
Clemmys muhlenbergii	Bog turtle	Threatened	Endangered
Clemmys insculpta	Wood turtle		Threatened
Elaphe guttata guttata	Corn snake		Endangered

TABLE 3.5 (CONTINUED)
Selected List of Endangered and Threatened Plant and Animal Species of the Mullica River Watershed

Scientific Name	Common Name	Federal Status	State Status
Pituophis melanoleucus melanoleucus	Northern pine snake		Threatened
Crotalus horridus horridus	Timber rattlesnake		Endangered
Rynchops niger	Black skimmer		Endangered
Sterna antillarum	Least tern		Endangered
Sterna dougallii	Roseate tern	Endangered	Endangered
Haliaeetus leucocephalus	Bald eagle	Threatened	Endangered
Falco peregrinus	Peregrine falcon		Endangered
Charadrius melodus	Piping plover	Threatened	Endangered
Bartramia longicauda	Upland sandpiper		Endangered
Calidris cantus	Red knot		Threatened
Cistothorus platensis	Sedge wren		Endangered

Source: The Natural Heritage Program, Trenton, NJ.

species can be seen over a range of habitats in the JCNERR from coastal dunes and marshlands to forest bogs and ponds. Some predatory birds (e.g., broad-winged hawk, great horned owl, and eastern screech owl) are largely confined to upland forested habitat. The raptors aggressively hunt for fish, mammals, reptiles (lizards and snakes), small birds, insects, and other animals. For example, the osprey preys on fish, whereas the peregrine falcon consumes small birds. The northern harrier, in turn, feeds on a wide range of prey such as small mammals and birds, as well as rodents.

The Mullica River watershed supports numerous species of songbirds and rails that feed, breed, and nest in various habitats. For example, Virginia rails (*Rallus limicola*), clapper rails (*Rallus longirostris*), and sora (*Porzana carolina*) feed and breed in freshwater and brackish marshes. They are commonly seen foraging in marsh habitat along the Bass, Wading, and Mullica rivers.

An array of songbirds also inhabits marshlands as well as other lowland areas along swamps, bogs, and lands surrounding lakes and streams. The marsh wren (*Cistothorus palustris*), sharp-tailed sparrow (*Ammospiza caudacuta*), and seaside sparrow (*Ammospiza maritima*) construct nests in tidal and freshwater marshes. While the marsh wren builds nests above ground level in emergent vegetation (i.e., cattails, common reed, and cordgrass), the sharp-tailed sparrow and seaside sparrow construct nests at or near ground level in cordgrass and salt meadow marshes (Kroodsma and Verner, 1997). The marsh wren is observed rather infrequently in freshwater marsh habitats.

Many insectivores frequent swamp, lake, and bog habitats in search of food. Among these species are the song sparrow (*Melospiza melodia*), yellow warbler

(*Dendroica petechia*), northern parula warbler (*Parula americana*), white-eyed vireo (*Vireo griseus*), tree swallow (*Iridoprocne bicolor*), purple martin (*Progne subis*), eastern wood pewee (*Contopus virens*), redstart (*Setophaga ruticilla*), yellowthroat (*Geothlypis trichas*), wood thrush (*Hylocichla mustilina*), and gray catbird (*Dumetella carolinensis*). Other insectivorous species concentrate in upland oak–pine or pine–oak woodlands. For example, the pine warbler (*Dendroica pinus*) occupies pine–oak stands, whereas the ovenbird (*Seiurus aurocapillus*), black-and-white warbler (*Mniotilta varia*), and red-eyed vireo (*Vireo olivaceus*) prefer oak-dominated stands. Some species (e.g., gray catbird, *Dumetella carolinensis* and rufous-sided towhee, *Pipilo erythrophthalmus*) are particularly abundant in scrubby undergrowth and dense thickets, which afford greater protection from predators (U.S. Fish and Wildlife Service, 1996; Leck, 1998).

Various neotropical migrant species utilize scrub–shrub and forest habitats of the JCNERR system. Some representative groups are the hummingbirds, swifts, swallows, flycatchers, grosbeaks, buntings, tanagers, and nightjars (De Graaf and Rappole, 1995). Southern bird species common in upland areas of the watershed are the mockingbird (*Mimus polyglottus*), Carolina wren (*Thryothorus ludovicianus*), and Carolina chickadee (*Parus carolinensis*).

Fish

Zampella et al. (2001) documented 22 species of fish in the streams and impoundments of the Mullica River Basin, consisting of both native and nonnative forms (Table 3.6). Based on their surveys, the banded sunfish (*Enneacanthus obesus*), swamp darter (*Etheostoma fusiforme*), chain pickerel (*Esox niger*), and eastern mudminnow (*Umbra pygmaea*) are the most frequently encountered native species in the Pinelands streams. Commonly occurring nonnative species include the golden shiner (*Notemigonus crysoleucas*), yellow perch (*Perca flavescens*), bluegill sunfish (*Lepomis macrochirus*), pumpkinseed (*L. gibbosus*), tesselated darter (*Etheostoma olmetedi*), brown bullhead (*Ameiurus nebulosus*), and largemouth bass (*Micropterus salmoides*). Three native species (i.e., blackbanded sunfish, *Enneacanthus chaetodon*; bluespotted sunfish, *E. gloriosus*; and banded sunfish, *E. obesus*) dominate the impoundment assemblages. Fish inhabiting streams in the Mullica River watershed are primarily acid-tolerant, sedentary forms that prefer areas with considerable vegetation (Hastings, 1998).

The acidic waters of the Mullica River watershed create inhospitable conditions for many freshwater fish. Hastings (1998) identified only 16 species of fish indigenous to the acidic waters of the region. He also divided Pine Barrens fish into five groups:

1. Characteristic species
2. Peripheral species
3. Introduced species
4. Anadromous species
5. Marine species

TABLE 3.6
Taxonomic List of Fish Collected in the Mullica River Basin

Common Name	Scientific Name
Mud sunfish	*Acantharchus pomotis*
Yellow bullhead	*Ameiurus natalis*
Brown bullhead	*Ameiurus nebulosus*
American eel	*Anguilla rostrata*
Pirate perch	*Aphredoderus sayanus*
Blackbanded sunfish	*Enneacanthus chaetodon*
Bluespotted sunfish	*Enneacanthus gloriosus*
Banded sunfish	*Enneacanthus obesus*
Creek chubsucker	*Erimyzon oblongus*
Redfin pickerel	*Esox americanus*
Chain pickerel	*Esox niger*
Swamp darter	*Etheostoma fusiforme*
Tessellated darter	*Etheostoma olmstedi*
Banded killifish	*Fundulus diaphanus*
Pumpkinseed	*Lepomis gibbosus*
Bluegill	*Lepomis macrochirus*
Largemouth bass	*Micropterus salmoides*
Golden shiner	*Notemigonus crysoleucas*
Tadpole madtom	*Noturus gyrinus*
Yellow perch	*Perca flavescens*
Black crappie	*Pomoxis nigromaculatus*
Eastern mudminnow	*Umbra pygmaea*

Source: Zampella, R.A., J.F. Bunnell, K.J. Laidig, and C.L. Dow. 2001. The Mullica River Basin. Technical Report, New Jersey Pinelands Commission, New Lisbon, NJ.

Included among the characteristic species are the yellow bullhead (*Ameiurus natalis*), pirate perch (*Aphredoderus sayanus*), ironcolor shiner (*Notropis chalybaeus*), swamp darter (*Etheostoma fusiforme*), mud sunfish (*Acantharchus pomotis*), banded sunfish (*Enneacanthus obesus*), bluespotted sunfish (*E. gloriosus*), and blackbanded sunfish (*E. chaetodon*). Other, more widespread species comprising this group are the brown bullhead (*Ameiurus nebulosus*), creek chubsucker (*Erimyzon oblongus*), tadpole madtom (*Noturus gyrinus*), tessellated darter (*Etheostoma olmstedi*), eastern mudminnow (*Umbra pygmaea*), chain pickerel (*Esox niger*), redfin pickerel (*Esox americanus*), and American eel (*Anguilla rostrata*). The distribution of these characteristic forms, as well as the species constituting the other finfish groups in the Pine Barrens, depends on their tolerance to the acidic waters, the competition from similar or related species, and the requirement of some species for sluggish streams or standing water with dense vegetation (Hastings, 1998).

Some species occupy nonacidic or weakly acidic waters in marginal areas of the Pine Barrens and avoid acidic waters. Pineland species found in nonacidic waters of the Pine Barrens are the bluntnose minnow (*Pimephales notatus*), silvery minnow (*Hybognathus nuchalis*), creek chub (*Semotilus atromaculatus*), fallfish (*S. corporalis*), margined madtom (*Noturus insignis*), blacknose dace (*Rhinichthys atratulus*), gizzard shad (*Dorosoma cepedianum*), common shiner (*Notropis cornutus*), comely shiner (*N. amoenus*), bridled shiner (*N. bifrenatus*), satinfin shiner (*N. analostanus*), and American brook lamprey (*Lampetra lamottei*). Peripheral species restricted to the lower reaches of the Mullica River and other influent systems of the JCNERR coastal bays include the mummichog (*Fundulus heteroclitus*), banded killifish (*Fundulus diaphanus*), golden shiner (*Notemigonus crysoleucas*), spotted shiner (*Notropis hudsonius*), redbreasted sunfish (*Lepomis auritus*), white sucker (*Catostomus commersoni*), white perch (*Morone americana*), and yellow perch (*Perca flavescens*) (U.S. Fish and Wildife Service, 1996; Hastings, 1998).

A number of introduced species have adapted to streams in the Mullica River watershed. Some of these species are stocked fish that now occur in peripheral areas of the Pine Barrens. Among notable introduced forms are the brook trout (*Salmo fontinalis*), rainbow trout (*Salmo trutta*), and brown trout (*Salmo gairdneri*). In addition to these salmonids, the flathead minnow (*Pimephales promelas*), black bullhead (*Ictalurus melas*), and channel catfish (*Ictalurus punctatus*) are introduced forms. Carp (*Cyprinus carpio*), goldfish (*Carassius auratus*), largemouth bass (*Micropterus salmoides*), black crappie (*Pomoxis nigromaculatus*), and bluegill (*Lepomis macrochirus*) likewise are introduced species occupying peripheral areas of the Pine Barrens (U.S. Fish and Wildlife Service, 1996).

Several anadromous species utilize the Mullica River and its tributaries. Species that migrate into these influent systems, being largely confined to their tidal reaches, are the alewife (*Alosa pseudoharengus*), blueback herring (*A. aestivalis*), hickory shad (*A. mediocris*), American shad (*A. sapidissima*), and striped bass (*Morone saxatilis*). The construction of dams in past decades has created obstructions to the upstream movement of anadromous fish, thereby hindering spawning runs. These obstructions have restricted anadromous fish spawning to the lower reaches of rivers and streams in the watershed.

Aside from the anadromous fish, a number of marine species also occur in rivers and streams of the Pine Barrens. Among this group of marine species are the Atlantic croaker (*Micropogonias undulatus*), Atlantic menhaden (*Brevoortia tyrannus*), weakfish (*Cynoscion regalis*), kingfish (*Menticirrhus saxatilis*), hogchoker (*Trinectes maculatus*), three-spined stickleback (*Gasterosteus aculeatus*), four-spined stickleback (*Apeltes quadracus*), and bay anchovy (*Anchoa mitchilli*). They are typically found in the lower tidal reaches of the influent systems.

ESTUARINE BIOTIC COMMUNITIES

PLANT COMMUNITIES

Benthic Flora

Submerged aquatic vegetation (SAV) in the estuarine waters of the JCNERR consists of rooted macrophytes, both marine angiosperms (i.e., true seagrasses) and freshwater macrophytes, that have colonized areas of the coastal bays. Eelgrass (*Zostera marina*) is the dominant species, with widgeon grass (*Ruppia maritima*) also found in some areas. In addition, an array of benthic macroalgae occurs in the coastal bays either attached to substrates or as part of a drift community. Eelgrass beds in the JCNERR are restricted to Little Egg Harbor and lower Barnegat Bay. Benthic macroalgae, in turn, are more widely distributed; they have been observed in lower Barnegat Bay, Little Egg Harbor, Great Bay, and the coastal bays to the south (i.e., Little Bay, Reeds Bay, and Absecon Bay).

The most abundant benthic macroalgae in the Barnegat Bay–Little Egg Harbor Estuary are *Agardhiella subulata, Ceramium fastigiatum, Chaetomorpha* spp., *Codium fragile, Gracilaria tikvahiae*, and *Ulva lactuca* (Loveland and Vouglitois, 1984; Kennish et al., 2001c). In the Barnegat Bay–Little Egg Harbor Estuary as well as Great Bay, *Ulva lactuca* exhibits the greatest biomass. When present in high abundances, benthic macroalgae can attenuate light and hinder the growth and survival of seagrasses. However, benthic macroalgae can provide habitat for some organisms, such as amphipods, crabs, and shrimp (Wilson et al., 1990; Sogard and Able, 1991; Jivoff and Able, 2001).

The distribution of eelgrass beds appears to have decreased significantly during the past 25 years in lower Barnegat Bay and Little Egg Harbor. For example, in a survey of SAV beds in Little Egg Harbor from July to October 1999, Bologna et al. (2000) found 1298.9 ha of *Zostera marina* and 6.8 ha of *Ruppia maritima*. This compares to a total SAV coverage of 3448 ha reported in Little Egg Harbor by Macomber and Allen (1979) in the 1970s. Bologna et al. (2000) noted, therefore, that SAV coverage in Little Egg Harbor declined by ~62% between the mid-1970s and 1999 due to the loss of *Zostera marina* beds. A substantial loss of eelgrass also appears to have occurred in Barnegat Bay during this interval (Figure 3.10). Despite the obvious loss of eelgrass in much of Little Egg Harbor since the 1970s, evidence exists of some recent recolonization of habitat in the southern part of the estuary.

Several factors other than benthic macroalgal infestation may be contributing to the loss of SAV in the Barnegat Bay–Little Egg Harbor Estuary. Noteworthy in this respect is the input of excessive amounts of nutrients, particularly nitrogen, leading to phytoplankton blooms and excessive growth of epiphytes that cause shading impacts on SAV. In addition, wasting disease by the protist, *Labyrinthula zosterae*, also decimates eelgrass beds. For example, wasting disease eliminated 400 ha of eelgrass beds in Barnegat Bay during 1995, and it was evident in up to 50% of the eelgrass leaves examined in 1996 (McClain and McHale, 1997).

Eelgrass has historically occurred in relatively dense concentrations along the perimeter of the Barnegat Bay–Little Egg Harbor Estuary. It is most abundant where sandy sediments predominate along the eastern margin of the estuary (Figure 3.10) (Lathrop et al., 2001). Bologna et al. (2000) recorded a peak eelgrass biomass of 230 g FDW/m^2 in Little Egg Harbor. Wootton and Zimmerman (2001) found above-ground and below-ground biomass values of eelgrass beds in Barnegat Bay ranging from 8.7 to 270.6 g/m^2.

Great Bay and the back-bays to the south (i.e., Little Bay, Reeds Bay, and Absecon Bay) are essentially devoid of *Zostera marina* and other seagrasses.

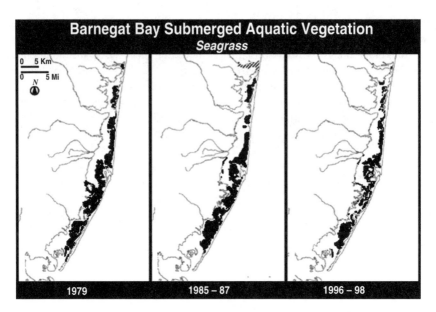

FIGURE 3.10 Seagrass distribution in the Barnegat Bay–Little Egg Harbor Estuary over the 1979–1998 period. (From Lathrop, R.G., J.A. Bognar, A.C. Hendrickson, and P.D. Bowers. 2001. In: Kennish, M.J. (Ed.). *The Scientific Characterization of the Barnegat Bay–Little Egg Harbor Estuary and Watershed.* Technical Report, Barnegat Bay National Estuary Program, Toms River, NJ, Appendix 3.)

Eelgrass beds once flourished in these bays but were eradicated by the wasting disease pandemic of the 1930s, which devastated eelgrass beds along most of the East Coast (Cottam and Munro, 1954). Benthic macroalgae (sea lettuce, *Ulva lactuca*; hollow green weed, *Enteromorpha* spp.; and rockweed, *Fucus* sp.) replaced *Z. marina* as the dominant benthic macroflora in these systems after the 1930s (Sogard and Able, 1991). They remain the dominant forms today.

The contraction of eelgrass beds in the JCNERR is a growing concern because of the multiple ecological roles of SAV in the system. Aside from its importance in primary production, SAV provides food and habitat for a multitude of organisms. Waterfowl (e.g., American brant, canvasbacks, and green-winged teal), turtles, some fish (e.g., Acanthuridae and Scaridae), and sea urchins consume SAV. Seagrasses serve as nursery and protective habitat for fish, crabs, and many other estuarine fauna. They also trap sediment, stabilize the substrate, and mitigate erosion. These plants act as nutrient pumps and thus may be significant in nutrient cycling in the estuary.

Phytoplankton

Olsen and Mahoney (2001) conducted the most extensive phytoplankton studies in Great Bay, Little Egg Harbor, and Barnegat Bay during the past 20 years. They surveyed phytoplankton communities in these estuaries from May through September during 1987, and in 1995 to 2000. These surveys were prompted by the recurrence of intense picoplankton blooms dominated by *Aureococcus anophagefferens* (Pelagophyceae) and *Nannochloris atomus* (Chlorophyceae). Intense picoplankton blooms have caused brownish water discoloration each summer since 1985; these blooms have been most intense and widespread in Little Egg Harbor. Phytoplankton counts in these blooms typically exceed 10^6 cells/ml.

Based on the surveys of Olsen (1989) and Olsen and Mahoney (2001), as well as previous investigations of Martin (1929) and Mountford (1965, 1967, 1969, 1971), a total of 242 phytoplankton species have been identified in Barnegat Bay, Little Egg Harbor, and Great Bay. Olsen and Mahoney (2001) determined that 41% of the 132 phytoplankton species they identified were dinoflagellates and 31% were diatoms. Aside from *Aureococcus anophagefferens* and *Nannochloris atomus*, other cosmopolitan species that are widely distributed both spatially and temporally in these coastal bays include *Calycomonas ovalis*, *Chlorella* sp., *Chroomonas vectensis*, *C. amphioxiea*, *C. minuta*, *Cyclotella* sp., *Cylindrotheca closterium*, *Euglena/Eutreptia* spp., *Heterosigma carterae*, *Katodinium rotundatum*, *Nitzschia* spp., *Prorocentrum minimum*, *Pyramimonas* spp., and *Skeletonema costatum*. In proximity to inlet sites (i.e., Little Egg Inlet and Barnegat Inlet), however, neritic or coastal ocean species dominate the phytoplankton communities. These consist mainly of centric diatoms (e.g., *Cerataulina*, *Chaetoceros*, and *Thalassiosira* spp.) and thecate dinoflagellates (e.g., *Ceratium*, *Dinophysis*, *Prorocentrum*, and *Protoperidinium* spp.).

Picoplankton, bacteria-sized forms 1 to 5 μm in size, are an important component of the phytoplankton community, responsible for a significant fraction of the total phytoplankton production. Mountford (1971) ascertained that *Nannochloris atomus* dominated the phytoplankton community in surveys he conducted in Barnegat Bay during the 1960s. Olsen and Mahoney (2001) corroborated the numerical importance of *N. atomus* and also confirmed the significance of the brown-tide alga, *Aureococcus anophagefferens*. The more frequent and widespread occurrence of brown-tide blooms, particularly in Little Egg Harbor, may signal shifts in water quality of the coastal bays favoring the bloom species and promoting more eutrophic conditions. Such a change is problematic because the picoplankton blooms also appear to displace the normal phytoplankton community, thereby potentially altering bottom-up controls and upper-trophic-level organisms in the bays.

Phytoplankton productivity in the Barnegat Bay–Little Egg Harbor Estuary averages ~480 g C /m²/yr (Moser, 1997). In this system, the total nitrogen concentrations range from ~20–80 μM, and phosphate levels < 1 μM (Seitzinger et al., 2001). Durand (1984), investigating phytoplankton productivity along a salinity gradient of the Mullica River–Great Bay Estuary, recorded the following productivity values:

1. 422–1081 mg C/m²/day at a lower Mullica River site
2. 485–985 mg C/m²/day at the head of the bay
3. 1362 mg C/m²/day at a down-bay site

Phytoplankton production in Great Bay occurs thorughout the water column, while it is limited to a portion of the water column (the upper 1.5 m) in the Mullica

River due to higher turbidity levels. Thus, Great Bay has greater abundances of phytoplankton than does the Mullica River.

ANIMAL COMMUNITIES

Zooplankton

Few zooplankton studies have been conducted in the back-bay waters of the JCNERR. Durand and Nadeau (1972) examined the zooplankton community in greatest detail, focusing on the Mullica River–Great Bay Estuary. Zooplankton in Little Egg Harbor and the back-bays to the south (i.e., Little Bay, Reeds Bay, and Absecon Bay), however, remain largely uncharacterized.

Along a salinity gradient of the Mullica River–Great Bay Estuary, Durand and Nadeau (1972) collected zooplankton samples at four sites:

1. Lower Bank (freshwater–saltwater interface)
2. French Point
3. Graveling Point (head of the bay)
4. Rutgers Marine Field Station (polyhaline waters)

Maximum zooplankton abundance occurred during the March through September period, with a mean count of 64 organisms/l being recorded at this time (Table 3.7). Minimum zooplankton counts (10 organisms/l or less) were registered in winter (December and January), and intermediate counts in October, November, and February. Zooplankton abundance was higher at the bay and lower Mullica River sampling sites.

Cladocerans, rotifers, and copepods dominated the microzooplankton in the estuary. Cladocerans and rotifers were less abundant at the Lower Bank site ~25 km upriver. At this location, the density of cladocerans and rotifers amounted to 190 and 40 organisms/l, respectively.

Calanoid and harpacticoid copepods were both important components of the microzooplankton. *Acartia tonsa*, *Eurytemora affinis*, and *Oithona similis* were the most abundant copepod species. *Eurytemora affinis* dominated the microzooplankton in the Mullica River, and *A. tonsa* was the dominant form in Great Bay. *Oithona similis*, a coastal ocean species, predominated in polyhaline waters near Little Egg Inlet. Other important members of the zooplankton community in Great Bay were *Centropages hamatus*, *C. typicus*, *Labidocera aestiva*, *Paracalanus crassirostris*, *P. parva*, *Pseudocalanus minutus*, *Pseudodiaptomus coronatus*, *Temora longicornis*, and *Tortanus discaudatus*.

Sandine (1984) and Kennish (2001b), reviewing the macrozooplankton of the Barnegat Bay–Little Egg Harbor Estuary, reported that *Crangon septemspinosa*, *Jassa falcata*, *Neomysis americana*, *Neopanope texana*, *Panopeus herbstii*, *Rathkea octopunctata*, *Sagitta* spp., and *Sarsia* spp. were the dominant macrozooplankton species. As in the case of microzooplankton in the Barnegat Bay–Little Egg Harbor Estuary and Great Bay, macrozooplankton attained peak abundance during the spring and summer months. Arrow worms (*Sagitta* spp.) and ctenophores (*Mnemiopsis leidyi*) are important forms that prey heavily on microzooplankton in the coastal

TABLE 3.7
Monthly Mean Abundance of Zooplankton in the Mullica River–Great Bay Estuary[a]

Month	Lower Bank[b]	French Point[c]	Graveling Point[d]	Rutgers Marine Field Station[e]
October	—	4.59	2.71	5.27
November	—	—	3.12	0.85
December	12.29	4.08	0.18	0.78
January	1.46	2.47	1.27	4.50
February	5.17	3.69	6.78	—
March	57.87	10.93	34.49	13.49
April	25.65	335.44	48.94	24.50
May	39.02	54.06	35.68	—
June	126.28	118.78	33.60	32.07
July	82.42	64.70	41.72	30.47
August	214.52	56.87	36.08	32.84
September	15.31	46.60	96.47	45.28
October	5.47	19.39	3.64	39.00
November	5.67	7.62	18.11	—
December	11.06	1.42	0.35	—
January	2.46	1.95	55.85	—
February	1.90	3.17	23.20	—
March	43.55	14.18	171.94	—
April	53.72	35.61	102.56	26.51
May	517.45	18.71	42.77	—
June	103.40	9.51	25.44	4.73

[a] Number/liter.
[b] 25 km upriver.
[c] Mouth of the Mullica River.
[d] Head of Great Bay.
[e] Near Little Egg Inlet.

Source: Durand, J.B. and R.J. Nadeau. 1972. Water Resources Development in the Mullica River Basin. Part I. Biological Evaluation of the Mullica River–Great Bay Estuary. Technical Report, New Jersey Water Resources Research Institute, Rutgers University, New Brunswick, NJ.

bays and appear to play a significant role in regulating the population sizes of these organisms (Mountford, 1980).

Meroplankton and ichthyoplankton are also important elements of the zooplankton communities in JCNERR estuarine waters. Pulses of barnacle, bivalve, gastropod, polychaete, and cyphonaute larvae contribute to microzooplankton maxima in the spring and summer. Monthly densities of each of these meroplanktonic groups may exceed 10,000 individuals/m^3 during the spring and summer seasons (Kennish, 2001b).

Among the most abundant ichthyoplankton in Great Bay and other JCNERR estuarine waters are the bay anchovy (*Anchoa mitchilli*), Atlantic silverside (*Menidia menidia*), mummichog (*Fundulus heteroclitus*), gobies (*Gobiosoma* spp.), Atlantic menhaden (*Brevoortia tyrannus*), American eel (*Anguilla rostrata*), northern pipefish (*Syngnathus fuscus*), fourspine stickleback (*Apeltes quadracus*), sand lance (*Ammodytes* sp.), winter flounder (*Pseudopleuronectes americanus*), cunner (*Tautogolabrus adspersus*), and hogchoker (*Trinectes maculatus*). Ichthyoplankton of atherinids and blennids are also abundant. Most of the aforementioned ichthyoplankton species attain peak abundance in the estuaries during the June through September period, exceptions being the sand lance and winter flounder whose ichthyoplankton numbers are greatest from January through April. The maximum monthly densities of the ichthyoplankton (eggs and larvae) generally range from ~20 to 75 individuals/m^3 (Sandine, 1984).

Benthic Fauna

No recent detailed investigations of benthic faunal communities have been conducted in JCNERR estuaries. Durand and Nadeau (1972) used a Petersen dredge and modified oyster dredge to collect system-wide benthic faunal samples in the Mullica River–Great Bay Estuary. This work represents the most comprehensive investigation to date on the benthic faunal communities of the estuary. Moser (1997) studied benthic infauna at one site near Westecunk Creek on the western side of Little Egg Harbor. Hales et al. (1995) and Viscido et al. (1997) examined epibenthic invertebrate assemblages on Beach Haven Ridge at the site of the Long-term Ecosystem Observatory of Rutgers University, an area within the JCNERR on the inner continental shelf. The following discussion provides an overview of the benthic invertebrate communities in the estuarine and coastal marine waters of the JCNERR based on the aforementioned studies.

More than 200 benthic invertebrate species have been identified in the Barnegat Bay–Little Egg Harbor Estuary (Loveland and Vouglitois, 1984; Kennish, 2001c), and nearly 150 benthic invertebrate species have been recorded in Great Bay (Durand and Nadeau, 1972). Taxonomic studies of benthic faunal communities in the Barnegat Bay–Little Egg Harbor Estuary have been mainly conducted in the central zone between Stouts Creek and Oyster Creek. However, investigations of the benthic faunal communities in Great Bay have covered a larger part of the bay.

Annelids, mollusks, and arthropods dominate the benthic invertebrate community of Great Bay (Table 3.8). Durand and Nadeau (1972) found that the tube-building amphipod, *Ampelisca abdita*, was the most abundant species in the bay. In some areas of the central bay, it exceeded densities of 5000 individuals/m^2 (Figure 3.11). *Ampelisca abdita* is not only abundant but also widely distributed. In addition to *A. abdita*, six other species have been identified as true estuarine forms, occupying the lower Mullica River and Great Bay. These include *Corophium cylindricum, Cyathura polita, Notomastus laterus, Polydori ligni, Scoloplos robustus,* and *Turbonilla* sp. All are also abundant. Still other species are more spatially restricted, being classified as river, bay, or lower bay forms (Table 3.9).

TABLE 3.8
Taxonomic List of Benthic Invertebrates Collected by Petersen Dredge in the Mullica River–Great Bay Estuarine System

Phylum Porifera

Class Demospongiae
Cliona sp.   *Microciona prolifera* (Ellis and Solander)

Phylum Coelenterata

Class Anthozoa
Stylactis hooperi (Sigerfoos)   *Sagartia modesta* (Verrill)

Phylum Platyhelminthes

Class Turbellaria
Euplana gracilis (Girard)

Phylum Nemertinea

Class Anopla
Carinoma tremaphoros (Leidy)   *Zygeupolia rubens* (Coe)
Cerebratulus lacteus (Leidy)

Phylum Annelida

Class Polychaeta
Amphitrite affinis (Malmgren)   *Hypaniola grayi* (Pettibone)
Amphitrite cirrata (O.F. Müller)   *Lumbrineris tenuis* (Verrill)
Amphitrite johnstoni (Malmgren)   *Maldinopsis elongata* (Verrill)
Aricidea jeffreysii (McIntosh)   *Nephtys bucera* (Ehlers)
Brania clavata (Claparede)   *Nephtys incisa* (Malmgren)
Chone infundibuliformis (Kroyer)   *Nephtys picta* (Ehlers)
Cirratulus grandis (Verrill)   *Nereis arenaceodonta* (Moore)
Clymenella torquata (Leidy)   *Nereis grayi* (Pettibone)
Diopatra cuprea (Bosc)   *Nereis succinea* (Frey and Leuckart)
Dispio uncinata (Hartman)   *Nerinides agilis* (Verrill)
Drilonereis longa (Webster)   *Notomastus latereus* (Sara)
Drilonereis magna (Webster and Benedict)   *Paranaitis speciosa* (Webster)
Eteone heteropoda (Hartman)   *Pectinaria gouldii* (Verrill)
Eteone longa (Fabricius)   *Phyllodoce arenae* (Webster)
Eumida sanguinea (Cersted)   *Pista palmata* (Verrill)
Exogone dispar (Webster)   *Polycirrus eximius* (Leidy)
Glycera americana (Leidy)   *Polydora ligni* (Webster)
Glycera dibranchiata (Ehlers)   *Pygospio elegans* (Verrill)
Glycinde solitaria (Webster)   *Sabella microphthalma* (Verrill)
Harmothoe imbricata (L.)   *Scolecolepides viridis* (Verrill)
Hydroides dianthus (Verrill)   *Scoloplos fragilis* (Verrill)
Scoloplos robustus (Verrill)   *Streblospio benedicti* (Webster)

(continued)

TABLE 3.8 (CONTINUED)
Taxonomic List of Benthic Invertebrates Collected by Petersen Dredge in the Mullica River–Great Bay Estuarine System

Sphaerosyllis hystrix (Claparède)
Spio filicornis (O.F. Müller)

Tharyx acutus (Webster and Benedict)

Phylum Mollusca

Class Gastropoda
Acteocina canaliculata (Say)
Acteon punctostriatus (C.B. Adams)
Anachis avara (Say)
Bittium alternatum (Say)
Busycon canaliculatum (L.)
Crepidula convexa (Say)
Crepidula fornicata (L.)
Crepidula plana (Say)
Cylichna alba (Brown)
Epitonium lineatum (Say)

Eupleura caudata (Say)
Littorina littorea (L.)
Nitrella lunata (Say)
Nassarius obsoletus (Say)
Nassarius vibex (Say)
Odostomia impressa (Say)
Triphora nigrocincta (C.B. Adams)
Trophon truncatus (Say)
Turbonilla sp.
Urosalpinx cinera (Say)

Class Bivalvia
Arca pexata (Say)
Crassostrea virginica (Gmelin)
Ensis directus (Conrad)
Gemma gemma (Totten)
Lyonsia hyalina (Conrad)
Macoma tenta (Say)
Mercenaria mercenaria (L.)

Mulinia lateralis (Say)
Mya arenaria (L.)
Mytilus edulis (L.)
Nucula sp.
Spisula solidissima (Dillwyn)
Tagelus divisus (Spengler)
Tellina agilis (Stimpson)

Phylum Arthropoda

Class Crustacea
Aeginella longicornis (Kröyer)
Ampelisca abdita (Mills)
Ampelisca verrilli (Mills)
Amphithoe longimana (Smith)
Amphithoe rubricata (Montagu)
Anoplodactylus lentus (Wilson)
Batea secunda (Holmes)
Caprella geometrica (Say)
Carinogammarus mucronatus (Say)
Chiridotea almyra (Bowman)
Corophium cylindricum (Say)
Crangon septemspinosus (Say)
Cyathura polita (Stimpson)
Edotea triloba (Say)
Elasmopus laevis (Smith)
Neomysis americana (S.I. Smith)
Neopanope texana (Smith)

Erichsonella attenuata (Harger)
Ericthonius minax (Smith)
Eurypanopeus depressus (Smith)
Gammarus locusta (L.)
Grubia compta (Smith)
Haustorius arenarius (Slabber)
Heteromysis formosa (S.I. Smith)
Hippolyte zostericolor (Smith)
Idotea balthica (Pallas)
Labidocera aestiva (Wheeler)
Leucon americanus (Zimmer)
Lysianopsis alba (Holmes)
Melita nitida (Smith)
Microdeutopus gryllotalpa (Costa)
Monoculodes edwardsi (S.I. Smith)
Ptilocheirus pinquis (Stimpson)
Rithropanopeus harrisii (Gould)

TABLE 3.8 (CONTINUED)
Taxonomic List of Benthic Invertebrates Collected by Petersen Dredge in the Mullica River–Great Bay Estuarine System

Oxyurostylis smithi (Calman)
Pagurus longicarpus (Say)
Palaemonetes vulgaris (Say)
Paraphosux spinosus (Holmes)

Stenothoe cypris (Say)
Sympleustes glaber (Boeck)
Unciola irrorata (Say)

Phylum Ecotoprocta

Class Gymnolaemata
Electra crustulenta (Pallas)
Electra hastingsae (Marcus)

Membranipora sp.
Schizoporella unicornis (Johnston)

Phylum Chordata

Class Ascidiacea
Molgula manhattensis (De Kay)

Source: Durand, J.B. and R.J. Nadeau. 1972. Water Resources Development in the Mullica River Basin. Part I. Biological Evaluation of the Mullica River–Great Bay Estuary. Technical Report, New Jersey Water Resources Research Institute, Rutgers University, New Brunswick, NJ.

The composition of bottom sediments, notably the silt–clay fraction, greatly affects the spatial distribution of benthic invertebrates in the estuary. Some species (e.g., *Ampelisca verrilli, Ensis directus, Haustorius arenarius, Pygospio elegans,* and *Oxyurostylis smithi*) occupy coarser sediments, where the percentage of silt and clay is less than 20%. Other species (e.g., *Acteocina canaliculata, Lumbrineris tenuis, Maldinopsis elongata, Tellina agilis, Turbonilla* sp., and *Unciola irrorata*) prefer finer sediments, where the amount of silt and clay exceeds 38%. When proceeding seaward along the length of the estuary, *Ptilocherirus pinquis* commonly occurs in Mullica River sediments and is nearly absent in Great Bay. *Nassarius obsoletus* predominates near the mouth of the Mullica River and along the western perimeter of the bay. While *Acteocina canaliculata* and *Glycinde solitaria* are most numerous on the southwestern side of the bay, *Ampelisca abdita* and *Polydora ligni* attain high abundances in the center of the bay. *Ampelisca verrilli* and *Gemma gemma*, in turn, reach peak numbers in the southern and eastern reaches (Durand and Nadeau, 1972).

Species richness is greatest along the western and southern areas of Great Bay. More than 40 species of benthic invertebrates have been found at some sampling sites in these areas. Mullica River exhibits lower species richness, with 10 to 20 species registered at most sampling sites (Durand and Nadeau, 1972).

Moser (1997), analyzing sediment cores collected in 1993 and 1994 on the western side of Little Egg Harbor near Westecunk Creek (~12 km north of Little Egg Inlet), reported high abundances of *Cossura* sp., amounting to more than 4000 individuals/m^2 in 1993 and more than 3000 individuals/m^2 in 1994. *Sphaerosyllis* spp. (*Sphaerosyllis* sp., *S. longicauda,* and *S. taylori*) were also numerous, exceeding 4000 individuals/m^2 in 1994. Other less abundant forms included *Ampelisca abdita,*

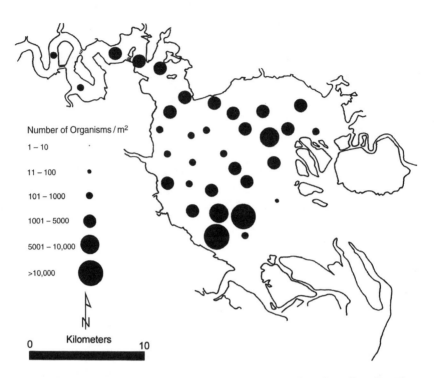

FIGURE 3.11 Density of benthic macroinvertebrates in the Mullica River–Great Bay Estuary. (From Durand, J.B. and R.J. Nadeau. 1972. Water Resources Development in the Mullica River Basin. Part 1. Biological Evaluation of the Mullica River–Great Bay Estuary. Technical Report, New Jersey Water Resources Research Institute, Rutgers University, New Brunswick, NJ.)

Brania clavata, Eumida sanguinea, Exogone dispar, Leitoscoloplos robustus, Mediomastus sp., *Melinna cristata, Microphthalmus sezelkowii, Nucula* sp., *Paraphosux* sp., and *Polycirrus hematodes.*

Along the Beach Haven Ridge approximately 9 km east of Little Egg Inlet (39°28′18″N, 74°15′10″W), Hales et al. (1995) documented significant numbers of bivalves (*Spisula solidissima*), gastropods (*Busycon* spp., *Euspira heros,* and *Nevirita duplicata*), and polychaetes (*Diopatra cuprea*). Echinoderms (*Asterias forbesi, Arbacia punctulata,* and *Echinarachnius parma*) dominated deeper water areas in proximity to the ridge. Epibenthic decapods numerically abundant on the ridge included the Atlantic rock crab (*Cancer irroratus*), spider crab (*Libinia emarginata*), lady crab (*Ovalipes ocellatus*), and sand shrimp (*Crangon septemspinosa*). Abundance of these species was seasonally variable, with *C. septemspinosa* and *L. emarginata* being most numerous in the spring and fall and *Cancer irroratus* and *O. ocellatus* attaining peak numbers in the summer (Viscido et al., 1997).

Finfish

Little Egg Harbor, Great Bay, and the back-bays to the south (i.e., Little Bay, Reeds Bay, and Absecon Bay) provide valuable habitat for numerous species of fish that

TABLE 3.9
Spatial Distribution of Benthic Invertebrates along a Salinity Gradient of the Mullica River–Great Bay Estuary

River-Dominant Forms

Cerebratulus lacteus	*Nereis succinea*
Chiridotea almyra	*Ptilocherirus pinquis*
Gammarus locusta	*Sagartia modesta*
Glycera dibranchiata	*Scolecolepides viridis*
Hypaniola grayi	*Scoloplos fragilis*
Lyonsia hyalina	*Streblospio benedicti*
Melita nitida	*Sympleustes glaber*
Neopanope texana	

Bay-Dominant Forms

Acteocina canaliculata	*Leucon americanus*
Ampelisca verrilli	*Lumbrinereis tenuis*
Amphitrite cirrata	*Maldinopsis elongata*
Arca pexata	*Mulinia lateralis*
Brania clavata	*Oxyurostylis smithi*
Crepidula convexa	*Polycirrus eximus*
Elasmopus laevis	*Tellina agilis*
Glycera americana	*Triphora nigrocincta*
Glycinde solitaria	*Unciola irrorata*

Lower-Bay–Dominant Forms

Caprella geometrica	*Idotea balthica*
Cirratulus grandis	*Nassarius vibex*
Crangon septemspinosa	*Nephtys picta*
Cylichna alba	*Pagurus longicarpus*
Ensis directus	*Pygospio elegans*
Haustorius arenarius	*Stenothoe cypris*

Estuarine Forms (Entire Range)

Ampelisca abdita	*Polydora ligni*
Corophium cylindricum	*Scoloplos robustus*
Cyathura polita	*Turbonilla* sp.
Notomastus latereus	

Source: Durand, J.B. and R.J. Nadeau. 1972. Water Resources Development in the Mullica River Basin. Part I. Biological Evaluation of the Mullica River–Great Bay Estuary. Technical Report, New Jersey Water Resources Research Institute, Rutgers University, New Brunswick, NJ.

utilize these embayments as spawning, nursery, and feeding areas. The fish fauna of these coastal bays has been the subject of several detailed investigations (see Able et al., 1996; Szedlmayer and Able, 1996; U.S. Fish and Wildlife Service, 1996; Jivoff and Able, 2001). Based on this work, more than 60 species of fish are known to occur in the coastal bays. A taxonomic list of fish compiled for the Barnegat Bay–Little Egg Harbor Estuary by Jivoff and Able (2001) also includes those forms encountered in the JCNERR bays to the south (Table 3.10). The fish assemblages observed in these estuarine waters are highly diverse, and they vary considerably in numerical abundance. Forage species, such as the bay anchovy (*Anchoa mitchilli*) and Atlantic silverside (*Menidia menidia*), are typically most abundant.

Durand and Nadeau (1972) determined that 20 fish species accounted for more than 99% of all fish collected in surveys of Great Bay during the 1970s. The top ten species in decreasing order of abundance were as follows: the bay anchovy (*Anchoa mitchilli*), Atlantic silverside (*Menidia menidia*), silver perch (*Bairdiella chrysoura*), alewife (*Alosa pseudoharengus*), striped killifish (*Fundulus majalis*), sea herring (*Clupea harengus*), white perch (*Morone americana*), northern puffer (*Sphoeroides maculatus*), oyster toadfish (*Opsanus tau*), and striped anchovy (*Anchoa hepsetus*). In a later field survey of Great Bay and Little Egg Harbor from June 1988 through October 1989, Szedlmayer and Able (1996) recorded 69 fish species; the most abundant forms were the bay anchovy (*Anchoa mitchilli*) (50.5% of the total catch), spot (*Leiostomus xanthurus*) (10%), Atlantic silverside (*Menidia menidia*) (9.7%), fourspine stickleback (*Apeltes quadracus*) (5.9%), and northern pipefish (*Syngnathus fuscus*) (4.2%). At least a dozen fish species are of recreational or commercial importance in Great Bay and Little Egg Harbor: *Anguilla rostrata*, *Brevoortia tyrannus*, *Morone americana*, *Centropristis striata*, *Pomatomus saltatrix*, *Leiostomus xanthurus*, *Menticirrhus saxatilis*, *Micropogonias undulatus*, *Tautoga onitis*, *Paralichthys dentatus*, *Pseudopleuronectes americanus*, and *Sphoeroides maculatus*.

Fish populations in the JCNERR coastal bays are characterized by wide seasonal variations in abundance; some species exhibit annual variations in abundance of 50 to 100%. This is due in large part to the seasonal occurrence of warm-water migrants, as well as reproduction by seasonal and year-round residents during the spring and summer. Fish abundance, therefore, peaks during the May through November period when warm-water migrants and the young of residents coexist in the bays. As warm-water migrants (e.g., black sea bass, *Centropristis striata*; northern pipefish, *Syngnathus fuscus*; striped searobin, *Prionotus evolans*; and summer flounder, *Paralichthys dentatus*) exit the bays during the late summer and fall, the overall species abundance of fish declines appreciably. Relatively few species inhabit the coastal bays in winter. The winter flounder, *Pseudopleuronectes americanus*, is an important cool-water migrant.

Of 70,000 fish collected along estuarine shores in Great Bay and Little Egg Harbor, Able et al. (1996) determined that 57% consisted of migratory species dominated by forms that overwinter on the continental shelf. These species primarily occur as young-of-the-year or older juveniles (38%) that utilize the estuaries as a nursery and those that appear as seasonal residents (19%), with all life-history stages occupying the embayments during part of the year but migrating to ocean waters

TABLE 3.10
Taxonomic List of Fishes and Selected Decapods Caught in Little Egg Harbor

Species	Common Name	Gear Used
Fish		
Alosa aestivalis	Blueback herring	S, T, W
Alosa pseudoharengus	Alewife	S, T, W
Alosa sapidissima	American shad	S, T
Aluterus scriptus	Scrawled filefish	T
Ammodytes americanus	American sand lance	S
Anchoa mitchilli	Bay anchovy	S, T, W, SS, TT
Anchoa hepsetus	Striped anchovy	S, T, W
Anguilla rostrata	American eel	S, T, W, SS, TT
Apeltes quadracus	Four-spine stickleback	S, T, SS, TT
Astroscopus guttatus	Northern stargazer	S
Bairdiella chrysoura	Silver perch	S, T
Brevoortia tyrannus	Atlantic menhaden	S, T, W
Caranx hippos	Crevalle jack	S, T, W
Centropristis striata	Black sea bass	T, SS, TT
Chasmodes bosquianus	Striped blenny	S, T
Chilomycterus schoepfi	Striped burrfish	S, T
Clupea harengus	Atlantic herring	S, T, W
Conger oceanicus	Conger eel	S, T
Cynoscion regalis	Weakfish	S, T
Cynoscion nebulosus	Spotted seatrout	
Cyprinodon variegatus	Sheepshead minnow	S, T, W, TT
Dasyatis sp.	Stingray	S
Dorosoma cepedianum	Gizzard shad	S, T
Engraulis eurystole	Silver anchovy	S, T
Epinephelus gorio	Red grouper	S
Esox niger	Chain pickerel	S
Etropus microstomus	Smallmouth flounder	S, T, SS, TT
Fundulus diaphanus	Banded killifish	S, T
Fundulus heteroclitus	Mummichog	S, T, W, SS, TT
Fundulus luciae	Spotfin killifish	W
Fundulus majalis	Striped killifish	S, T, W, SS, TT
Gasterosteus aculeatus	Three-spined stickleback	S, T, W
Gobiosoma bosc	Naked goby	S, T, W, SS, TT
Gobiosoma ginsburgi	Starboard goby	SS, TT
Hippocampus erectus	Lined seahorse	S, T, SS, TT
Hyporhamphus unifasciatus	Halfbeak	S
Hypsoblennius hentzi	Feather blenny	S, T
Ictalurus punctatus	Channel catfish	
Lagodon rhomboids	Pinfish	S
Leiostomus xanthurus	Spot	S, T, W, TT

(*continued*)

TABLE 3.10 (CONTINUED)
Taxonomic List of Fishes and Selected Decapods Caught in Little Egg Harbor

Species	Common Name	Gear Used
Lucania parva	Rainwater killifish	S, T, W, SS, TT
Lutjanus griseus	Grey snapper	S, SS, TT
Menidia beryllina	Inland silverside	S, W
Menidia menidia	Atlantic silverside	S, T, W, SS, TT
Menticirrhus saxatilis	Northern kingfish	S, T
Merluccius bilinearis	Silver hake	
Micropogonias undulatus	Atlantic croaker	T
Monacanthus hispidus	Planehead filefish	S
Morone americana	White perch	S, T
Morone saxatilis	Striped bass	S
Mugil cephalus	Striped mullet	S, T
Mugil curema	White mullet	S, T, W
Mustelus canis	Smooth dogfish	S, T, W
Myoxocephalus aenaeus	Grubby	S, T
Notemigonus crysoleucas	Golden shiner	S
Ophidion marginatum	Cusk-eel	T
Opsanus tau	Oyster toadfish	S, T, W, SS, TT
Paralichthys dentatus	Summer flounder	S, T, W, SS, TT
Peprilus triacanthus	Butterfish	S, T
Perca flavescens	Yellow perch	
Pogonias cromis	Black drum	S
Pollachius virens	Pollack	S, W, TT
Pomatomus saltatrix	Bluefish	S, T, W
Priacanthus arenatus	Bigeye	S
Prionotus carolinus	Northern searobin	S, T, TT
Prionotus evolans	Striped searobin	S, T
Pseudopleuonectes americanus	Winter flounder	S, T, W, SS, TT
Raja elanteria	Skate	
Rhinoptera bonansus	Cownose ray	
Sardinella aurita	Spanish sardine	S, W
Sciaenops ocellatus	Red drum	
Scomber scombrus	Atlantic mackerel	
Scophthalmus aquosus	Windowpane	S, T
Selene vomer	Lookdown	S, T
Sphoeroides maculatus	Northern puffer	S, T, TT
Sphyraena borealis	Northern sennet	S, T, W
Stenotomus chrysops	Scup	T
Strongylura marina	Atlantic needlefish	S, T, W
Syngnathus fuscus	Pipefish	S, T, W, SS, TT
Synodus foetens	Lizardfish	T
Tautoga onitis	Tautog	S, T, W, SS, TT
Tautogolabrus adspersus	Cunner	S

TABLE 3.10 (CONTINUED)
Taxonomic List of Fishes and Selected Decapods Caught in Little Egg Harbor

Species	Common Name	Gear Used
Trachinotus falcatus	Permit	S, T
Trinectes maculatus	Hogchoker	S, T
Urophycis chuss	Red hake	T
Urophycis regia	Spotted hake	S, T, W
Urophycis tenuis	White hake	S
Vomer setapinnis	Atlantic moonfish	S
Selected Decapods		
Callinectes sapidus	Blue crab	W, TT
Callinectes similis	Lesser blue crab	TT
Cancer irroratus	Rock crab	TT
Libinia dubia	Six-spined spider crab	TT
Libinia emarginata	Common spider crab	TT
Limulus polyphemus	Horseshoe crab	
Ovalipes ocellatus	Lady crab	TT
Portunus gibbesii	Swimming crab	

Note: S = seine; T = trawl; W = weir; SS = suction sampling; TT = throw trap.

Source: Jivoff, P. and K.W. Able. 2001. *Journal of Coastal Research* Special Issue 32, pp. 178–196.

for the winter. Other species found in the estuaries are classified as strays (22%) or true residents (21%).

Examples of resident species are the Atlantic silverside (*Menidia menidia*), mummichog (*Fundulus heteroclitus*), fourspine stickleback (*Apeltes quadracus*), and oyster toadfish (*Opsanus tau*). Marine strays include the windowpane (*Scophthalmus aquosus*) and blackcheek tonguefish (*Symphurus plaguisa*). Some of the species that use the bays as nursery areas are the Atlantic menhaden (*Brevoortia tyrannus*), bluefish (*Pomatomus saltatrix*), spot (*Leiostomus xanthurus*), and weakfish (*Cynoscion regalis*). Summer spawners in the bays are exemplified by the Atlantic silverside, bay anchovy (*Anchoa mitchilli*), northern pipefish (*Syngnathus fuscus*), and gobies (*Gobiosoma* spp.); winter spawners, by the winter flounder (*Pseudopleuronectes americanus*) and sand lance (*Ammodytes americanus*) (Tatham et al., 1984; U.S. Fish and Wildlife Service, 1996).

While some species (i.e., cosmopolitan forms) have a widespread distribution in the coastal bays and have been found in all habitats sampled (e.g., *Anchoa mitchilli, Menidia menidia, Leiostomus xanthurus,* and *Pseudopleurnectes americanus*), other species are habitat specific. For example, Jivoff and Able (2001) showed that the small-mouth flounder (*Etropus microstomus*), windowpane (*Scophthalmus aquosus*), skate (*Raja eglanteria*), and hakes (*Urophycis* spp.) prefer the deep channels of Little Egg Harbor. The silver perch (*Bairdiella chrysoura*), lizardfish (*Synodus foetens*), and fourspine stickleback (*Apeltes quadracus*) are associated with eelgrass (*Zostera marina*) beds. Szedlmayer and Able (1996) linked *Leiostomus xanthurus* with marsh channels. They also coupled clupeids with upper estuary subtidal creeks. Able et al. (1996) observed many species only in subtidal habitats of Great Bay and Little Egg Harbor (e.g., *Hippocampus erectus, Opsanus tau,* and *Syngnathus fuscus*).

In conclusion, fish assemblages in JCNERR estuaries are characterized by the numerical dominance of a few species, most notably forage fishes and juveniles. There is a conspicuous seasonal occurrence of warm- and cool-water migrants. In addition, there is considerable reproduction in the estuaries by seasonal and year-round residents. The net effect is a large flux in absolute abundance of fish populations in the system.

SUMMARY AND CONCLUSIONS

The JCNERR consists of more than 45,000 ha of upland, wetland, and open water habitats. The reserve is remarkably pristine, largely because of the federally protected New Jersey Pinelands, state and federal managed lands surrounding the coastal bays, and only 553 ha of developed landscape (<2% of the total area). Upland vegetation consists of pine–oak forests, which are replaced seaward by freshwater-, brackish-, and salt (*Spartina*) marshes. Marsh habitat covers more than 13,000 ha (> 28%) of the reserve. Rich and diverse plant and animal communities inhabit watershed areas of the JCNERR. Numerous species of amphibians, reptiles, mammals, birds, fish, and invertebrates occur in the JCNERR watershed.

The most extensive area of the JCNERR consists of open water habitat that covers more than 27,000 ha. Included here are the open waters of the Mullica River–Great Bay Estuary, Little Egg Harbor, lower Barnegat Bay, Little Bay, Reeds Bay, and Absecon Bay to the south. These estuarine waters support an array of planktonic, nektonic, and benthic organisms. A number of finfish (e.g., bluefish, *Brevoortia tyrannus*; weakfish, *Cynoscion regalis*; summer flounder, *Paralichthys dentatus*; and winter flounder, *Pseudopleuronectes americanus*) and shellfish (blue crab, *Callinectes sapidus*; and hard clam, *Mercenaria mercenaria*) populations are of recreational and commercial importance.

Submerged aquatic vegetation (SAV), notably eelgrass (*Zostera marina*) and widgeon grass (*Ruppia maritima*), are critically important habitat in lower Barnegat Bay and Little Egg Harbor. SAV not only generates a significant amount of primary production but also provides habitat for benthic epifauna and infauna as well as spawning, nursery, and feeding grounds for an array of finfish populations. Some fish (e.g., Acanthuridae and Scaridae), turtles, waterfowl (e.g., American brant, canvasbacks, and green-winged teal), and sea urchins consume SAV. In addition, these vascular plants baffle waves and currents and mitigate substrate erosion, thereby stabilizing bottom sediments. Eelgrass and widgeon grass are confined to the Barnegat Bay–Little Egg Harbor Estuary. In Great Bay and coastal bays to Absecon Bay, benthic macroalgae (e.g., *Ulva lactuca* and *Enteromorpha* spp.) proliferate, but eelgrass and widgeon grass are absent.

The Barnegat Bay–Little Egg Harbor Estuary is a moderately eutrophic system. Picoplankton blooms dominated by *Nannochloris atomus* and *Aureococcus anophagefferens* commonly occur in this estuary. During the bloom events, the phytoplankton cell counts often exceed 10^6 cells/ml. Brown tides composed of *A. anophagefferens* have been most intense and widespread in Little Egg Harbor, but they have also been documented in Barnegat Bay and Great Bay. These phytoplankton blooms are problematic because they cause brownish water discoloration and shading effects that can be detrimental to SAV.

Phytoplankton productivity in JCNERR coastal bays rivals or exceeds that of many other coastal bays in the U.S. and abroad (Seitzinger et al., 2001). Phytoplankton directly support zooplankton and benthic invertebrate populations in the system. Calanoid and harpacticoid copepods are major components of the zooplankton community in JCNERR estuaries. Meroplankton and ichthyoplankton are also important constituents because they are critical life-history stages for the proliferation of benthic invertebrate and finfish populations.

More than 150 benthic invertebrate species have been recorded in Great Bay, and more than 200 benthic invertebrate species have been documented in the Barnegat Bay–Little Egg Harbor Estuary. The composition of bottom sediments, particularly the grain size, strongly influences the distribution and abundance of the benthic organisms. The amount of silt and clay appears to be a controlling factor in this regard.

Finfish in JCNERR estuaries can be divided into several major groups. These include resident species, warm-water migrants, cool-water migrants, and stray species. Forage species, such as the bay anchovy (*Anchoa mitchilli*) and Atlantic silverside (*Menidia menidia*), typically dominate in numerical abundance. The occurrence and abundance of finfish populations in the JCNERR estuaries are highly variable due to seasonal migrations and the reproductive flux by seasonal and year-round residents. Annual variations in abundance of finfish populations commonly range from 50 to 100%. Many species found in the estuaries exhibit a preference for specific habitats (e.g., tidal creeks, eelgrass beds, and deep channels). Surveys conducted in these systems by various investigators underscore the significance of myriad habitats to the success of fishery resources in the JCNERR.

REFERENCES

Able, K.W. and S.M. Hagan. 2000. Effects of common reed (*Phragmites australis*) invasion on marsh surface macrofauna: response of fishes and decapod crustaceans. *Estuaries* 23: 633–646.

Able, K.W., D.A. Witting, R.S. McBride, R.A. Rountree, and K.J. Smith. 1996. Fishes of polyhaline estuarine shores in Great Bay–Little Egg Harbor, New Jersey: a case study of seasonal and habitat influences. In: Nordstrom, K.F. and C.T. Roman (Eds.). *Estuarine Shores: Evolution, Environments, and Human Alterations.* John Wiley & Sons, New York, pp. 335–353.

Bologna, P.A.X., R. Lathrop, P.D. Bowers, and K.W. Able. 2000. Assessment of the health and distribution of submerged aquatic vegetation from Little Egg Harbor, New Jersey. Technical Report, Institute of Marine and Coastal Sciences, Rutgers University, New Brunswick, NJ.

Burger, J., C.D. Jenkins, Jr., F. Lesser, and M. Gochfeld. 2001. Status and trends of colonially nesting birds in Barnegat Bay. *Journal of Coastal Research* Special Issue 32, pp. 197–211.

Chant, R.J. 2001. Tidal and subtidal motion in a shallow bar-built multiple inlet/bay system. *Journal of Coastal Research* Special Issue No. 32, pp. 102–114.

Conant, R. 1998. A zoogeographical review of the amphibians and reptiles of southern New Jersey, with emphasis on the Pine Barrens. In: Forman, R.T.T. (Ed.). *Pine Barrens: Ecosystem and Landscape*. Rutgers University Press, New Brunswick, NJ, pp. 467–488.

Cottam, C. and D.A. Munro. 1954. Eelgrass status and environmental relations. *Journal of Wildlife Management* 18: 449–460.

De Graaf, R.M. and J.H. Rappole. 1995. *Neotropical Migrating Birds: Natural History, Distribution, and Population Change*. Cornell University Press, Ithaca, NY.

Dow, C.L. and R.A. Zampella. 2000. Specific conductance and pH as watershed disturbance indicators in streams of the New Jersey Pinelands, U.S.A. *Environmental Management* 26: 437–445.

Durand, J.B. and R.J. Nadeau. 1972. Water Resources Development in the Mullica River Basin. Part I. Biological Evaluation of the Mullica River–Great Bay Estuary. Technical Report, New Jersey Water Resources Research Institute, Rutgers University, New Brunswick, NJ.

Durand, J.B. 1988. Field Studies in the Mullica River–Great Bay Estuarine System, Vol. 1. Technical Report, Center for Coastal and Estuarine Studies, Rutgers University, New Brunswick, NJ.

Durand, J.B. 1998. Nutrient and hydrological effects of the Pine Barrens on neighboring estuaries. In: Forman, R.T.T. (Ed.). *Pine Barrens: Ecosystem and Landscape*. Rutgers University Press, New Brunswick, NJ, pp. 195–227.

Forman, R.T.T. (Ed.). 1998. *Pine Barrens: Ecosystem and Landscape*. Rutgers University Press, New Brunswick, NJ.

Good, R.E., N.F. Good, and J.W. Andreasen. 1998. Pine Barrens plains. In: Forman, R.T.T. (Ed.). *Pine Barrens: Ecosystem and Landscape*. Rutgers University Press, New Brunswick, NJ, pp. 283–295.

Hales, L.S., Jr., R.S. McBride, E.A. Bender, R.L. Hoden, and K.W. Able. 1995. Characterization of Nontarget Invertebrates and Substrates from Trawl Collections During 1991–1992 at Beach Haven Ridge (LEO-15) and Adjacent Sites in Great Bay and on the Inner Continental Shelf off New Jersey. Technical Report, Institute of Marine and Coastal Sciences, Rutgers University, New Brunswick, NJ.

Hastings, R.W. 1998. Fish of the Pine Barrens. In: Forman, R.T.T. (Ed.). *Pine Barrens: Ecosystem and Landscape*. Rutgers University Press, New Brunswick, NJ, pp. 489–504.

Hunchak-Kariouk, K., R.S. Nicholson, D.E. Rice, and T.I. Ivahnenko. 2001. A synthesis of currently available information on freshwater quality conditions and nonpoint source pollution in the Barnegat Bay watershed. In: Kennish, M.J. (Ed.). *Barnegat Bay–Little Egg Harbor Estuary and Watershed*. Characterization Report, Barnegat Bay Estuary Program, Toms River, NJ.

Jacques Cousteau National Estuarine Research Reserve. 1999. Jacques Cousteau National Estuarine Research Reserve at Mullica River–Great Bay, New Jersey. Final Management Plan, Institute of Marine and Coastal Sciences, Rutgers University, New Brunswick, NJ.

Jivoff, P. and K.W. Able. 2001. Characterization of the fish and selected decapods in Little Egg Harbor. *Journal of Coastal Research* Special Issue 32, pp. 178–196.

Kennish, M.J. (Ed.). 2001a. Characterization Report of the Barnegat Bay–Little Egg Harbor Estuary and Watershed. Technical Report, Barnegat Bay Estuary Program, Toms River, NJ.

Kennish, M.J. 2001b. Zooplankton of the Barnegat Bay–Little Egg Harbor Estuary. *Journal of Coastal Research* Special Issue 32, pp. 163–166.

Kennish, M.J. 2001c. Benthic communities of the Barnegat Bay–Little Egg Harbor Estuary. *Journal of Coastal Research* Special Issue 32, pp. 167–177.

Kennish, M.J. and S. O'Donnell. 2002. Water quality monitoring in the Jacques Cousteau National Estuarine Research Reserve. *Bulletin of the New Jersey Academy of Science* 47: 1–13.

Kroodsma, D.E. and J. Verner. 1997. Marsh wren (*Cistothorus palustris*). In: Poole, A. and F. Gill (Eds.). *The Birds of North America*. Publication No. 127, The Academy of Natural Sciences, Philadelphia and the American Ornithologists Union, Washington, D.C.

Lathrop, R.G. and T.M. Conway. 2001. A Build-out Analysis of the Barnegat Bay Watershed. CRSSA Technical Report, Rutgers University, New Brunswick, NJ.

Lathrop R.G., J.A. Bognar, A.C. Hendrickson, and P.D. Bowers. 2001. Data synthesis report for the Barnegat Bay Estuary Program: habitat loss and alteration. In: Kennish, M.J. (Ed.). *The Scientific Characterization of the Barnegat Bay–Little Egg Harbor Estuary and Watershed*. Technical Report, Barnegat Bay National Estuary Program, Toms River, NJ, Appendix 3.

Leck, C.F. 1998. Birds of the Pine Barrens. In: Forman, R.T.T. (Ed.). *Pine Barrens: Ecosystem and Landscape*. Rutgers University Press, New Brunswick, NJ, pp. 457–466.

Loveland, R.E. and J.J. Vouglitois. 1984. Benthic fauna. In: Kennish, M.J. and R.A. Lutz (Eds.). *Ecology of Barnegat Bay, New Jersey*. Springer-Verlag, New York, pp. 135–170.

Macomber, R. and D. Allen. 1979. The New Jersey Submerged Aquatic Vegetation Distribution Atlas Final Report. Technical Report, Earth Satellite Corporation, Washington, D.C.

Martin, G.W. 1929. Dinoflagellates from Marine and Brackish Waters in New Jersey. University of Iowa Studies in Natural Waters in New Jersey, XII, Ames, IA.

McCormick, J. 1998. The vegetation of the New Jersey Pine Barrens. In: Forman, R.T.T. (Ed.). *Pine Barrens: Ecosystem and Landscape*. Rutgers University Press, New Brunswick, NJ, pp. 229–263.

McLain, P. and M. McHale. 1997. Barnegat Bay eelgrass investigation 1995–1996. In: Flimlin, G.E. and M.J. Kennish (Eds.). *Proceedings of the Barnegat Bay Ecosystem Workshop*. Rutgers Cooperative Extension of Ocean County, Toms River, NJ, pp. 165–171.

Moser, F.C. 1997. Sources and Sinks of Nitrogen and Trace Metals and Benthic Macrofauna Assemblages in Barnegat Bay, New Jersey. Ph.D. thesis, Rutgers University, New Brunswick, NJ.

Moul, E.T. and H.F. Buell, 1998. Algae of the Pine Barrens. In: Forman, R.T.T. (Ed.). *Pine Barrens: Ecosystem and Landscape*. Rutgers University Press, New Brunswick, NJ, pp. 425–440.

Mountford, K. 1965. A late summer red tide in Barnegat Bay, New Jersey. *Underwater Naturalist* 3: 32–34.

Mountford, K. 1967. The occurrence of Pyrrophyta in a brackish cove — Barnegat Bay, New Jersey at Mantoloking, May through December, 1966. *Bulletin of the New Jersey Academy of Science* 12: 9–12.

Mountford, K. 1969. A Seasonal Plankton Cycle for Barnegat Bay, New Jersey. M.S. thesis, Rutgers University, New Brunswick, NJ.

Mountford, K. 1971. Plankton Studies in Barnegat Bay. Ph.D. thesis, Rutgers University, New Brunswick, NJ.

Mountford, K. 1980. Occurrence and predation by *Mnemiopsis leidyi* in Barnegat Bay, NJ. *Estuarine, Coastal and Shelf Science* 10: 393–402.

Nichols, T.C. and P.M. Castelli. 1997. Waterfowl hunting in Barnegat Bay and Little Egg Harbor Bay, New Jersey. In: Flimlin, G.E. and M.J. Kennish (Eds.). *Proceedings of the Barnegat Bay Estuary Program*. Rutgers Cooperative Extension of Ocean County, Toms River, NJ, pp. 309–327.

Olsen, P.S. 1989. Development and distribution of a brown-water algal bloom in Barnegat Bay, New Jersey with perspective on resources and other red tides in the region. In: Cosper, E.M., E.J. Carpenter, and V.M. Bricelj (Eds.). *Novel Phytoplankton Blooms: Causes and Impacts of Recurrent Brown Tides and Other Unusual Blooms*. Springer-Verlag, Berlin, pp. 189–212.

Olsen, P.S. and J.B. Mahoney. 2001. Phytoplankton in the Barnegat Bay–Little Egg Harbor estuarine system: species composition and picoplankton bloom development. *Journal of Coastal Research* Special Issue 32, pp. 115–143.

Rhodehamel, E.C. 1998. Hydrology of the New Jersey Pine Barrens. In: Forman, R.T.T. (Ed.). *Pine Barrens: Ecosystem and Landscape*. Rutgers University Press, New Brunswick, NJ, pp. 147–167.

Sandine, P.H. 1984. Zooplankton. In: Kennish, M.J. and R.A. Lutz (Eds.). *Ecology of Barnegat Bay, New Jersey*. Springer-Verlag, New York, pp. 95–134.

Seitzinger, S.P., R.M. Styles, and I.E. Pillig. 2001. Benthic microalgal and phytoplankton production in Barnegat Bay, New Jersey (U.S.A.): microcosm experiments and data synthesis. *Journal of Coastal Research* Special Issue 32, pp. 144–162.

Smith, K.J. and K.W. Able. 1994. Salt-marsh tide pools as winter refuges for the mummichog, *Fundulus heteroclitus*, in New Jersey. *Estuaries* 17: 226–234.

Sogard, S.M. and K.W. Able. 1991. A comparison of eelgrass, sea lettuce macroalgae, and marsh creeks as habitats for epibenthic fishes and decapods. *Estuarine, Coastal and Shelf Science* 33: 501–519.

Szedlmayer, S.T. and K.W. Able. 1996. Patterns of seasonal availability and habitat use by fishes and decapod crustaceans in a southern New Jersey estuary. *Estuaries* 19: 697–709.

Tatham, T.R., D.L. Thomas, and D.D. Danila. 1984. Fishes of Barnegat Bay, New Jersey. In: Kennish, M.J. and R.A. Lutz (Eds.). *Ecology of Barnegat Bay, New Jersey*. Springer-Verlag, New York, pp. 241–281.

U.S. Fish and Wildlife Service. 1996. Significant Habitats and Habitat Complexes of the New York Bight Watershed. Technical Report, U.S. Fish and Wildlife Service, Southern New England–New York Bight Coastal Ecosystem Program, Charlestown, RI.

Viscido, S.V., D.E. Stearns, and K.W. Able. 1997. Seasonal and spatial patterns of an epibenthic decapod crustacean assemblage in northwest Atlantic continental shelf waters. *Estuarine, Coastal and Shelf Science* 45: 377–392.

Wilson, K.A., K.W. Able, and K.L. Heck, Jr. 1990. Habitat use by juvenile blue crabs: a comparison among habitats in southern New Jersey. *Bulletin of Marine Science* 46: 105–114.

Windom, L. and R. Lathrop. 1999. Effects of *Phragmites australis* (common reed) invasion on above-ground biomass and soil properties in brackish tidal marsh of the Mullica River, New Jersey. *Estuaries* 22: 927–935.

Wolgast, L.J. 1998. Mammals of the New Jersey Pine Barrens. In: Forman, R.T.T. (Ed.). *Pine Barrens: Ecosystem and Landscape*. Rutgers University Press, New Brunswick, NJ, pp. 443–455.

Wootton, L. and R.W. Zimmerman. 2001. Water Quality and Biomass of *Zostera marina* (Eelgrass) Beds in Barnegat Bay. Technical Report, Georgian Court College, Lakewood, NJ.

Zampella, R.A. 1994. Characterization of surface water quality along a watershed disturbance gradient. *Water Resources Bulletin* 30: 605–611.

Zampella, R.A. and K.J. Laidig. 1997. Effect of watershed disturbance on Pinelands stream vegetation. *Journal of the Torrey Botanical Society* 124: 52–66.

Zampella, R.A. and J.F. Bunnell. 1998. Use of reference-site fish assemblages to assess aquatic degradation in Pinelands streams. *Ecological Applications* 8: 645–658.

Zampella, R.A., J.F. Bunnell, K.J. Laidig, and C.L. Dow. 2001. The Mullica River Basin. Technical Report, New Jersey Pinelands Commission, New Lisbon, NJ.

Zimmer, B.J. 1981. Nitrogen Dynamics in the Surface Water of the New Jersey Pine Barrens. Ph.D. thesis, Rutgers University, New Brunswick, NJ.

Case Study 3

4 Delaware National Estuarine Research Reserve

INTRODUCTION

The National Oceanic and Atmospheric Administration (NOAA) designated the Delaware National Estuarine Research Reserve (DNERR) as a National Estuarine Research Reserve System (NERRS) program site on July 21, 1993. The DNERR consists of two well-defined component sites about 32 km apart (see Figure 4.1):

1. The Lower St. Jones River Reserve site located south of Dover in east-central Kent County, Delaware
2. The Upper Blackbird Creek Reserve site located between Odessa and Smyrna in southern New Castle County, Delaware

Both the Lower St. Jones River Reserve site and the Upper Blackbird Creek Reserve site are subestuaries of the Delaware River estuary. Tidal marshes and tidal streams comprise the primary habitats of both reserve sites.

The Lower St. Jones River Reserve site covers a 1518-ha area along the lower 8.8-km portion of the St. Jones River watershed. Here, agricultural land use predominates in the watershed. The St. Jones River stretches for 16.8 km across the Delmarva Peninsula, and it discharges to the mid–Delaware Bay zone. The Trunk Ditch, Beaver Branch, and Cypress Branch are the largest tributaries of the Lower St. Jones River, which is characterized by mesohaline salinity conditions. The lower boundary of the reserve site extends 3.2 km into the open waters of Delaware Bay; it encompasses a 1036-ha area of subtidal bottom.

The Upper Blackbird Creek Reserve site, which covers an area of 477 ha, lies ~9.0 to 18.5 km upstream of the Blackbird Creek mouth. This site is characterized by low salinity brackish or freshwater tidal creek habitat. Woodlots, croplands, and upland fields also occur within the designated boundaries of the reserve. Forested wetlands with coastal plain ponds blanket much of the land area upstream of the reserve site, notably in the Blackbird State Forest. Forested and agricultural land cover dominates much of the Blackbird Creek watershed. Extensive tidal mud flats and *Spartina* marshes border Blackbird Creek bayward of the Beaver Branch tributary in the upper creek segment.

This chapter provides an overview of the DNERR based in large part on published reports of the Delaware Department of Natural Resources and Environmental Control (1993, 1994, 1995, 1999) and the U.S. Environmental Protection Agency (Dove and Nyman, 1995; Sutton et al., 1996). The estuarine profile of the DNERR is a particularly important source of information on the reserve.

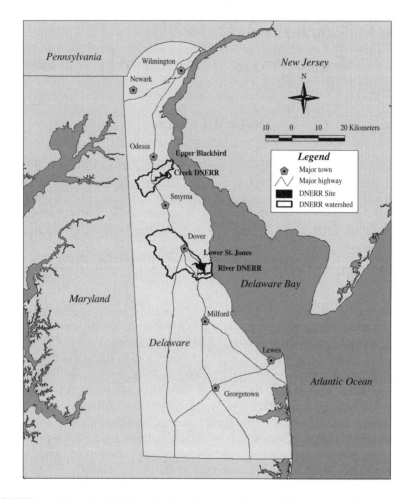

FIGURE 4.1 Map of the Delaware National Estuarine Research Reserve showing the location of the Upper Blackbird Creek and the Lower St. Jones River DNERR sites. (From the Delaware Department of Natural Resources and Environmental Control. 1999. Delaware National Estuarine Research Reserve: Estuarine Profiles. Technical Report, Delaware Department of Natural Resources and Environmental Control, Dover, DE.)

LOWER ST. JONES RIVER RESERVE SITE

WATERSHED

More than 50,000 people reside in the 233-km^2 area of the St. Jones River watershed. Development is greatest in the middle and upper watershed, being highly concentrated in the urbanized area of Dover, Delaware. Considering the entire St. Jones River watershed, approximately 48% of the land use cover is agriculture, 25% developed, 14% wetlands, 10% forested land, and 3% open water. Impervious land cover in the watershed amounts to nearly 25%. Within the reserve area, most of the land remains in private ownership despite the purchase of nearly 300 ha of wetland and upland habitat by NERRS in 1991–1992.

The Lower St. Jones River Reserve site and Upper Blackbird Creek Reserve site occur in the Atlantic Coastal Plain and are underlain by thick layers of unconsolidated sediments and semi-consolidated sedimentary rocks. The terrane is gently sloping; relief in the Lower St. Jones River Reserve ranges from sea level to 22 m. Soils in the upland areas of both reserve sites consist of well-drained or moderately drained sandy loams to poorly drained sandy–clay loams rich in organic matter. Because tidal wetlands are extensive, tidal marsh soils predominate in large areas of the reserve. These soils are composed of clay and sand layers mixed in many places with mucky peat. They attain a thickness of nearly 30 m in the marsh habitat at the mouth of the St. Jones River (DNERR, 1999).

Upland Vegetation

The Lower St. Jones River Reserve site supports two types of forest communities: upland forest and tidal marsh forest. Principal species of the upland forest community are the white oak (*Quercus alba*), southern red oak (*Q. falcata*), sassafras (*Sassafras albidum*), black cherry (*Prunus serotina*), American beech (*Fagus grandifolia*), American holly (*Ilex opaca*), black haw (*Viburnum prunifolium*), and tulip tree (*Liriodendron tulipifera*). The tidal swamp forest community includes red cedar (*Juniperus virginiana*), red maple (*Acer rubrum*), black gum (*Nyssa sylvatica*), sweet gum (*Liquidambar styraciflua*), green ash (*Fraxinus pennsylvanica*), and willows (*Salix* spp.). Farmland, old fields, and mixed deciduous hardwood forests comprise most of the upland land cover.

Wetland Vegetation

Forested wetland vegetation, scrub forest, and scrub–marsh mixes define the marsh habitat along the St. Jones River, with 66 species of plants reported (Wetlands Research Associates and Environmental Consulting Services, 1995) (Table 4.1). Mixed associations of emergent vegetation typify tidal wetland habitat of the Lower St. Jones River Reserve site. Emergent vegetation of high marsh areas exhibits greater diversity of plant assemblages than that of low marsh areas. The smooth cordgrass (*Spartina alterniflora*) is the dominant species of the tidal marsh, covering 62.2% of the Lower St. Jones River Reserve wetlands area (Table 4.2). Tall-form *S. alterniflora* grows along tide-channel banks, and short-form *S. alterniflora* spreads across broad expanses of intertidal habitat. The big cordgrass (*S. cynosuroides*) is also found along channel edges. Marsh edaphic algae (diatoms) in the top few millimeters of marsh sediments constitute a valuable food source for fish and other fauna of the salt marsh system. Pickerel weed (*Ponderia cordata*), marshpepper smartweed (*Polygonum hydroviper*), and swamp rose mallow (*Hibiscus palustris*) are subdominants within the cordgrass communities. Above mean high water (MHW), salt meadow cordgrass (*Spartina patens*) and salt grass (*Distichlis spicata*) concentrate in patches, most conspicuously immediately below the border areas of the upper marsh.

TABLE 4.1
Taxonomic List of Plants Identified in St. Jones River Marshes

Common Name	Scientific Name
Red maple	*Acer rubrum*
Ground nut	*Apios americana*
Orach	*Atriplex patula*
Grounsel bush	*Baccharis halimifolia*
Hedge bindweed	*Calystegia sepium*
Winged sedge	*Carex alata*
Greenish-white sedge	*Carex albolutescens*
Lone sedge	*Carex lonchocarpa*
Lurid sedge	*Carex lurida*
Uptight sedge	*Carex stricta*
Bitternut hickory	*Carya cordiformis*
Buttonbush	*Cephalanthus occidentalis*
Water hemlock or spotted cowbane	*Cicuta maculata*
Sweet pepperbush	*Clethra alnifolia*
Canker root	*Coptis trifolia*
Persimmon tree	*Diospyros virginiana*
Salt grass	*Distichlis spicata*
Spikerush	*Eleocharis ambigens*
American beech	*Fagus grandifolia*
Green ash	*Fraxinus pennsylvanicus*
Common madder	*Galium tinctorium*
Manna grass	*Glyceria stricta*
Swamp rose mallow	*Hibiscus palustris*
Many-flowered pennywort	*Hydrocotyle umbellata*
Inkberry	*Ilex glabra*
American holly	*Ilex opaca*
Winterberry	*Ilex verticillata*
Jewel weed	*Impatiens capensis*
Marsh elder	*Iva frutescens*
Black walnut	*Juglans nigra*
Red cedar	*Juniperus virginiana*
Rice cutgrass	*Leersia oryzoides*
Sweet gum	*Liquidambar styraciflua*
Japanese honeysuckle	*Lonicera japonica*
Sweet bay magnolia	*Magnolia virginiana*
Yellow pond lily	*Nuphar lutea*
Black gum	*Nyssa sylvatica*
Cinnamon fern	*Osmunda cinnamomea*
Royal fern	*Osmunda regalis*
Arrow arum	*Peltandra virginica*
Common reed	*Phragmites australis*
Black bindweed	*Polygonum convolvulus*

TABLE 4.1 (CONTINUED)
Taxonomic List of Plants Identified in St. Jones River Marshes

Common Name	Scientific Name
Black cherry	*Prunus serotina*
Mock bishopweed	*Ptilimnium capillaceum*
Willow oak	*Quercus phellos*
Swamp honeysuckle	*Rhododron viscosum*
Multiflora rose	*Rosa multiflora*
Swamp rose	*Rosa palustris*
Swampdock	*Rumex verticillatus*
Saltwort	*Salsola kali*
Common elderberry	*Sambucus canadensis*
American three-square	*Scirpus americanus*
Saltmarsh bulrush	*Scirpus robustus*
Giant bulrush	*Scirpus validus*
Sawbriar	*Smilax glauca*
Smooth cordgrass	*Spartina alterniflora*
Big cordgrass	*Spartina cynosuroides*
Saltmeadow cordgrass	*Spartina patens*
Skunk cabbage	*Symplocarpus foetides*
Tall meadow-rue	*Thalictrum pubescens*
Poison ivy	*Toxicodendron radicans*
Narrow-leaf cattail	*Typha angustifolia*
Broad-leaf cattail	*Typha latifolia*
Highbush blueberry	*Vaccinium corumbosum*
Northern arrowwood	*Viburnum recognitum*
Black haw	*Viburnum prunifloloium*

Source: From the Delaware Department of Natural Resources and Environmental Control. 1999. Delaware National Estuarine Research Reserve: Estuarine Profiles. Technical Report, Delaware Department of Natural Resources and Environmental Control, Dover, DE.

Marsh shrub communities proliferate in higher marsh areas. Two types of shrub communities are evident:

1. Fresher, lower salinity tidal communities
2. Brackish, higher salinity tidal communities

Woody plants dominate the lower salinity tidal communities; smooth alder (*Alnus serrulata*), winterberry (*Ilex verticillata*), buttonbush (*Cephalanthus occidentalis*), sweet pepperbush (*Clethra alnifolia*), and dogwoods (*Cornus* spp.) are plant species commonly observed here. In higher salinity tidal communities, the predominant marsh shrubs include the marsh elder (*Iva frutescens*) and groundsel bush

TABLE 4.2
Vegetation Cover in Wetlands of the Delaware National Estuarine Research Reserve

Upper Blackbird Creek		Lower St. Jones River	
Map Unit	Percent Cover	Map Unit	Percent Cover
Spartina alterniflora	28.64	*Spartina alterniflora*	62.23
Tidal flat	26.44	*Phragmites australis*	13.38
Open water	14.21	Impoundment	7.54
Phragmites australis	11.05	Open water	7.03
Spartina alterniflora mix	4.74	Marsh shrub	3.63
Tidal swamp forest	4.20	Salt hay	2.37
Marsh shrub	3.64	*Spartina cynosuroides*	1.68
Spartina cynosuroides	2.72	Tidal swamp forest	0.92
Typha spp.	1.53	Tidal flat	0.72
Zizania aquatica	1.09	Marsh shrub	0.43
Peltandra virginica	0.96	*Typha (latifolia/angustifolia)*	0.04
Ponderia cordata	0.44	*Scirpus americanus*	0.02
Impoundment	0.30	*Atriplex triangularis*	0.01
Nuphar lutea	0.04	*Peltandra virginica*	0.01
Scirpus americanus	0.01		

Source: From the Delaware Department of Natural Resources and Environmental Control. 1999. Delaware National Estuarine Research Reserve: Estuarine Profiles. Technical Report, Delaware Department of Natural Resources and Environmental Control, Dover, DE.

(*Baccharis halimifolia*). Red cedar (*Juniperus virginiana*) is also evident in the brackish tidal communities.

Deep-water emergents form a low-marsh mixed association along the edges of creeks and ponds. Comprising this plant association are arrow arum (*Peltandra virginica*), yellow pondweed (*Nuphar lutea*), marshpepper smartweed (*Polygonum hydropiper*), and pickerel weed (*Ponderia cordata*). Dense monospecific stands of yellow pondweed occur in some areas.

The common reed (*Phragmites australis*), an invasive species, inhabits fresh and brackish marshes along marsh upland borders. This nuisance species has spread most rapidly in the Upper Blackbird Creek marshes but also has been documented in more restricted areas along the upland edge and major river banks of the Lower St. Jones River Reserve site. Increasing distribution of the common reed in reserve marshes is a growing concern because this species generally degrades coastal wetland habitat values for wildlife. Efforts to control the spread of *Phragmites* in the DNERR have consisted of aerial herbicide spraying followed by prescribed burning. However, this species is resilient, and its persistent monotypic stands remain a target for various remedial programs. Monotypic stands of *Phragmites* currently cover about 10 to 15% of Delaware's tidal wetlands (DNREC, 1993).

Three other emergent wetland plant communities exist in the DNERR: *Typha* spp., *Scirpus americanus*, and *Zizania aquatica* communities. Although relatively minor emergent wetland components, these communities provide important habitat for a number of animal populations such as muskrats (*Ondatra zibethicus*) and an array of bird species. Two cattail species (*Typha angustifolia* and *T. latifolia*) have been documented in *Typha* communities of the reserve. The cattails may be present as monospecific stands or a mixed community with a number of co-dominants (i.e., smooth cordgrass, *Spartina alterniflora*; rice cutgrass, *Leersia oryzoides*; salt marsh water hemp, *Acnida cannabina*; and nodding bur-marigold, *Bidens cernua*). Emergent plants of the *Scirpus* community grow along brackish shorelines as either monotypic stands of the American three-square (*Scirpus americanus*) or a mixed community with a few other common marsh plants (i.e., *Spartina alterniflora*, *S. patens,* and *Distichlis spicata*). Extensive monospecific stands of wild rice comprise the *Zizania aquatica* community, which proliferates in fresh to slightly brackish water areas.

Aquatic Habitat

The lower 8.8-km section of the St. Jones River site is a medium-salinity tidal river subjected to semidiurnal tides. The mean tidal range at the mouth of the river amounts to about 1.5 m; at spring tide, however, the mean tidal range averages 1.7 m. Significant tidal range attenuation occurs upriver.

The channel width of the Lower St. Jones River at the site of the reserve ranges from ~40 to 90 m. Mid-channel depths at low tide along this stretch of the river range from ~2.4 to 5.5 m. The highest current velocities are recorded in the lower segment of the river. Here, maximum current velocities observed during spring tides and neap tides are ~30 to 40 and ~20 to 30 cm/sec, respectively. The water column is relatively well mixed, with little evidence of two-layered estuarine flow. Hence, flow is mainly unidirectional from surface to bottom in the lower river with slight differences (10 to 20%) in current velocity observed throughout the water column.

Water Quality

A YSI Model 6000 data logger deployed at Scotton Landing in the middle reach of the Lower St. Jones River during 1996 recorded physical–chemical data semicontinuously (every 30 min) year-round. The water quality parameters monitored were temperature, salinity, dissolved oxygen, pH, turbidity, and depth (Figure 4.2 and Figure 4.3). The absolute temperature over the annual period at the Scotton Landing site ranged from less than 0 to 30°C. The monthly mean water temperature, in turn, ranged from less than 0°C in February to 25.5°C in August. The annual salinity range in the middle reach of the river ranged from ~1‰ to more than 20‰. Mean monthly salinity values ranged from ~3‰ in December to more than 12‰ in September. Waters in the Lower St. Jones River are generally classified as mesohaline.

Annual dissolved oxygen values (% saturation) varied from less than 20% to more than 120% saturation. The monthly mean dissolved oxygen, however, ranged from more than 40% in July to more than 80% in March. Hypoxic events were also

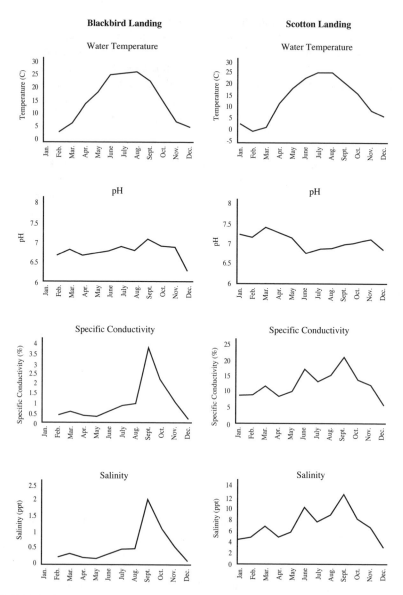

FIGURE 4.2 Comparison of monthly mean water temperature, pH, specific conductivity, and salinity for Blackbird Landing and Scotton Landing in 1996. (From the Delaware Department of Natural Resources and Environmental Control. 1999. Delaware National Estuarine Research Reserve: Estuarine Profiles. Technical Report, Delaware Department of Natural Resources and Environmental Control, Dover, DE.)

documented during the summer months. Absolute dissolved oxygen measurements varied from 0 to 14.9 mg/l. The monthly mean absolute dissolved oxygen values ranged from ~4 mg/l (July) to 10 mg/l (March). The mean annual dissolved oxygen value was 6.45 mg/l.

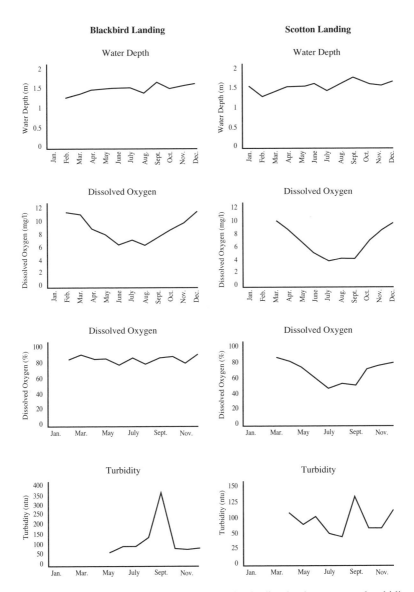

FIGURE 4.3 Comparison of monthly mean water depth, dissolved oxygen, and turbidity for Blackbird Landing and Scotton Landing in 1996. (From the Delaware Department of Natural Resources and Environmental Control. 1999. Delaware National Estuarine Research Reserve: estuarine profiles. Technical Report, Delaware Department of Natural Resources and Environmental Control, Dover, DE.)

The pH measurements at Scotton Landing for 1996 ranged from 6.01 to 8.87. The mean pH value for the year was 7.02. Highest pH levels were observed in March, and lowest pH levels were noted in June and July.

Turbidity generally ranged from 50 to 125 NTU, with highest levels (mean ~125 NTU) registered in September. However, spiked events of more than 500 NTU occasionally interrupted periods of relatively stable, low turbidity conditions. These episodic events typically resulted from storms and elevated stormwater runoff, which transported large concentrations of sediments and other particulate matter into the system. The roiling of bottom sediments by high winds and other factors also contributed to higher turbidity levels.

Water depth can influence the amount of turbidity in the water column because bottom agitation and erosion of sediments may be substantially less in deeper waters. At the Scotton Landing site, mean monthly water depths varied from about 1.3 to 1.7 m. The shallowest depths were reported in February and the deepest depths in September.

An extensive water quality database on the DNERR for the period from 1996 through 2002 can be obtained over the Internet from the NERRS Centralized Data Management Office (CDMO). The CDMO database can be accessed at the following Internet address: http://cdmo.baruch.sc.edu.

ANTHROPOGENIC IMPACTS

Pollution

The Delaware River estuary and its watershed have historically experienced significant alteration due to heavy industrialization and other human activities. A wide array of anthropogenic problems, including excessive watershed development, point and nonpoint source runoff, habitat loss and alteration, toxic chemical contaminants, and degraded water quality (Sutton et al., 1996), potentially threatens the environmental integrity of the system. The Delaware River watershed drains an area of ~33,000 km^2, and more than 6 million people live in areas surrounding the estuary. Poor development planning in the watershed has led to considerable population pressure on habitats and resources in the system.

Water quality has been compromised in some areas of the Delaware Estuary due to substantial loadings of nutrients, trace metals, volatile organics, polycyclic aromatic hydrocarbons (PAHs), and some chlorinated hydrocarbon compounds (Kennish, 2000) (Tables 4.3 to 4.5). Water quality degradation in the estuary peaked during the period from the 1940s to the 1960s as a consequence of rapid industrialization of the Delaware River basin, accelerated growth of major cities, and the expansion of urban water and sewer systems. Water quality has generally improved in the estuary since the 1970s through the application of environmental remediation programs, as evidenced by increased dissolved oxygen and pH levels observed over the past three decades.

Nutrient loading in the Delaware Estuary is higher than that in many other major U.S. estuaries, such as the Chesapeake Bay; nitrogen loading in the estuary amounts to ~7500 mmol N/m^2/yr, and phosphorus loading, ~600 mmol P/m^2/yr (Kennish, 2000). The total nitrogen concentration in the estuary (1.5 to 3 mg N/l) far exceeds the phosphate concentration (~0.02 to 0.12 mg P/l) (Sutton et al., 1996). Heavy metal concentrations are also elevated relative to those of other major estuarine systems. Arsenic, chromium, copper, and lead loadings in the estuary are ~1 × 10^5 kg/yr, while mercury loading is ~1 × 10^4 kg/yr (Sutton et al., 1996). The loading of

TABLE 4.3
Representative Toxic Substances of Concern in the Delaware Estuary

Metals

Aluminum	Chromium[a]	Nickel
Arsenic[a]	Copper[a]	Selenium
Beryllium	Lead[a]	Silver[a]
Cadmium	Mercury[a]	Zinc[a]

Volatile Organics

Acrolein	Chloroform	Trichloroethene
Acrylonitrile	1,2-Dichloroethane[a]	Vinyl chloride
Benzene	Tetrchloroethene[a]	
Carbon tetrachloride	Toluene	

Nonvolatile Organics

Polycyclic Aromatic Hydrocarbons (PAHs)[a]

Acenaphthene	Benzopyrenes	Fluorenes
Acenaphthylene	Biphenyl	Indeno [1,2,3-c,d] pyrene
Anthracene	Chrysene	Naphthalene
Benzo [a] anthracene	Dibenzo [a,h] anthracene	Perylene
Benzofluoranthenes	Dibenzothiophene	Phenanthrene
Benzo [g,h,i] perylene	Fluoranthene	Pyrene

Organochlorines

Chlorinated Pesticides

Aldrin	Endosulfan	Pentachlorophenol
Chlordane[a]	Heptachlor	Toxaphene
DDT and its metabolites[a]	Hexachlorobenzene	Endrin
Dieldrin[a]	Mirex	

Polychlorinated Biphenyls (PCBs)

Others

Dinitrophenol	Nitrophenol
Nitrobenzene	Phenol

[a] These substances were named to the preliminary list of toxic pollutants of concern by the Delaware Estuary Program's Toxics Task Force.

Source: Sutton, C.C., J.C. O'Herron, III, and R.T. Zappalorti. 1996. The Scientific Characterization of the Delaware Estuary. Technical Report (DRBC Project No. 321; HA File No. 93.21), Delaware Estuary Program, U.S. Environmental Protection Agency, New York.

TABLE 4.4
Loading of Toxic Substances to the Delaware Estuary

Contaminant	Source				Percent of Total Loading
	PS[a]	UR[a]	AR[a]	AD[a]	
As	43.8/7.3	8.9/3.2	46.6/92	0.7/2.0	10.4
Cr	87.4/20.1	11.6/5.8	—	1.0/3.8	14.3
Cu	82.1/18.7	15.6/7.7	—	2.3/9.0	14.2
Pb	70.3/13.2	24.5/10.0	—	5.2/16.9	11.7
Hg	10.1/0.2	10.1/0.3	—	79.8/20.2	0.9
Ag	100.0/2.2	—	—	—	1.4
Zn	52.6/33.4	43.5/59.8	—	4.0/43.6	39.6
PAH	—	95.1/10.6	—	4.9/4.4	3.2
Chlorinated pesticides	39.5/0.4	2.6/0.1	57.9/7.8	—	0.7
PCBs	66.7/<0.01	—	—	33.3/0.1	<0.01
Volatile organics	79.0/4.5	21.0/2.6	—	—	3.5
Percent of total loading by source	62.3	28.8	5.2	3.6	99.9/99.9

Note: PS = point source; UR = urban source; AR = agricultural runoff; AD = atmospheric deposition.

[a] Percent loading of a substance by source/percent contribution of a substance to loading from a source.

Source: Sutton, C.C., J.C. O'Herron, III, and R.T. Zappalorti. 1996. The Scientific Characterization of the Delaware Estuary. Technical Report (DRBC Project No. 321; HA File No. 93.21), Delaware Estuary Program, U.S. Environmental Protection Agency, New York.

TABLE 4.5
Dissolved Trace Metal Concentrations in the Delaware River and Other East Coast Rivers

River	Trace Metal (µg/l)							
	Cd	Co	Cu	Fe	Mn	Ni	Pb	Zn
Delaware	0.17	0.42	2.36	32.9	155	3.86	0.27	12.1
Susquehanna	0.089	1.0	1.21	57.3	655	5.75	0.21	2.62
Hudson	0.25	—	3.24	31.9	10.7	2.41	—	8.83
Connecticut	0.10	—	4.17	113	45.9	—	—	0.98
Potomac	—	—	—	—	—	—	—	0.55
Southeastern U.S. (Mean)	0.078	—	0.56	30.7	18	0.26	—	0.64

Source: Sutton, C.C., J.C. O'Herron, III, and R.T. Zappalorti. 1996. The Scientific Characterization of the Delaware Estuary. Technical Report (DRBC Project No. 321; HA File No. 93.21), Delaware Estuary Program, U.S. Environmental Protection Agency, New York.

PAHs equals 3.28×10^4 kg/yr (Frithsen et al., 1995). Among organochlorine contaminants, PCBs and DDTs continue to be problematic. The loading of PCBs and DDTs in the estuary is estimated to be 89 kg/yr and 7900 kg/yr, respectively (Frithsen et al., 1995). Bottom sediments are a repository for the largest fraction of chemical contaminants that enter the estuary. In some areas, the contaminants may pose a significant health threat to some biota, particularly upper-trophic-level organisms (Kennish, 2000).

Point and nonpoint source pollution contributes to the same water quality problems in the Lower St. Jones River as observed in the Delaware River estuary, including elevated levels of nutrients (nitrogen and phosphorus) and chemical contaminants (heavy metals, hydrocarbons, and PCBs). For example, several industrial facilities exist in the St. Jones River watershed where point source wastewater discharges are regulated by the National Pollution Discharge Elimination System (NPDES) of Delaware's Department of Natural Resources and Environmental Control. The central sewer system in Dover, Delaware periodically releases sewage waste in combined sewer overflows. However, the most persistent water quality problems in the St. Jones River are ascribable to nonpoint source pollution from both urban and rural areas in the St. Jones River watershed. Escalating urban land use in the Dover area has increased pollutant export to the river via accelerated stormwater runoff from impervious surfaces or residential landscapes. Nonpoint source pollutant loads from Dover and surrounding areas originate largely from construction sites, high-density commercial zones, and industrial centers; they consist of heavy metals, oil and grease, organochlorine compounds, and other contaminants. In more rural areas, agricultural runoff mainly associated with corn and soybean production or animal feedlots transports nutrients and sediments to the St. Jones River. In addition to nitrogen, phosphorus, and sediment inputs, constituent loads of concern from farmlands include oxygen-demanding compounds and pesticides. Aside from urban and agricultural runoff, the effects of silviculture, land disposal, leaching of nutrients and coliform bacteria from septic fields, and the atmospheric deposition of an array of contaminants augment pollutant inputs. Nutrient loading is of particular concern because of its link to eutrophication of estuarine waters (Kennish, 1997; Livingston, 2001, 2003).

The accumulation of PCBs in the St. Jones River watershed is a resource management problem. A health advisory was issued on March 18, 1993 for all tidal and several nontidal reaches of the St. Jones River watershed because of elevated PCB levels in aquatic sediments and the food web. This advisory recommended limited consumption (i.e., no more than two 226.8-g meals a year) of catfish (*Ameiurus catus, A. nebulosus,* and *Ictalurus punctatus*), white perch (*Morone americana*), carp (*Cyprinus carpio*), and largemouth bass (*Micropterus salmoides*) taken in the upper portions of the St. Jones River downstream to Bowers Beach (DNERR, 1999). The exact source of the PCBs remains undetermined.

Some biota throughout the Delaware Estuary have also accumulated high levels of PCBs and other toxins (Sutton et al., 1996; Kennish, 2000). Owing to widespread contamination by PCBs, DDTs, chlordane, dioxin, and mercury, consumption advisories have been issued by government agencies for a number of fish species in the estuary, notably bluefish (*Pomatomus saltatrix*), striped bass (*Morone saxatilis*), white perch (*Morone americana*), American eel (*Anguilla rostrata*), white catfish (*Ameiurus catus*), channel catfish (*Ictalurus punctatus*), and chain pickerel (*Esox niger*). Apart from these finfish species, other fauna exhibiting high concentrations of certain toxins in the estuary are mussels (*Mytilus edulis*), oysters (*Crassostrea virginica*), and osprey (*Pandion haliaetus*) (Kennish, 2000).

Habitat Alteration

The most profound alteration of upland habitat in the DNERR region is the conversion of natural forested land cover to population centers and farmlands. These land use conversions have contributed to various levels of habitat destruction and nonpoint source pollution in the watershed. However, other land use conversions have also impacted habitat and water quality in watershed areas. Included here are the development of exurban residential subdivisions, installation of septic systems in environmentally sensitive areas, construction of highways, operation of a major airbase, creation of borrow pits for sand and gravel mining, and nonselective marsh ditching for mosquito control (DNERR, 1999).

Dredging of the Delaware River main shipping channel deepens the waterway, resulting in improved circulation of the estuary. Sharp et al. (1994) showed that, soon after dredging, changes in salinity, dissolved oxygen, turbidity, and water quality occur throughout the estuary. Upper estuary locations experience increased tidal amplitudes. These changes may also influence water quality conditions in subestuaries such as the St. Jones River. However, because the lower reserve site is a considerable distance from the channel dredging areas, the dredging effects are likely to be small.

Shoreline erosion is an escalating problem along the Delaware River and Bay. Rising sea level and wave erosion are threatening the wetland shoreline habitat in the system. Principal shoreline protective measures implemented to control shoreline erosion include the installation of permanent engineering structures such as bulkheads and seawalls. However, these shoreline structures alter or even destroy habitat for turtles, horseshoe crabs, shorebirds, and various wildlife populations. In addition, toxins (e.g., wood preservatives) that leach from the treated wooden structures can contaminate adjacent waters and bottom sediments, thereby posing a potential danger to organisms inhabiting these areas.

Three federal Superfund sites exist in the St. Jones watershed:

1. Dover Air Force Base
2. Wildcat Landfill
3. Dover Gas Light Company

All have serious chemical contamination problems that can cause the degradation of water quality offsite if leachates are not effectively controlled. Dover Air Force Base has been responsible for significant groundwater contamination due to volatile organic compounds (solvents and gasoline) and heavy metals derived from aircraft operations on site. Tributaries of the St. Jones River have received groundwater contaminated with pollutants from the base. However, no serious environmental impacts have been attributed to this water contamination.

The Wildcat Landfill was a privately owned and operated industrial and municipal waste disposal facility located about 3.7 km upstream of the Lower St. Jones River Reserve site. The facility, which closed in 1973 due to permit violations associated with illegal disposal of waste materials, caused contamination of surface water and sediments along the St. Jones River as a result of inputs of PCBs and other toxins (DNREC, 1994). Since its closing, the landfill has been reclaimed for wildlife purposes. However, during its operation the landfill may have been a significant source of PCBs, which are now stored in bottom sediments of the St. Jones River.

The Dover Gas Light Company was another source of PCB contamination in the St. Jones River watershed during the 20th century. It also caused coal tar contamination of soils and groundwater in the watershed. This site is located in Dover, Delaware, and its impact may have been more problematic for the Lower St. Jones River than for the upper reaches of the system (DNERR, 1999).

Biotic Communities

Phytoplankton

Phytoplankton surveys conducted at three stations in the Lower St. Jones River during 1995 and 1996 identified 44 taxa, with most belonging to the Bacillariophyta (diatoms) (N = 24) and the Chlorophycota (green algae) (N = 10). The numerically dominant taxa in decreasing order of abundance were *Melosira* spp., *Guinardia* spp., *Ceratium* spp., and *Biddulphia* spp. Three of these taxa (*Melosira*, *Guinardia*, and *Biddulphia*) are diatoms. *Volvox* spp., *Ankistrodesmus* spp., *Scenedesmus* spp., *Chlamydomonas* spp., *Hydrodictyon* spp., and *Chlorella* spp. were the most abundant green algae, and *Anabaena* spp., *Microcystis* spp., and *Oscillatoria* spp. were the dominant blue-green algae. Table 4.6 provides a taxonomic list of phytoplankton collected in both the Lower St. Jones River and Upper Blackbird Creek (DNERR, 1999).

A distinct seasonal pattern of phytoplankton abundance and diversity occurs in the Lower St. Jones River. Phytoplankton abundance here peaks in the summer and drops to a minimum in the winter. Diatoms rank among the most abundant taxa during all seasons. Maximum diversity takes place in the summer and minimum diversity in the fall.

The phytoplankton community is much more diverse in the Delaware River estuary, where more than 250 species and over 100 genera have been registered (Marshall, 1992). Watling et al. (1979) documented 113 phytoplankton species in Delaware Bay. Diatoms predominate from fall through spring, with several species (*Skeletonema costatum, Thalassiosira nordenskioldii, Asterionella glacialis, Chaetoceras* sp., and *Rhizosolenia* sp.) acting as the principal constituents of the spring bloom (Watling et al., 1979; Marshall, 1992). Phytoplankton biomass peaks in the lower estuary during March and in the upper estuary during July (Pennock and Sharp, 1986). Phytoplankton diversity is highest in the summer and fall when small flagellates are most abundant in the estuary (Marshall, 1992, 1995; Kennish, 2000).

TABLE 4.6
Taxonomic List of Phytoplankton Collected in the Blackbird Creek and St. Jones River Areas of the Delaware National Estuarine Research Reserve

Division Cyanophycota (Blue-Green Algae)
Anabaena spp.
Anacystis spp.
Microcystis spp.
Oscillatoria spp.
Sphaerocystis spp.
Unidentified Cyanophycota

Division Bacillariophyta (Diatoms)
Class Coscinodiscophyceae (Centric Diatoms)
Biddulphia spp.
Chaetoceros spp.
Coscinodiscus spp.
Cyclotella spp.
Ditylum spp.
Guinardia spp.
Leptocylindrus spp.
Lithodesmium spp.
Melosira spp.
Rhizosolenia spp.
Skeletonema spp.
Unidentified Coscinodiscophyceae

Class Fragilariophyceae (Araphid, Pennate Diatoms)
Asterionella spp.
Fragilaria spp.
Synedra spp.
Tabellaria spp.
Thalassionema spp.
Thalassiothrix spp.

Class Bacillariophyceae (Raphid, Pennate Diatoms)
Achnanthes spp.

Gyrosigma spp.
Hantzschia spp.
Navicula spp.
Nitzschia spp.
Pinnularia spp.
Surirella spp.

Division Chlorophycota (Green Algae)
Actinastrum spp.
Ankistrodesmus spp.
Chlamydomonas spp.
Chlorella spp.
Hydrodictyon spp.
Pediastrum spp.
Scenedesmus spp.
Tetraedron spp.
Ulothrix spp.
Volvox spp.
Unidentified Chlorphycota

Division Cryptophycota
Class Cryptophyceae
Cryptomonas spp.

Division Pyrrhophycota (Dinoflagellates)
Unidentified Pyrrhophycota
Class Dinophyceae
Ceratium spp.
Gymnodinium spp.
Noctiluca spp.
Peridinium spp.

Unidentified Phytoflagellates
Unidentified Phytoplankton

Source: Delaware Department of Natural Resources and Environmental Control. 1999. Delaware National Estuarine Research Reserve: Estuarine Profiles. Technical Report, Delaware Department of Natural Resources and Environmental Control, Dover, DE.

Zooplankton

Zooplankton sampling in the Lower St. Jones River during the 1995–1996 period collected 39 microzooplankton taxa and 53 mesozooplankton taxa. Among the most common microzooplankton (<64 μm) taxa observed in plankton collections were copepod nauplii, rotifers (*Brachionus* spp., *Filinia* spp., *Keratella* spp., *Notholca* spp., and unidentified forms), protozoans (*Tintinnidium* spp., Arcellinida, Peritrichia,

Zoomastigophora), cladocerans (*Daphnia* spp.), bivalve larvae, Gastropoda, and Polychaeta. The Ascidiacea were also abundant. Copepod nauplii dominated the microzooplankton during the spring, summer, and fall; unidentified Rotifera dominated during the winter. Cladocerans and polychaete larvae were also abundant in the Lower St. Jones River during a given season. Microzooplankton diversity was highest in the summer and lowest in the fall (DNERR, 1999).

Of the 53 mesozooplankton (64 to 250 μm) taxa found in the Lower St. Jones River, polychaete larvae dominated the collections and were most abundant in the spring and summer. Nematodes predominated in the winter, and the copepod, *Eurytemora affinis*, was most numerous in the fall. Several other copepod taxa were also abundant, notably copepod nauplii, *Acartia tonsa, Acartia* spp., copepodites, Calanoida, *Leptastacus* spp., Cyclopoida, *Cyclops* spp., *Halicyclops fosteri, Pseudodiaptomus pelagicus,* and Harpacticoida. In addition, crab larvae (*Uca* spp. and *Rhithropanopeus* spp.), mysid shrimp (*Neomysis americana*), cladocerans (*Bosmina* spp., *Daphnia* spp., and *Diaphanosoma* spp.), rotifers (*Brachionus* spp., *Notholca* spp., and unidentified Rotifera), Ascidiacea, Cirripedia, Cnidaria medusa, Gastropoda, and Tardigrada were numerically important mesozooplankton.

Zooplankton attained highest densities in the upper reaches of both the St. Jones River and Blackbird Creek. While phytoplankton densities likewise peaked in the upper reaches of the St. Jones River, they were highest in the lower reaches of Blackbird Creek. Densities of plankton were generally greatest in the summer, an exception being phytoplankton in the St. Jones River, which exhibited maximum densities in the spring (Table 4.7).

TABLE 4.7
Net Plankton Density Recorded in the Blackbird Creek and St. Jones River

Season	Phytoplankton (n/ml)	Microzooplankton (n/m^3)	Mesozooplankton (n/m^3)
Blackbird Creek			
Fall 1995	2,662	16,235	2,977
Winter 1995–96	1,732	5,241	788
Spring 1996	4,972	14,525	6,476
Summer 1996	7,587	27,901	11,305
Annual Mean, 1995–96	6,450	25,608	10,940
St. Jones River			
Fall 1995	2,392	7,824	2,086
Winter 1995–96	1,721	8,641	1,199
Spring 1996	11,120	4,949	2,472
Summer 1996	3,978	15,223	8,235
Annual Mean, 1995–96	8,797	14,518	6,162

Source: Delaware Department of Natural Resources and Environmental Control. 1999. Delaware National Estuarine Research Reserve: Estuarine Profiles. Technical Report, Delaware Department of Natural Resources and Environmental Control, Dover, DE.

In the Delaware Estuary, copepods account for ~85% of the total zooplankton biomass (Herman, 1988). Among the numerically dominant copepod species in the estuary are *Acartia hudsonica, Acartia tonsa, Eurytemora affinis, Halicyclops fosteri, Oithona colcarva,* and *Pseudodiaptomus pelagicus.* Of these species, *A. tonsa* is most abundant, attaining peak numbers in the summer (Stearns, 1995; Kennish, 2000). *Oithona colcarva* and *P. pelagicus* are also abundant at this time, and along with *H. fosteri,* persist into the fall. Abundant forms in the winter and spring include *A. hudsonica, E. affinis, O. colcarva,* and *P. pelagicus.* Zooplankton attain peak numbers in Delaware Bay in the summer, often exceeding 0.5×10^5 individuals/m^3 (Herman et al., 1983).

Sutton et al. (1996) reported that cladocerans, cyclopoid copepods, and gammarid amphipods dominate the zooplankton community in the tidal waters upestuary. Estuarine and marine species (e.g., calanoid copepods) predominate in Delaware Bay. Salinity is a major factor influencing the spatial distribution of zooplankton species in the system (Stearns, 1995).

Benthic Fauna

More than 30 macroinvertebrate taxa were collected in benthic surveys conducted in the Lower St. Jones River during the 1994–1995 sampling period. These benthic fauna belong to five phyla, notably the Annelida, Arthropoda, Mollusca, Platyhelminthes, and Nemertea (Table 4.8). Ten of the taxa comprised more than 90% of all the organisms collected at eight sampling sites. These ten taxa are listed here in order of decreasing abundance: Oligochaeta (58% of the total), Chironomidae (9%), *Corophium* sp. (5%), Polychaeta (3%), *Neomysis americana* (3%), *Edotea triloba* (3%), *Streblospio benedicti* (3%), *Gammarus* sp. (3%), *Ilyanassa* sp. (2%), and turbellarians (2%). While oligochaetes were by far the most abundant benthic macroinvertebrate taxa in the Lower St. Jones River, the opossum shrimp (*Neomysis americana*) was the overwhelming dominant member of the parabenthic community there, constituting more than 92% of all parabenthic organisms collected (DNERR, 1999).

The seasonal densities of the benthic macroinvertebrates ranged from 3850 to 4573 individuals/m^2, with maximum numbers recorded during spring. The densities of parabenthic organisms ranged from 578 to 21,210 individuals/m^2. Peak densities of these organisms were found in the fall.

Bivalves and polychaetes dominate the soft-bottom benthic community in Delaware Bay. In polyhaline waters near the mouth of the bay, surf clams (*Spisula solidissima*) and sand dollars (*Echinarachnius parma*) predominate in sandy sediments, and polychaetes (*Nucula proxima* and *Nephtys* spp.) are most abundant in silty sediments. Other numerically important macroinvertebrate species found in the bay are the bivalves *Crassostrea virginica* and *Ensis directus* as well as the polychaetes *Glycera dibranchiata* and *Heteromastus filiformis.* In mesohaline silts and fine sands, the bivalves *Gemma gemma, Mulinia lateralis,* and *Mya arenaria* are likewise abundant (Maurer et al., 1978; Steimle, 1995).

Proceeding to mesohaline salt marsh habitats adjacent to the St. Jones River, the most common members of the macroinvertebrate community are fiddler crabs

TABLE 4.8
Mean Density of Benthic Macroinvertebrates in the Tidal River and Channels of the St. Jones River during 1994 (Summer and Fall) and 1995 (Spring)[a]

Taxon	Upper	Lower	Mean	% Total
	Tidal River			
Oligochaeta	1302	1335	1319	34.9
Chironomidae	1765	118	942	24.9
Corophium sp.	1103	11	557	14.7
Edotea triloba	226	135	180	4.8
Neomysis americana	318	27	172	4.6
Ilyanassa sp.	0	312	156	4.1
Gammarus sp.	151	16	83	2.2
Gammaridae	145	0	73	1.9
Polychaeta	0	135	67	1.8
Bivalvia	5	102	54	1.4
Streblospio benedicti	5	91	48	1.3
Scolecolepides viridis	11	54	32	0.9
Polydora ligni	27	22	24	0.6
Cyathura polita	38	0	19	0.5
Nemertea	32	0	16	0.4
Nereis succinea	22	11	16	0.4
Eurypanopeus depressus	0	16	8	0.2
Coecidotea sp.	11	0	5	0.1
Isopoda	0	5	3	0.1
Spionidae	5	0	3	0.1
Idotea sp.	0	5	3	0.1
Xanthidae	5	0	3	0.1
	Channels			
Oligochaeta	5839	737	3288	74.7
Turbellaria	581	59	320	7.3
Neomysis americana	22	301	161	3.7
Gammarus sp.	22	237	129	2.9
Streblospio benedicti	11	183	97	2.2
Corophium sp.	—	172	86	2.0
Chironomidae	140	11	75	1.7
Nereis succinea	22	108	65	1.5
Polydora ligni	38	65	51	1.2
Edotea triloba	81	16	48	1.1
Scolecolepides viridis	32	0	16	0.4
Sphaeriidae	32	—	16	0.4
Gammaridae	16	5	11	0.2
Xanthidae	5	11	8	0.2
Hypaniola florida	5	5	5	0.1
Nassarius sp.	0	11	5	0.1

(*continued*)

TABLE 4.8 (CONTINUED)
Mean Density of Benthic Macroinvertebrates in the Tidal River and Channels of the St. Jones River during 1994 (Summer and Fall) and 1995 (Spring)[a]

Taxon	Upper	Lower	Mean	% Total
Nemertea	11	0	5	0.1
Bivalvia	0	5	3	0.1
Hirudinea	0	5	3	0.1
Limulus polyphemus	0	5	3	0.1
Melita nitida	0	5	3	0.1

[a] Density = number/m^2.

Source: Delaware Department of Natural Resources and Environmental Control. 1999. Delaware National Estuarine Research Reserve: Estuarine Profiles. Technical Report, Delaware Department of Natural Resources and Environmental Control, Dover, DE..

(*Uca* spp.), salt marsh snails (*Melampus bidentatus*), mud snails (*Ilyanassa obsoleta*), grass shrimp (*Palaemonetes* spp.), marsh crabs (*Sesarma reticulatum*), blue crabs (*Callinectes sapidus*), ribbed mussels (*Geukensia demissa*), amphipods (*Orchestia grillus* and *Gammarus* spp.) and isopods (*Edotea triloba*). Quadrat sampling revealed a mean density of marsh surface macroinvertebrates amounting to 44 individuals/m^2, with the most numerous species being the salt marsh snail *Melampus bidentatus* (mean density = 37.6 individuals/m^2) followed in decreasing order of abundance by *Uca minax, Orchestia grillus, U. pugnax, Geukensia demissa, Palaemonetes pugio,* and *Sesarma reticulatum*. Of all salt marsh areas sampled, *Spartina alterniflora* habitat had the highest mean density of macroinvertebrates (135.4 individuals/m^2) (DNERR, 1999).

Finfish

Nektonic surveys conducted in the summer and fall of 1994 and the spring of 1995 documented 25 species of fish in the Lower St. Jones River. Estuarine species dominated the assemblage. For example, the Atlantic silverside (*Menidia menidia*), mummichog (*Fundulus heteroclitus*), spot (*Leiostomus xanthurus*), sheepshead minnow (*Cyprinodon variegatus*), white perch (*Morone americana*), and bay anchovy (*Anchoa mitchilli*) — in decreasing order of abundance — were the most abundant forms, comprising nearly 95% of the total finfish catch (Table 4.9). Finfish abundance was significant in secondary tributaries (DNERR, 1999).

The finfish community is much more diverse in the Delaware Estuary, where more than 200 species of fish have been recorded. O'Herron et al. (1994) identified the following priority species in the estuary: alewife (*Alosa pseudoharengus*), American shad (*A. sapidissima*), blueback herring (*A. aestivalis*), American eel (*Anguilla rostrata*), Atlantic menhaden (*Brevoortia tyrannus*), Atlantic sturgeon (*Acipenser*

TABLE 4.9
Finfish Abundance in the Upper Blackbird Creek and Lower St. Jones River during 1994 (Summer and Fall) and 1995 (Spring)

Species	Blackbird Creek	St. Jones River	Percent of Total
Leiostomus xanthurus (spot)	2319	829	32.80
Fundulus heteroclitus (mummichog)	523	1143	17.36
Brevoortia tyrannus (Atlantic menhaden)	1241	—	12.93
Menidia menidia (Atlantic silverside)	18	1198	12.67
Morone americana (white perch)	585	342	9.66
Cyprinodon variegatus (sheepshead minnow)	—	634	6.60
Anchoa mitchilli (bay anchovy)	3	281	2.96
Trinectes maculatus (hogchoker)	78	65	1.49
Cynoscion regalis (weakfish)	19	51	0.73
Morone saxatilis (striped bass)	36	29	0.68
Anguilla rostrata (American eel)	20	32	0.54
Urophycis regia (spotted hake)	—	31	0.32
Pogonias cromis (black drum)	5	21	0.27
Ameiurus gibbosus (brown bullhead)	16	—	0.17
Hybognathus nuchalis (silvery minnow)	15	—	0.16
Clupea harengus (Atlantic herring)	—	12	0.13
Micropogonias undulatus (Atlantic croaker)	—	10	0.10
Ictalurus punctatus (channel catfish)	4	4	0.08
Bairdiella chrysoura (silver perch)	—	5	0.05
Fundulus majalis (striped killifish)	—	4	0.04
Mugal cephalus (striped mullet)	—	4	0.04
Dorosoma cepedianum (gizzard shad)	3	—	0.03
Menidia beryllina (inland silverside)	1	2	0.03
Centropristis striata (black sea bass)	—	2	0.02
Pomatomus saltatrix (bluefish)	—	2	0.02
Prionotus carolinus (northern searobin)	—	2	0.02
Opsanus tau (oyster toadfish)	—	2	0.02
Perca flavescens (yellow perch)	2	—	0.02
Fundulus diaphanus (banded killifish)	—	1	0.01
Pomoxis nigromaculatus (black crappie)	1	—	0.01
Alosa aestivalis (blueback herring)	1	—	0.01
Lepomis macrochirus (bluegill)	1	—	0.01
Lepomis gibbosus (pumpkinseed)	1	—	0.01
Paralichthys dentatus (summer flounder)	—	—	0.01

Note: Density = number/m^2.

Source: Delaware Department of Natural Resources and Environmental Control. 1999. Delaware National Estuarine Research Reserve: Estuarine Profiles. Technical Report, Delaware Department of Natural Resources and Environmental Control, Dover, DE.

oxyrhynchus), white perch (*Morone americana*), striped bass (*M. saxatilis*), weakfish (*Cynoscion regalis*), bluefish (*Pomatomus saltatrix*), spot (*Leiostomus xanthurus*), scup (*Stenotomus versicolor*), Atlantic croaker (*Micropogonias undulatus*), black drum (*Pogonias cromis*), channel catfish (*Ictalurus punctatus*), white catfish (*Ameiurus catus*), summer flounder (*Paralichthys dentatus*), windowpane flounder (*Scophthalmus aquosus*), and carp (*Cyprinus carpio*). The bay anchovy (*Anchoa mitchilli*) and Atlantic silverside (*Menidia menidia*) are important forage species in the system. The species of commercial importance primarily include the weakfish (*Cynoscion regalis*), bluefish (*P. saltatrix*), Atlantic menhaden (*B. tyrannus*), summer flounder (*P. dentatus*), and spot (*L. xanthurus*). They have largely replaced the prominent upriver forms (alewife, *A. pseudoharengus*; American shad, *A. sapidissima*; blueback herring, *A. aestivalis*; and Atlantic sturgeon, *A. oxyrhynchus*) in the fishery (Price and Beck, 1988; Kennish, 2000).

Amphibians and Reptiles

Frogs and salamanders are common in the DNERR, especially at the Upper Blackbird Creek Reserve site. Among the most important frog species in the DNERR are the green frog (*Rana clamitans melanota*), bullfrog (*R. catesbeiana*), wood frog (*R. sylvatica*), northern spring peeper (*Pseudacris crucifer crucifer*), and southern leopard frog (*R. utricularia*). Salamanders of significance include the northern two-lined salamander (*Eurycea bislineata*) and red-backed salamander (*Plethodon cinereus*). The greater areal coverage and diversity of wetland habitats along the Upper Blackbird Creek provide more suitable conditions for amphibian populations than those along the Lower St. Jones River.

Four species of turtles occupy wetland habitats of both DNERR component sites, specifically the snapping turtle (*Chelydra serpentina*), northern diamondback terrapin (*Malaclemys terrapin terrapin*), eastern mud turtle (*Kinosternon subrubrum subrubrum*), and red-bellied turtle (*Chrysemys rubriventris*). Marine turtles observed in Delaware Bay include the green sea turtle (*Chelonia mydas*), Kemp's Ridley turtle (*Lepidochelys kempii*), and loggerhead turtle (*Caretta caretta*).

Two species of snakes primarily inhabit the wetlands and uplands of the DNERR. These are the black rat snake (*Elaphe obsoleta*) and northern water snake (*Nerodia sipedon*). A few other species of snakes may range into reserve areas but are less common than the aforementioned forms (DNERR, 1999).

Birds

The Delaware Estuary and surrounding areas are havens for rich and diverse groups of avifauna. Waterbirds, raptors, and passerines are well represented in the St. Jones River watershed. An extensive list of shorebirds, wading birds, and waterfowl has been compiled for the reserve site based on surveys conducted between May 1994 and June 1995 (Table 4.10). Among the common shorebird species identified in the St. Jones River survey are dunlin (*Calidris alpina*), sanderling (*C. alba*), red knot (*C. canutus*), semipalmated sandpiper (*C. pusilla*), least sandpiper (*C. minutilla*), western sandpiper (*C. mauri*), black-bellied plover

TABLE 4.10
Bird Species Recorded in the Blackbird Creek and St. Jones River DNERR Sites during Field Surveys Conducted from May 1994 through June 1995

Species	Blackbird Creek	St. Jones River
American black duck		X
American coot		X
American crow	X	X
American goldfinch	X	X
American robin	X	X
Bald eagle	X	
Barn swallow	X	X
Belted kingfisher	X	X
Black-bellied plover		X
Black skimmer		X
Black vulture	X	X
Blue jay	X	
Blue-winged teal		X
Boat-tailed grackle	X	X
Canada goose	X	X
Carolina chickadee	X	
Carolina wren	X	X
Cattle egret	X	
Chimney swift		X
Clapper rail		X
Common grackle	X	X
Common rail	X	X
Common merganser		X
Common snipe		X
Common tern	X	X
Common yellowthroat	X	X
Double-crested cormorant	X	X
Downy woodpecker	X	X
Dunlin		X
Eastern bluebird	X	
Eastern kingbird	X	X
Eastern meadowlark	X	
Eastern wood pewee	X	
Fish crow	X	X
Forster's tern	X	X
Glossy ibis	X	X
Great black-backed gull		X
Great blue heron	X	X

(*continued*)

TABLE 4.10 (CONTINUED)
Bird Species Recorded in the Blackbird Creek and St. Jones River DNERR Sites during Field Surveys Conducted from May 1994 through June 1995

Species	Blackbird Creek	St. Jones River
Great crested flycatcher	X	X
Great egret	X	X
Great horned owl	X	
Greater yellowlegs	X	X
Green-backed heron	X	X
Green-winged teal	X	X
Grey catbird	X	X
Hairy woodpecker	X	
Herring gull	X	X
House wren		X
Indigo bunting	X	X
Killdeer	X	X
Laughing gull	X	X
Least tern		X
Mallard	X	X
Marsh wren	X	X
Mockingbird	X	
Mourning dove	X	X
Night heron	X	
Northern bobwhite	X	X
Northern cardinal	X	X
Northern flicker	X	X
Northern harrier	X	X
Northern pintail		X
Northern shoveler		X
Orchard oriole	X	
Peep	X	X
Pileated woodpecker	X	
Purple martin	X	
Red head		X
Red knot		X
Red-bellied woodpecker	X	X
Red-breasted merganser		X
Red-eyed vireo	X	
Red-tailed hawk	X	X
Red-winged blackbird	X	X
Ring-billed gull	X	X
Ruby-throated hummingbird		X
Ruddy duck		X

TABLE 4.10 (CONTINUED)
Bird Species Recorded in the Blackbird Creek and St. Jones River DNERR Sites during Field Surveys Conducted from May 1994 through June 1995

Species	Blackbird Creek	St. Jones River
Ruddy turnstone		X
Rufous-sided towhee	X	X
Sanderling		X
Savannah sparrow	X	
Scarlet tanager	X	
Seaside sparrow	X	X
Sharp-shinned hawk	X	
Sharp-tailed sparrow		X
Shore birds — mixed flocks	X	X
Short-billed downcatcher		X
Snow goose	X	X
Snowy egret	X	X
Song sparrow	X	X
Spotted sandpiper	X	
Swamp sparrow		X
Tree swallow	X	X
Tufted titmouse	X	
Turkey vulture	X	X
White-crowned sparrow		X
White-eyed vireo	X	X
White-throated sparrow	X	
Willet		X
Wood duck	X	
Wood thrush	X	
Yellow-billed cuckoo	X	X
Yellow-rumped warbler	X	
Yellow warbler	X	X

Source: Delaware Department of Natural Resources and Environmental Control. 1999. Delaware National Estuarine Research Reserve: Estuarine Profiles. Technical Report, Delaware Department of Natural Resources and Environmental Control, Dover, DE.

(*Pluvialis squatarola*), short-billed dowitcher (*Limnodromus griseus*), killdeer (*Charadrius vociferus*), greater yellowlegs (*Tringa melanoleuca*), willet (*Catoptrophorus semipalmatus*), and common snipe (*Gallinago gallinago*). Shorebirds are most numerous in spring and early summer, with peak abundance recorded in May. Each spring more than a million shorebirds utilize the beach and marsh habitats of the Delaware Estuary (Clark, 1988; Kennish, 2000). Some species inhabit the mudflats and tidal marshes year-round (Sutton et al., 1996).

Jenkins and Gelvin-Innvaer (1995) identified 10 species of wading birds in the Delaware Estuary. They are the glossy ibis (*Plegadis falcinellus*), cattle egret (*Bubulcus ibis*), yellow-crowned night heron (*Nycticorax violaceus*), black-crowned night heron (*Nycticorax nycticorax*), green-backed heron (*Butorides virescens*), great egret (*Casmerodius albus*), snowy egret (*Egretta thula*), tricolored heron (*E. tricolor*), little blue heron (*E. caerulea*), and great blue heron (*Ardea herodias*). Five of these species were recorded in surveys of the Lower St. Jones River, including the snowy egret, great egret, great blue heron, green-backed heron, and glossy ibis. Of these species, the snowy egret and great blue heron were most common at the reserve site.

Many species of waterbirds have been observed in the open waters and tidal wetlands of the Lower St. Jones River. Colonial nesting birds, such as the black scoter (*Melanitta nigra*), American black duck (*Anas rubripes*), and bufflehead (*Bucephala albeola*), are well represented. Other waterbirds of note frequenting the DNERR sites include the least tern (*Sterna antillarum*), Forster's tern (*S. fosteri*), common tern (*S. hirundo*), laughing gull (*Larus atricilla*), herring gull (*L. argentatus*), ring-billed gull (*L. delawarensis*), greater black-backed gull (*L. marinus*), surf scoter (*Melanitta perspicillata*), lesser scaup (*Aythya affinis*), oldsquaw (*Clangula hyemalis*), common merganser (*Mergus merganser*), red-breasted merganser (*M. serrator*), northern gannet (*Morus bassanus*), double-crested cormorant (*Phalacrocorax auritus*), horned grebe (*Podiceps auritus*), and red-throated loon (*Gavia stellata*). Rails (e.g., king rail, *Rallus elegans*; Virginia rail, *R. limicola*; and clapper rail, *R. longirostris*) are also often seen in the reserve. The black rail (*Laterallus jamaicensis*), yellow rail (*Coturnicops noveboracensis*), and sora rail (*Porzana carolina*) have been reported in other areas of the Delaware Estuary as well (Kerlinger and Widjeskog, 1995).

Various passerines occupy tidal wetlands habitat in the DNERR. The most numerous species are the red-winged blackbird (*Agelaius phoeniceus*), seaside sparrow (*Ammospiza maritima*), sharp-tailed sparrow (*A. caudacuta*), common yellowthroat (*Geothlypis trichas*), marsh wren (*Cistothorus palustris*), and boat-tailed grackle (*Quiscalus major*). The red-winged blackbird, present year-round, is perhaps the most common passerine in the tidal wetlands.

Predatory birds (eagles, hawks, falcons, and owls) are widely dispersed in the Delaware Estuary region and occur at both DNERR sites. The bald eagle (*Haliaeetus leucocephalus*), osprey (*Pandion haliaetus*), and peregrine falcon (*Falco peregrinus*) nest along the shores of the Delaware. The red-tailed hawk (*Buteo jamaicensis*), Cooper's hawk (*Accipiter cooperii*), and great horned owl (*Bubo virginianus*) prefer woodlots, forest edges, and upland forests. The red-shouldered hawk (*Buteo lineatus*) and barred owl (*Strix varia*) nest and feed in hardwood swamps (Sutton et al., 1996; Kennish, 2000). An estimated 80,000 raptors fly through the mouth of the estuary each year (Niles and Sutton, 1995).

The northern harrier (*Circus cyaneus*) is the most frequently seen raptor in the tidal wetlands of the DNERR. Four species of hawks (rough-legged hawk, *Buteo lagopus*; sharp-shinned hawk, *Accipiter striatus*; red-shouldered hawk,

Buteo lineatus; and red-tailed hawk, *B. jamaicensis*) are also observed in DNERR habitats. Other raptors documented in the DNERR system include the osprey (*Pandion haliaetus*), short-eared owl (*Asio flammeus*), great horned owl (*Bubo virginianus*), Eastern screech owl (*Otus asio*), bald eagle (*Haliaeetus leucocephalus*), and turkey vulture (*Cathartes aura*).

Upland avifaunal species are quite diverse in woodlands and thickets, as well as open or semiopen habitats. Some species which may occur in woodlands and shrubby thickets are the blue jay (*Cyanocitta cristata*), red-bellied woodpecker (*Melanerpes carolinus*), downy woodpecker (*Picoides pubescens*), Carolina chickadee (*Parus carolinensis*), rufous-sided towhee (*Pipilo erythrophthalmus*), gray catbird (*Dumetella carolinensis*), white-eyed vireo (*Vireo griseus*), red-eyed vireo (*V. olivaceus*), yellow-rumped warbler (*Dendroica coronata*), ovenbird (*Seiurus aurocapillus*), wood thrush (*Hylocichla mustelina*), tufted titmouse (*Parus bicolor*), and white-throated sparrow (*Zonotrichia albicollus*). In open or semiopen habitats, characteristic species may include the northern bobwhite (*Colinus virginianus*), mourning dove (*Zenaida macroura*), American crow (*Corvus brachyrhynchos*), northern mockingbird (*Mimus polyglottos*), prairie warbler (*Dendroica discolor*), American kestrel (*Falco sparverius*), brown thrasher (*Toxostoma rufum*), American robin (*Turdus migratorius*), yellow warbler (*Dendroica petechia*), common grackle (*Quiscalus quiscula*), northern oriole (*Icterus galbula*), northern cardinal (*Cardinalis cardinalis*), house sparrow (*Passer domesticus*), field sparrow (*Spizella pusilla*), chipping sparrow (*Spizella passerina*), song sparrow (*Melospiza melodia*), bobolink (*Dolichonyx oryzivorus*), blue grosbeak (*Guiraca caerulea*), indigo bunting (*Passerina cyanea*), and American goldfinch (*Carduelis tristis*) (DNERR, 1999).

Table 4.11 provides a list of bird species compiled for both the Lower St. Jones River Reserve site and the Upper Blackbird Creek Reserve site. This list includes the bird species known to occur or possibly found in reserve areas. All major groups are represented, notably shorebirds, wading birds, waterfowl, rails, raptors, and passerines.

Mammals

A number of harbor seals, dolphins, porpoises, and whales are reported in the lower Delaware Estuary each year. Harbor seals (*Phoca vitulina*) periodically appear in Delaware Bay; in addition, harp seals (*Pagophilus groenlandicus*) and gray seals (*Halichoerus grypus*) occasionally occur in the system. The bottlenose porpoise (*Tursiops truncatus*) is relatively common in the lower estuarine waters. The humpback whale (*Megaptera novaeangliae*) is the most frequently observed cetacean species in Delaware Bay, but the northern right whale (*Balaena glacialis*) and finback whale (*Balaenoptera physalus*) have also been found here (Kennish, 2000).

In fresh and brackish marshes of the DNERR, the beaver (*Castor canadensis*), river otter (*Lutra canadensis*), and muskrat (*Ondatra zibethicus*) are important mammalian species. The muskrat, a fur-bearing animal, is most abundant and remains commercially valuable for its pelts. In upland forested areas, an array of mammalian species has been documented, such as the eastern cottontail rabbit (*Sylvilagus floridanus*), white-footed mouse (*Peromyscus leucopus*), meadow jumping mouse (*Zapus hudsonius*), pine vole (*Microtus pinetorum*), meadow vole (*M. pennsylvanicus*), gray

TABLE 4.11
Species of Birds Known to Occur or Likely to Occur in the Delaware National Estuarine Research Reserve

Common Name	Scientific Name
American black duck	*Anas rubripes*
Gadwall	*Anas strepera*
Mallard	*Anas platyrhynchos*
American wigeon	*Anas americana*
Northern pintail	*Anas acuta*
Northern shoveler	*Anas clypeata*
Green-winged teal	*Anas crecca*
Blue-winged teal	*Anas discors*
Wood duck	*Aix sponsa*
Canvasback	*Aythya valisineria*
Redhead	*Aythya americana*
Greater scaup	*Aythya marila*
Lesser scaup	*Aythya affinis*
Ring-necked duck	*Aythya collaris*
Bufflehead	*Bucephala albeola*
Common goldeneye	*Bucephala clangula*
Oldsquaw	*Clangula hyemalis*
Surf scoter	*Melanitta perspicillata*
Black scoter	*Melanitta nigra*
White-winged scoter	*Melanitta fusca*
Ruddy duck	*Oxyura jamaicensis*
Common merganser	*Mergus merganser*
Red-breasted merganser	*Mergus serrator*
Hooded merganser	*Lophodytes cucullatus*
Tundra swan	*Cygnus columbianus*
Canada goose	*Branta canadensis*
Greater snow goose	*Anser caerulescens atlanticus*
Common loon	*Gavia immer*
Red-throated loon	*Gavia stellata*
Pied-billed grebe	*Podilymbus podiceps*
Horned grebe	*Podiceps auritus*
Double-crested cormorant	*Phalacrocorax auritus*
Northern gannet	*Morus bassanus*
Herring gull	*Larus argentatus*
Ring-billed gull	*Larus delawarensis*
Greater black-backed gull	*Larus marinus*
Laughing gull	*Larus atricilla*
Bonaparte's gull	*Larus philadelphia*
Common tern	*Sterna hirundo*
Forster's tern	*Sterna forsteri*
Least tern	*Sterna antillarum*
Gull-billed tern	*Sterna nilotica*

TABLE 4.11 (CONTINUED)
Species of Birds Known to Occur or Likely to Occur in the Delaware National Estuarine Research Reserve

Common Name	Scientific Name
Caspian tern	*Sterna caspia*
Royal tern	*Sterna maxima*
Black skimmer	*Rynchops niger*
Great egret	*Casmerodius albus*
Snowy egret	*Egretta thula*
Cattle egret	*Bubulcus ibis*
Great blue heron	*Ardea herodias*
Tricolored heron	*Egretta tricolor*
Little blue heron	*Egretta caerulea*
Black-crowned night heron	*Nycticorax nycticorax*
Yellow-crowned night heron	*Nycticorax violaceus*
Green-backed heron	*Butorides virescens*
American bittern	*Botaurus lentiginosus*
Least bittern	*Ixobrychus exilis*
Glossy ibis	*Plegadis falcinellus*
Clapper rail	*Rallus longirostris*
King rail	*Rallus elegans*
Virginia rail	*Rallus limicola*
Sora	*Porzana carolina*
American coot	*Fulica americana*
Common moorhen	*Gallinula chloropus*
American oystercatcher	*Haematopus palliatus*
American avocet	*Recurvirostra americana*
Black-necked stilt	*Himantopus mexicanus*
Black-bellied plover	*Pluvialis squatarola*
Ruddy turnstone	*Arenaria interpres*
Semipalmated plover	*Charadrius semipalmatus*
Killdeer	*Charadrius vociferus*
Semipalmated sandpiper	*Calidris pusilla*
Least sandpiper	*Calidris minutilla*
Western sandpiper	*Calidris mauri*
White-rumped sandpiper	*Calidris fuscicollis*
Sanderling	*Calidris alba*
Red knot	*Calidris canutus*
Dunlin	*Calidris alpina*
Long-billed dowitcher	*Limnodromus scolopaceus*
Short-billed dowitcher	*Limnodromus griseus*
Greater yellowlegs	*Tringa melanoleuca*
Lesser yellowlegs	*Tringa flavipes*
Willet	*Catoptrophorus semipalmatus*
Solitary sandpiper	*Tringa solitaria*

(*continued*)

TABLE 4.11 (CONTINUED)
Species of Birds Known to Occur or Likely to Occur in the Delaware National Estuarine Research Reserve

Common Name	Scientific Name
Upland sandpiper	*Bartramia loingicauda*
Pectoral sandpiper	*Calidris melanotos*
Spotted sandpiper	*Actitis macularia*
American woodcock	*Philohela minor*
Common snipe	*Gallinago gallinago*
Wild turkey	*Meleagris gallopavo*
Ring-necked pheasant	*Phasianus colchicus*
Northern bobwhite	*Colinus virginianus*
Bald eagle	*Haliaeetus leucocephalus*
Osprey	*Pandion haliaetus*
Peregrine falcon	*Falco peregrinus*
Merlin	*Falco columbarius*
American kestrel	*Falco sparverius*
Red-tailed hawk	*Buteo jamaicensis*
Red-shouldered hawk	*Buteo lineatus*
Broad-winged hawk	*Buteo platypterus*
Rough-legged hawk	*Buteo lagopus*
Sharp-shinned hawk	*Accipiter striatus*
Cooper's hawk	*Accipiter cooperii*
Northern harrier	*Circus cyaneus*
Black vulture	*Coragyps atratus*
Turkey vulture	*Cathartes aura*
Great horned owl	*Bubo virginianus*
Barred owl	*Strix varia*
Short-eared owl	*Asio flammeus*
Common barn owl	*Tyto alba*
Eastern screech owl	*Otus asio*
Rock dove	*Columba livia*
Mourning dove	*Zenaida macroura*
Yellow-billed cuckoo	*Coccyzus americanus*
Common nighthawk	*Chordeiles minor*
Whip-poor-will	*Caprimulgus vociferus*
Chimney swift	*Chaetura pelagica*
Ruby hummingbird	*Archilochus colubris*
Belted kingfisher	*Ceryle alcyon*
Red-bellied woodpecker	*Melanerpes carolinus*
Yellow-bellied sapsucker	*Sphyrapicus varius*
Downy woodpecker	*Picoides pubescens*
Hairy woodpecker	*Picoides villosus*
Northern flicker	*Colaptes auratus*
Eastern wood pewee	*Contopus virens*
Acadian flycatcher	*Empidonax virescens*

TABLE 4.11 (CONTINUED)
Species of Birds Known to Occur or Likely to Occur in the Delaware National Estuarine Research Reserve

Common Name	Scientific Name
Willow flycatcher	*Empidonax traillii*
Eastern phoebe	*Sayornis phoebe*
Great-crested flycatcher	*Myiarchus crinitus*
Eastern kingbird	*Tyrannus tyrannus*
Horned lark	*Eremophila alpestris*
Purple martin	*Progne subis*
Tree swallow	*Tachycineta bicolor*
Bank swallow	*Riparia riparia*
Barn swallow	*Hirundo rustica*
Blue jay	*Cyanocitta cristata*
American crow	*Corvus brachyrhynchos*
Fish crow	*Corvus ossifragus*
Carolina chickadee	*Parus carolinensis*
Tufted titmouse	*Parus bicolor*
Red-breasted nuthatch	*Sitta canadensis*
White-breasted nuthatch	*Sitta carolinensis*
Brown creeper	*Certhia americana*
Marsh wren	*Cistothorus palustris*
Sedge wren	*Cistothorus platensis*
Carolina wren	*Thryothorus ludovicianus*
House wren	*Troglodytes aedon*
Winter wren	*Troglodytes troglodytes*
Golden-crowned kinglet	*Regulus satrapa*
Ruby-crowned kinglet	*Regulus calendula*
Blue-gray gnatcatcher	*Polioptila caerulea*
Eastern bluebird	*Sialia sialis*
Veery	*Catharus fuscescens*
Swainson's thrush	*Catharus ustulatus*
Hermit thrush	*Catharus guttatus*
Wood thrush	*Hylocichla mustelina*
American robin	*Turdus migratorius*
Gray catbird	*Dumetella carolinensis*
Northern mockingbird	*Mimus polyglottos*
Brown thrasher	*Toxostoma rufum*
Water pipit	*Anthus spinoletta*
Cedar waxwing	*Bombycilla cedrorum*
Loggerhead shrike	*Lanius ludovicianus*
White-eyed vireo	*Vireo griseus*
Red-eyed vireo	*Vireo olivaceus*
Common yellowthroat	*Geothlypis trichas*
Northern parula warbler	*Parula americana*

(*continued*)

TABLE 4.11 (CONTINUED)
Species of Birds Known to Occur or Likely to Occur in the Delaware National Estuarine Research Reserve

Common Name	Scientific Name
Prothonotary warbler	*Protonotaria citrea*
Yellow warbler	*Dendroica petechia*
Chestnut-sided warbler	*Dendroica pensylvanica*
Magnolia warbler	*Dendroica magnolia*
Black-throated blue warbler	*Dendroica caerulescens*
Yellow-rumped warbler	*Dendroica coronata*
Black-throated green warbler	*Dendroica virens*
Pine warbler	*Dendroica pinus*
Prairie warbler	*Dendroica discolor*
Palm warbler	*Dendroica palmarum*
Blackpoll warbler	*Dendroica striata*
Black-and-white warbler	*Mniotilta varia*
Kentucky warbler	*Oporornis formosus*
Canada warbler	*Wilsonia canadensis*
American redstart	*Setophaga ruticilla*
Ovenbird	*Seiurus aurocapillus*
Northern waterthrush	*Seiurus noveboracensis*
Yellow-breasted chat	*Icteria viren*
Red-winged blackbird	*Agelaius phoeniceus*
Rusty blackbird	*Euphagus carolinus*
Brown-headed cowbird	*Molothrus ater*
Common grackle	*Quiscalus quiscula*
Boat-tailed grackle	*Quiscalus major*
Bobolink	*Dolichonyx oryzivorus*
Eastern meadowlark	*Sturnella magna*
European starling	*Sturnus vulgaris*
Orchard oriole	*Icterus spurius*
Northern oriole	*Icterus galbula*
Scarlet tanager	*Piranga olivacea*
Summer tanager	*Piranga rubra*
House sparrow	*Passer domesticus*
Dark-eyed junco	*Junco hyemalis*
Snow bunting	*Plectrophenax nivalis*
Northern cardinal	*Cardinalis cardinalis*
House finch	*Carpodacus mexicanus*
Purple finch	*Carpodacus purpureus*
American goldfinch	*Carduelis tristis*
Pine siskin	*Carduelis pinus*
Rose-breasted grosbeak	*Pheucticus ludovicianus*
Blue grosbeak	*Guiraca caerulea*
Indigo bunting	*Passerina cyanea*
Rufous-sided towhee	*Pipilo erythrophthalmus*

TABLE 4.11 (CONTINUED)
Species of Birds Known to Occur or Likely to Occur in the Delaware National Estuarine Research Reserve

Common Name	Scientific Name
House sparrow	*Passer domesticus*
American tree sparrow	*Spizella arborea*
Chipping sparrow	*Spizella passerina*
Field sparrow	*Spizella pustilla*
Savannah sparrow	*Passerculus sandwichensis*
Seaside sparrow	*Ammospiza maritima*
Sharp-tailed sparrow	*Ammospiza caudacuta*
Swamp sparrow	*Melospiza georgiana*
Song sparrow	*Melospiza melodia*
Fox sparrow	*Passerella iliaca*
White-throated sparrow	*Zonotrichia albicollus*
White-crowned sparrow	*Zonotrichia leucophrys*

Source: Delaware Department of Natural Resources and Environmental Control. 1999. Delaware National Estuarine Research Reserve: Estuarine Profiles. Technical Report, Delaware Department of Natural Resources and Environmental Control, Dover, DE.

squirrel (*Sciurus carolinensis*), red bat (*Lasiurus borealis*), red fox (*Vulpes vulpes*), gray fox (*Urocyon cinereoargenteus*), raccoon (*Procyon lotor*), long-tailed weasel (*Mustela frenata*), masked shrew (*Sorex cinereus*), least shrew (*Cryptotis parva*), short-tail shrew (*Blarina brevicauda*), striped skunk (*Mephitis mephitis*), and white-tailed deer (*Odocoileus virginianus*) (DNERR, 1999).

UPPER BLACKBIRD CREEK RESERVE SITE

WATERSHED

More than 4000 people inhabit the 80-km^2 area of the Blackbird Creek watershed. Land use cover in the watershed is as follows: 39% agriculture, 22% forested land, 25% wetlands, 10% developed, and 4% open water. Although the Upper Blackbird Creek Reserve site is less developed than the Lower St. Jones River Reserve site, low-density residential development continues to increase.

Upland Vegetation

Plant assemblages in upland habitats of the Upper Blackbird Creek Reserve site are very similar to those of the Lower St. Jones River Reserve site. Two types of forested areas are recognized in the reserve: upland forest and tidal swamp forest. The dominant trees in the upland forest of the reserve include the white oak (*Quercus alba*), red oak (*Q. rubra*), black cherry (*Prunus serotina*), sassafras (*Sassafras albidum*), American holly (*Ilex opaca*), and American beech (*Fagus grandifolia*). Principal species comprising the tidal swamp forest are the red cedar (*Juniperus virginiana*), black gum (*Nyssa sylvatica*), sweet gum (*Liquidambar styraciflua*), red maple (*Acer rubrum*), green ash (*Fraxinus pennsylvanica*), and willows (*Salix* spp.) (DNERR, 1999).

Wetland Vegetation

Coastal marsh vegetation in the Upper Blackbird Creek Reserve is very similar to that in the St. Jones River Reserve site, characterized by scrub/marsh mixes, scrub forest, and forested tidal wetlands. A mixed emergent marsh plant community proliferates along creek and tributary banks and the open marsh plain. *Spartina alterniflora* marsh, although a dominant habitat in the Upper Blackbird Creek Reserve site, is much less expansive than in the Lower St. Jones River Reserve site. It covers 28.6% of the wetlands area in the reserve site compared to 62.2% of the habitat in the Lower St. Jones River Reserve site. In addition, tidal flats are much more extensive in the Upper Blackbird Creek Reserve, amounting to 26.4% of the wetlands cover compared to only 0.7% of the wetlands cover in the Lower St. Jones River Reserve. Open water habitat is also greater in the Upper Blackbird Creek Reserve, covering a 14.2% area compared to a 7% area in the Lower St. Jones River Reserve.

The marsh plant communities are much more diverse in the Upper Blackbird Creek Reserve site (113 species) than in the Lower St. Jones River Reserve site (66 species) (Table 4.12). Salinities are lower in the Upper Blackbird Creek Reserve; this accounts for the higher species richness. However, the common reed (*Phragmites australis*) is also more densely distributed along the Lower Blackbird Creek. *Phragmites*-infested marsh remains a target for remedial management control programs of the DNERR.

The dominant species of the emergent tidal wetlands in the Upper Blackbird Creek Reserve site parallel those in the Lower St. Jones River Reserve site. Aside from the smooth cordgrass (*Spartina alterniflora*), which predominates as tall-form plants along tidal creek banks and as short-form plants on the marsh plain, other abundant plants are the big cordgrass (*S. cynosuroides*) along channel banks, salt grass (*Distichlis spicata*) and salt meadow cordgrass (*S. patens*) above MHW, and a low-marsh mixed association of arrow arum (*Peltandra virginica*), yellow pondweed (*Nuphar lutea*), marshpepper smartweed (*Polygonum hydropiper*), and pickerel weed (*Ponderia cordata*) distributed along channel edges. Arrow arum and yellow pondweed are especially prominent in the Upper Blackbird Creek Reserve site. Other emergent wetland assemblages include cattails (*Typha* spp.), American three-square (*Scirpus americanus*), and wild rice (*Zizania aquatica*). They provide valuable wildlife habitat. Marsh shrub communities are also well represented; they consist of woody plants (e.g., smooth alder, *Alnus serrulata*; winterberry, *Ilex verticillata*; buttonbush, *Cephalanthus occidentalis*; sweet pepperbush, *Clethra alnifolia*; dogwoods, *Cornus* spp.; and red cedar, *Juniperus virginiana*), as well as marsh shrubs (e.g., marsh elder, *Iva frutescens*; and

TABLE 4.12
Taxonomic List of Plants Identified in Upper Blackbird Creek Marshes

Common Name	Scientific Name
Red maple	*Acer rubrum*
Saltmarsh water hemp	*Acnida cannabina*
Sweet flag	*Acorus calamus*
Smooth alder	*Alnus serrulata*
Ground nut	*Apios americana*
Jack-in-the-pulpit	*Aarisaema triphyllum*
Chokeberry	*Aronia arbutifolia*
Swamp milkweed	*Asclepias incarnata*
White aster	*Aster vimineus*
Lady fern	*Athryium filix-femina*
Nodding bur-marigold	*Bidens cernua*
False nettle	*Boehmeria cylindrica*
Bluejoint	*Calamagrostis canadensis*
Marsh marigold	*Cartha palustris*
Trumpet creeper	*Campsis radicans*
Bearded carex	*Carex comosa*
Large sedge	*Carex gigantia*
Lurid carex	*Carex lurida*
Straw carex	*Carex straminea*
Uptight sedge	*Carex stricta*
Carex sedge	*Carex vinata*
Ironweed	*Carpinus caroliniana*
Mockernut hickory	*Carya tomentosa*
Buttonbush	*Cephalanthus occidentalis*
Water hemlock	*Cicuta maculata*
Virgin's bower	*Clematis virginiana*
Sweet pepperbush	*Clethra alnifolia*
Silky dogwood	*Cornus amomum*
Flowering dogwood	*Cornus florida*
Gray dogwood	*Cornus racemosa*
Umbrella sedge	*Cyperus strigosus*
Water willow	*Decodon verticillatus*
Water millet	*Echinochloa walteri*
Spike-rush	*Eleocharis ambigens*
Marsh spike-rush	*Eleocharis palustris*
Virginia wild-rye	*Elymus virginicus*
Joe-pye-weed	*Eupatorium fistulosum*
American beech	*Fagus grandifolia*
Wild strawberry	*Fragaria virginia*
Green ash	*Fraxinus pennsylvanica*
Common madder	*Galium tinctorium*

(continued)

TABLE 4.12 (CONTINUED)
Taxonomic List of Plants Identified in Upper Blackbird Creek Marshes

Common Name	Scientific Name
Rough avens	*Geum virginiana*
Witch hazel	*Hamamelis virginiana*
Swamp rose mallow	*Hibiscus palustris*
Water penny-wort	*Hydrocotyl sibthorpioides*
American holly	*Ilex opaca*
Winterberry	*Ilex verticillata*
Jewelweed	*Impatiens capensis*
Yellow flag	*Iris pseudacorus*
Blue flag	*Iris versicolor*
Virginia willow	*Itea virginica*
Black walnut	*Juglans nigra*
Red cedar	*Juniperus virginiana*
Seashore mallow	*Kosteletzya virginica*
Rice cutgrass	*Leersia oryzoides*
Canada lily	*Lilium canadense*
Spicebush	*Lindera benzoin*
Sweet gum	*Liquidambar styraciflua*
Tulip tree	*Liriodendron tulipifera*
Cardinal flower	*Lobelia cardinalis*
Climbing hempweed	*Mikania scandens*
Yellow pond lily	*Numphar luteum*
Sensitive fern	*Onoclea sensibilis*
Royal fern	*Osmunda regalis*
Panic grass	*Panicum laniterinum*
Virginia creeper	*Parthenocissus quinquefolia*
Arrow arum	*Peltandra virginica*
Common reed	*Phragmites australis*
Pokeweed	*Phytolacca americana*
Mayapple	*Podophyllum peltatum*
Halberd-leaved tearthumb	*Polygonum arifolium*
Marshpepper smartweed	*Polygonum hydropiper*
Arrow-leaved tearthumb	*Polygonum sagittatum*
Pickerel weed	*Ponderia cordata*
Mazzard	*Prunus* cf. *avium*
Black cherry	*Prunus serotina*
Mock bishopweed	*Ptilimnium capillaceum*
White oak	*Quercus alba*
Southern red oak	*Quercus falcata*
Red oak	*Quercus rubra*
White water-crowfoot	*Ranunculus longirostris*
Swamp honeysuckle	*Rhondodendron viscosum*
Winged sumac	*Rhus copallina*

**TABLE 4.12 (CONTINUED)
Taxonomic List of Plants Identified in Upper Blackbird Creek Marshes**

Common Name	Scientific Name
Marsh yellow cress	*Rorippa palustris*
Swamp rose	*Rosa palustris*
Wineberry	*Rubus phoenicolasius*
Blackberry	*Rubus* sp.
Swamp dock	*Rumex verticillatus*
Long-beaked arrowhead	*Sagittaria australis*
Sassafras	*Sassafras albidum*
Lizard's tail	*Saururus cernuus*
American three-square	*Scirpus americanus*
Leafy bulrush	*Scirpus polyphyllus*
Saltmarsh bulrush	*Scirpus robustus*
Giant bulrush	*Scirpus validus*
Blue-eyed grass	*Sisyrinchium graminoides*
Water parsnip	*Sium suave*
Sawbriar	*Smilax glauca*
Common greenbriar	*Smilax rotundifolia*
Saltmarsh cordgrass	*Spartina alterniflora*
Big cordgrass	*Spartina cynosuroides*
Saltmeadow cordgrass	*Spartina patens*
Smooth hedge nettle	*Stachys hispidus*
Skunk cabbage	*Symplocarpus foetidus*
Tall meadow-rue	*Thalictrum pubescens*
Poison ivy	*Toxicodendron radicans*
Narrow-leaf cattail	*Typha angustifolia*
Broad-leaf cattail	*Typha latifolia*
Slippery elm	*Ulmus rubra*
Highbush blueberry	*Vaccinium corymbosum*
Black haw	*Viburnum prunifolium*
Northern arrowwood	*Viburnum recognitum*
Rusty black haw	*Viburnum rufidulum*
Common cocklebur	*Xanthium strumarium*
Wild rice	*Zizania aquatica*

Source: Delaware Department of Natural Resources and Environmental Control. 1999. Delaware National Estuarine Research Reserve: Estuarine Profiles. Technical Report, Delaware Department of Natural Resources and Environmental Control, Dover, DE.

groundsel bush, *Baccharis halimifolia*). A tidal swamp forest assemblage is prominent in this reserve site, dominated by black gum (*Nyssa sylvatica*), red maple (*Acer rubrum*), green ash (*Fraxinus pennsylvanica*), and sweet gum (*Liquidambar styraciflua*). Wetland plant species of special concern in the Upper Blackbird Creek

Reserve are the Canada lily (*Lilium canadense*), rough avens (*Geum virginiana*), nodding bur-marigold (*Bidens cernua*), and marsh marigold (*Cartha palustris*) (DNERR, 1999).

Aquatic Habitat

The oligohaline conditions of the Upper Blackbird Creek contrast markedly with the mesohaline conditions of the Lower St. Jones River. However, semidiurnal tides characterize both locations. The mean tidal range at the mouth of the Blackbird Creek averages about 1.92 m, or 0.42 m greater than at the mouth of the St. Jones River.

The channel width of the Blackbird Creek at the site of the reserve ranges from ~75 to 110 m. The mid-channel depth of the creek, in turn, ranges from 0.5 to 5.7 m. Maximum currents observed at the mouth of the Blackbird Creek at mid-depth during neap and spring tides are 50 to 60 and 60 to 70 cm/sec, respectively. The water column is unstratified in the Upper Blackbird Creek, with flow essentially unidirectional from surface to bottom.

Water quality variables (temperature, salinity, dissolved oxygen, pH, turbidity, and depth) have been monitored in the tidal portions of the Blackbird Creek at Blackbird Landing using YSI data loggers. Figure 4.2 and Figure 4.3 show water quality data recorded by the DNERR at the Blackbird Landing site in 1996. The data are presented as monthly averages for the year.

The absolute temperature range at the Blackbird Landing site in 1996 was ~0 to 30°C. The monthly mean water temperature, in turn, ranged from ~2°C in February to more than 25°C in August. Absolute salinities in the Blackbird Creek during 1996 ranged from ~0.1 to 3.5‰, with the monthly mean salinity values ranging from ~0.1 to 2‰. Lowest salinities were recorded in winter (December) and highest salinities in summer (September).

The Blackbird Creek exhibits higher dissolved oxygen concentrations than the St. Jones River. Dissolved oxygen percent saturation at Blackbird Landing in 1996 was much less variable than at Scotton Landing in the St. Jones River, with values consistently exceeding 75% year-long. The annual mean percent saturation at this site was 83%. Dissolved oxygen levels ranged from near 6 mg/l in summer to more than 11 mg/l in winter. Despite depressed dissolved oxygen levels in summer, no hypoxia was observed at the site.

Waters in the Blackbird Creek are slightly more acidic than those in the St. Jones River. The pH measurements at the Blackbird Creek in 1996 ranged from 5.70 to 8.67, with lowest values in winter (December). The mean pH reading of 6.76 at Blackbird Landing was only slightly less than that at Scotton Landing (7.02) in the St. Jones River. The highest pH levels at Blackbird Landing were recorded in summer (September).

Monthly mean water depths at Blackbird Landing varied from a low of about 1.2 m (February) to a high of approximately 1.7 m (September). These values and seasonal patterns are very similar to those observed at Scotton Landing in the St. Jones River. Water depth measurements at both DNERR sites are consistent, showing relatively low variation from month to month.

ANTHROPOGENIC IMPACTS

Pollution and Habitat Alteration

At the present time, pollution and other anthropogenic impacts appear to be more serious at the Lower St. Jones River Reserve site than at the Upper Blackbird Creek Reserve site because of a wider array of water quality problems and other environmental stressors. However, the potential for habitat alteration and associated adverse human impacts may be greater in the Upper Blackbird Creek Reserve as a result of new highway construction (e.g., Route 13 Relief Route) and encroaching residential development in the Blackbird Creek watershed. With greater infrastructure complexity in the watershed due to more aggressive development, additional point and nonpoint source pollution problems and issues related to habitat modification are likely to arise (DNREC, 1995).

The Lower St. Jones River is experiencing more water quality problems than Blackbird Creek mainly because of urbanization of its middle and upper watershed areas. While there are a few sites in the St. Jones River watershed that require National Pollution Discharge Elimination System discharge permits, none exists in the Blackbird Creek watershed. Hence, point source pollution is not a problem at the Upper Blackbird Creek Reserve site. A growing concern in the Blackbird Creek watershed, however, is land use conversion to residential development in rural areas. In contrast to urban land use impacts in the St. Jones River watershed, rural land use effects are of overriding significance in the Blackbird Creek watershed.

Nonpoint source pollution in the Upper Blackbird Creek Reserve largely stems from silviculture activities, as well as land disposal operations and agricultural runoff (DNREC, 1995). These factors may contribute to elevated nutrient inputs, chemical contaminant influx, and sediment loading of waterways. An adverse effect of greater highway construction and residential development in the watershed will be the partitioning and loss of existing wildlife habitat.

BIOTIC COMMUNITIES

Phytoplankton

Phytoplankton surveys conducted in the Blackbird Creek in 1995 and 1996 recovered a total of 42 taxa. Diatoms dominated the community, with *Skeletonema* spp., *Melosira* spp., and *Nitzschia* spp. as the most abundant taxa. Blue-green algae (i.e., *Anabaena* spp. and *Microcystis* spp.) were of secondary numerical importance. *Actinastrum* spp., *Ankistrodesmus* spp., *Hydrodictyon* spp., *Scenedesmus* spp., *Tetraedron* spp., and *Volvox* spp. were the predominant green algae (DNERR, 1999).

Phytoplankton abundance in the Blackbird Creek peaked in the summer, as did phytoplankton diversity. The lowest phytoplankton abundance occurred in the winter. Phytoplankton diversity was lowest in the fall.

Zooplankton

Copepod nauplii and rotifers (*Filinia* spp. and *Notholca* spp.) dominated the microzooplankton assemblages in the Blackbird Creek during the 1995–1996 sampling period. Other rotifers as well as bivalve, gastropod, and polychaete larvae were also commonly occurring microzooplankton. During the sampling period, 36 microzooplankton taxa were recorded, with the greatest diversity of taxa observed in the summer. Microzooplankton diversity was lowest in the winter (DNERR, 1999).

The mesozooplankton in the Blackbird Creek consisted of 44 taxa. The dominant forms were the cladocerans *Diaphanosoma* spp., Copepoda nauplii, and *Acartia hudsonica*. Other numerically abundant taxa included the Bivalvia, Copepoda, Gastropoda, Polychaeta, Rotifera, Cladocera, Hydrozoa medusa, Cirripedia, and *Uca* spp. During spring, summer, and winter, copepods predominated. Polychaete larvae were most abundant in the fall. Among the important copepod taxa were nauplii, copepodites, *Acartia hudsonica, A. tonsa, Acartia* spp., *Cyclops* spp., *Ectinosoma* spp., *Eurytemora affinis, Halicyclops fosteri, Oithona* spp., and *Scottolana* spp. Mesozooplankton taxa were most diverse in the summer and least diverse in the winter (DNERR, 1999).

The density of both microzooplankton and mesozooplankton in the Blackbird Creek is greatest in the summer months (Table 4.7). The density of phytoplankton counts in the creek also peaks during the summer. Highest densities of microzooplankton and mesozooplankton occur in the upper reaches of the creek; progressively lower densities exist in the middle and lower reaches.

Benthic Fauna

The benthic macroinvertebrate community of the Upper Blackbird Creek is less diverse than that of the Lower St. Jones River; it is characterized by an assemblage of organisms that prefers lower salinity conditions. During the 1995–1996 sampling period, 21 benthic macroinvertebrate taxa were identified. They belonged to three phyla: (a) Annelida, (b) Arthropoda, and (c) Nemertinea (Table 4.13). Five taxa were by far most abundant, comprising more than 92% of all individuals collected. They were, in decreasing order of abundance, the oligochaeta, Chironomidae, amphipods (*Corophium* sp. and *Gammarus* sp.), and the isopod, *Cyathura polita*. Oligochaetes comprised 54% of all macroinvertebrates collected in the samples, and the chironomids, more than 20%. *Corophium* sp. and *Gammarus* sp. accounted for 11% and 4% of the individuals, respectively.

The parabenthic macroinvertebrate community of the Upper Blackbird Creek consisted of 11 taxa representing four phyla (i.e., Annelida, Arthropoda, Cnidaria, and Ctenophora). Crustacean taxa far outnumbered all others, comprising nearly 95% of the parabenthos collected. These taxa included, in order of decreasing abundance, the grass shrimp (*Palaemonetes* sp.), amphipod (*Gammarus* sp.), opossum shrimp (*Neomysis americana*), and scud (*Corophium* sp.).

The seasonal densities of benthic macroinvertebrates in the Upper Blackbird Creek ranged from 2040 to 4289 individuals/m^2. Minimum density occurred in the summer and maximum density in the spring. Oligochaetes were responsible for 41 to 62% of the density values calculated for the benthic samples, reflecting their numerical dominance of the benthic communities in the creek (DNERR, 1999).

TABLE 4.13
Mean Density of Benthic Macroinvertebrates in the Tidal River and Channels of Blackbird Creek during 1994 (Summer and Fall) and 1995 (Spring)[a]

Taxon	Upper	Lower	Mean	% Total
Tidal River				
Oligochaeta	646	2599	1623	42.3
Chironomidae	167	1668	918	23.9
Corophium sp.	22	1254	638	16.6
Gammarus sp.	129	269	199	5.2
Cyathura polita	81	274	178	4.6
Polydora ligni	75	54	65	1.7
Edotea triloba	0	102	51	1.3
Xanthidae	22	59	40	1.1
Gammaridae	0	65	32	0.8
Scolecolepides viridis	16	48	32	0.8
Parapluestes aestuarius	48	0	24	0.6
Nereis succinea	16	5	11	0.3
Hypaniola florida	5	11	8	0.2
Neomysis americana	5	11	8	0.2
Streblospio benedicti	11	0	5	0.1
Melita nitida	0	5	3	0.1
Nemertea	0	5	3	0.1
Rhithropanopeus harrisii	0	5	3	0.1
Channels				
Oligochaeta	1937	1222	1580	77.1
Chironomidae	452	118	285	13.9
Nereis succinea	172	16	94	4.6
Gammarus sp.	38	22	30	1.4
Scolecolepides viridis	0	43	22	1.1
Edotea triloba	11	5	8	0.4
Hypaniola florida	16	0	8	0.4
Nemertea	0	11	5	0.3
Ceratopogonidae	5	0	3	0.1
Chiridotea almyera	5	0	3	0.1
Corophium sp.	5	0	3	0.1
Diptera	5	0	3	0.1
Neomysis americana	5	0	3	0.1
Xanthidae	0	5	3	0.1

[a] Density = number/m^2

Source: Delaware Department of Natural Resources and Environmental Control. 1999. Delaware National Estuarine Research Reserve: Estuarine Profiles. Technical Report, Delaware Department of Natural Resources and Environmental Control, Dover, DE.

Finfish

Upper Blackbird Creek finfish surveys conducted in the summer and fall of 1994 and the spring of 1995 recorded 21 species of fish. Four of these species (spot, *Leiostomus xanthurus*; Atlantic menhaden, *Brevoortia tyrannus*; white perch, *Morone americana*; and mummichog, *Fundulus heteroclitus*) were overwhelmingly dominant, accounting for more than 95% of all individuals collected (Table 4.9). In decreasing order of abundance were the spot (47% of all individuals collected), Atlantic menhaden (25%), white perch (12%), and mummichog (11%). Most of the fish were found in secondary tributaries rather than the main channels. The spot and Atlantic menhaden are estuarine-dependent species, and the white perch and mummichog, estuarine-resident forms (DNERR, 1999).

Amphibians and Reptiles

Because freshwater wetland habitats are relatively extensive along the Upper Blackbird Creek, amphibians are more abundant here than along the Lower St. Jones River. For example, the red-backed salamander (*Plethodon cinereus*), northern two-lined salamander (*Eurycea bislineata*), green frog (*Rana clamitans melanota*), wood frog (*R. sylvatica*), bullfrog (*Rana catesbeiana*), and southern leopard frog (*Rana utricularia*) are more abundant at the Upper Blackbird Creek Reserve site than at the Lower St. Jones River Reserve site.

Several species of snakes and turtles are also commonly observed at the Upper Blackbird Creek Reserve site. For instance, the black rat snake (*Elaphe obsoleta*) and northern water snake (*Nerodia sipedon*) occur in both wetland and upland habitats of the upper reserve site. Among turtle species frequenting this location are the eastern mud turtle (*Kinosternon subrubrum subrubrum*), red-bellied turtle (*Pseudemys rubriventris*), and snapping turtle (*Chelydra serpentina*). The northern diamondback terrapin (*Malaclemys terrapin terrapin*) is also found here (DNERR, 1999).

Birds

Most avian species observed in the Lower St. Jones River Reserve site also occur in the Upper Blackbird Creek Reserve site (Table 4.10) (DNERR, 1999). The scrub/shrub wetlands, wooded wetlands, and uplands provide excellent habitat for passerines and waterbirds. The lower freshwater and brackish marshes support numerous wading birds, waterfowl, and shorebirds. Passerine species that may find tidal wetland habitats favorable include the song sparrow (*Melospiza melodia*), seaside sparrow (*Ammospiza maritima*), red-winged blackbird (*Agelaius phoeniceus*), common yellowthroat (*Geothlypis trichas*), prothonotary warbler (*Protonotaria citrea*), northern parula warbler (*Parula americana*), marsh wren (*Cistothorus palustris*), and boat-tailed grackle (*Quiscalus major*). Avifauna likely to occur in open habitat, thickets, or woodlands in upland areas are the American robin (*Turdus migratorius*), northern oriole (*Icterus galbula*), field sparrow (*Spizella pusilla*), song sparrow (*Melospiza melodia*), blue jay (*Cyanocitta cristata*), Carolina chickadee (*Parus carolinensis*), house wren (*Troglodytes aedon*), gray catbird (*Dumetella carolinensis*), wood thrush (*Hylocichla mustelina*), yellow warbler (*Dendroica petechia*), yellow-rumped warbler (*D. coronata*), ovenbird (*Seiurus aurocapillus*), bobolink (*Dolichonyx oryzivorus*), American kestrel (*Falco sparverius*), white-eyed vireo (*Vireo griseus*), red-eyed vireo (*V. olivaceus*), rufous-sided towhee (*Pipilo erythrophthalmus*), common grackle (*Quiscalus quiscula*), northern mockingbird (*Mimus polyglottos*), prairie warbler (*Dendroica discolor*), house wren (*Troglodytes aedon*), mourning dove (*Zenaida macroura*), eastern meadowlark (*Sturnella magna*), tree swallow (*Tachycineta bicolor*), barn swallow (*Hirundo rustica*), brown creeper (*Certhia americana*), and American goldfinch (*Carduelis tristis*).

Waterbirds also commonly occupy tidal wetland habitat of the Upper Blackbird Creek Reserve site. For example, wading birds documented at this site are the great blue heron (*Ardea herodias*), green-backed heron (*Butorides virescens*), snowy egret (*Egretta thula*), cattle egret (*Bubulcus ibis*), great egret (*Casmerodius albus*), and glossy ibis (*Plegadis falcinellus*). Several species of rails are also found here, such as the king rail (*Rallus elegans*), clapper rail (*Rallus longirostris*), and Virginia rail (*Rallus limicola*). Forster's tern (*Sterna forsteri*), laughing gull (*Larus atricilla*), and the willet (*Catoptrophorus semipalmatus*) are other waterbirds frequently seen in the Upper Blackbird Creek Reserve. Waterfowl of significance include the wood duck (*Aix sponsa*), green-winged teal (*Anas crecca*), and mallard (*A. platyrhynchos*), which overwinter in the tidal wetlands.

Shorebirds are less frequently observed in the Upper Blackbird Creek Reserve site than in the Lower St. Jones River Reserve site. Greater yellowlegs (*Tringa melanoleuca*) exhibited the greatest abundance at the northern site. Spotted sandpiper (*Actitis macularia*), killdeer (*Charadrius vociferus*), and unidentified peeps have also been recorded here. The Delaware Estuary in general is a major staging area for migrating shorebirds, although few species breed here.

Several raptors utilize Upper Blackbird Creek habitats. Among these are the northern harrier (*Circus cyaneus*), red-tailed hawk (*Buteo jamaicensis*), and sharp-shinned hawk (*Accipiter striatus*). The northern harrier is the most abundant raptor. The Upper Blackbird Creek Reserve site also supports a breeding pair of bald eagles (*Haliaeetus leucocephalus*).

Mammals

Terrestrial, aquatic, and marine mammals that utilize the Upper Blackbird Creek Reserve site are similar to those documented in the Lower St. Jones River Reserve site (DNERR, 1999). Among the mammalian species reported in upland forests, wooded fringes, or wetlands of the Upper Blackbird Creek Reserve site are the white-tailed deer (*Odocoileus virginianus*), gray fox (*Urocyon cinereoargenteus*), red fox (*Vulpes vulpes*), masked shrew (*Sorex cinereus*), least shrew (*Cryptotis parva*), short-tailed shrew (*Blarina brevicauda*), eastern cottontail rabbit (*Sylvilagus floridanus*), striped skunk (*Mephitis mephitis*), raccoon (*Procyon lotor*), gray squirrel (*Sciurus carolinensis*), long-tailed weasel (*Mustela frenata*), white-footed mouse (*Peromyscus leucopus*), meadow jumping mouse (*Zapus hudsonius*), pine vole (*Microtus pinetorum*), meadow vole (*Microtus pennsylvanicus*), and red bat

(*Lasiurus borealis*). Common aquatic forms include the beaver (*Castor canadensis*), river otter (*Lutra canadensis*), and muskrat (*Ondatra zibethicus*). The aforementioned mammals are upper-trophic-level organisms in the terrestrial and aquatic food webs of both reserve sites and hence are important in regulating the population sizes of prey in the system.

Commercially and Recreationally Important Species

Most commercial and recreational fishing, waterfowl hunting, and furbearer trapping occur along Delaware Bay, but some activities also extend into the subestuaries of the two reserve sites. The recreational fishery is of major economic importance in the Delaware Estuary and its tidal tributaries, with an estimated annual value of about $25 million (Sutton et al., 1996). The principal species of the recreational fishery include the bluefish (*Pomatomus saltatrix*), weakfish (*Cynoscion regalis*), striped bass (*Morone saxatilis*), and summer flounder (*Paralichthys dentatus*). In addition, the black drum (*Pogonias cromis*), scup (*Stenotomus versicolor*), and tautog (*Tautoga onitis*) are of considerable recreational importance.

Commercial landings data indicate a shift during the past century from the predominance of upriver species to the prevalence of estuarine and marine forms (Price and Beck, 1988; Kennish, 2000). The most valuable species landed in the commercial fishery during the past decade are the American eel (*Anguilla rostrata*), American shad (*Alosa sapidissima*), Atlantic menhaden (*Brevoortia tyrannus*), bluefish (*Pomatomus saltatrix*), and weakfish (*Cynoscion regalis*). Other species, such as the white perch (*Morone americana*) and spot (*Leiostomus xanthurus*), are also important in the commercial fishery. Overfishing has played a significant role in the historical decline of certain species (e.g., American shad, *A. sapidissima;* and Atlantic sturgeon, *Acipenser oxyrhynchus*).

The eastern oyster (*Crassostrea virginica*), blue crab (*Callinectes sapidus*), and horseshoe crab (*Limulus polyphemus*) are the most economically important shellfish species in the Delaware Estuary. The once thriving oyster industry has been decimated since the late 1950s by disease, bacterial contamination, and poor setting of seed oysters, which threaten the viability of the fishery. Two protozoan parasites, MSX (*Haplosporidium nelsoni*) and Dermo (*Perkinsus marinus*), have been particularly problematic, causing massive declines in oyster abundance. The oyster drill (*Urosalpinx cinerea*) is a major predator of the eastern oyster and has inflicted additional heavy losses on oyster beds in some areas.

The blue crab (*Callinectes sapidus*) is the most valuable shellfish species in terms of dollar value of harvest in the recreational and commercial fisheries (Sutton et al., 1996). It is widely distributed, utilizing the entire Delaware Estuary including tidal tributaries. Commercial fishermen harvest blue crabs principally by baited pots (April through October) and dredges (December through March). Recreational fishermen catch blue crabs primarily by baited traps. Blue crab harvests vary considerably from year to year due to the fluctuating abundance of crabs associated with the vagaries of environmental and biotic factors.

The horseshoe crab (*Limulus polyphemus*) is harvested for use as bait in eel, conch, and lobster trays. Horseshoe crab blood is also economically important in the medical and pharmaceutical industries because it is used to detect pyrogens (bacterial contamination) in injectable drugs and surgical implants. Horseshoe crabs are harvested by hand along sandy beaches and by dredges in subtidal waters of the estuary. State government agencies in Delaware and New Jersey regulate horseshoe crab harvests because in the past, excessive harvests have dramatically reduced population levels. Migratory shorebird species (e.g., red knot, *Calidris canutus*, and semipalmated sandpiper, *Calidris pusilla*) depend on horseshoe crab resources for survival; they consume large numbers of horseshoe crab eggs during their seasonal stopovers. The eggs provide the energy required by the birds to complete their migration to nesting grounds in far northern latitudes. The critical importance of this food resource for the birds has prompted state agencies to regulate harvests to protect and conserve horseshoe crabs in the estuarine system. For example, the New Jersey Department of Environmental Protection recently imposed a ban on harvesting of Delaware Bay horseshoe crabs from May 1 through the first week of June in 2003.

According to McConnell and Powers (1995), the muskrat (*Ondatra zibenthicus*) and, to a lesser extent, the river otter (*Lutra canadensis*) and mink (*Mustela vison*) are harvested by fur trappers in or near the two DNERR sites. Most trapping of furbearers takes place in tidal wetlands. The trend has been for decreasing furbearer harvest in recent years, largely because of the decreasing demand for furs.

The northern diamondback terrapin (*Malaclemys terrapin terrapin*) and the snapping turtle (*Chelydra serpentina*) are harvested along the Delaware Estuary. The open season on the diamondback terrapin lasts less than 90 days, while that on the snapping turtle lasts more than 300 days. Although no harvest limits exist for either species in Delaware, government regulations in New Jersey impose more restrictive harvesting of these species.

Waterfowl hunting is a major recreational pursuit in the wetlands habitat along the Delaware Estuary and its tributary systems. Several species of ducks are popular to hunt in the region. These include dabblers, most notably the black duck (*Anas rubripes*), wigeon (*A. americana*), gadwall (*A. strepera*), pintail (*A. acuta*), blue-winged teal (*A. discors*), green-winged teal (*A. crecca*), and wood duck (*Aix sponsa*), as well as divers, such as the red-breasted merganser (*Mergus serrator*), bufflehead (*Bucephala albeola*), canvasback (*Aythya valisineria*), redhead (*A. americana*), lesser scaup (*A. affinis*), and greater scaup (*A. marila*). In recent years, declining population sizes of certain species (e.g., *Anas rubripes, A. acuta,* and *Aythya valisineria*) have resulted in more restrictive harvests. The black duck, in particular, has exhibited significant reductions in abundance since the mid-1960s (Sutton et al., 1996; Kennish, 2000).

SUMMARY AND CONCLUSIONS

The Lower St. Jones River Reserve and the Upper Blackbird Creek Reserve are the two component sites of the Delaware National Estuarine Research Reserve (DNERR). The DNERR program primarily focuses on resource protection and conservation, estuarine research and monitoring, and environmental education and interpretation. A major goal is to better manage resources in the reserve sites and to engender a greater sense of responsibility and stewardship among residents in the two watersheds of the DNERR. Both reserve sites are subestuaries of the Delaware River estuary and hence are greatly affected by biotic and abiotic processes occurring there.

The Lower St. Jones River Reserve site encompasses an area of 1518 ha, and the Upper Blackbird Creek Reserve an area of 477 ha. The Lower St. Jones River, a mesohaline tidal system, flows into mid-Delaware Bay. Lower salinities characterize the Upper Blackbird Creek, which empties into the lower Delaware River. The St. Jones River watershed is more heavily developed (25% of the reserve site area) than the Blackbird Creek watershed (10%), and consequently water quality in the Lower St. Jones River is more heavily impacted than that in the Upper Blackbird Creek.

The watershed areas of both reserve sites consist of luxuriant emergent tidal wetlands as well as rich assemblages of upland vegetation. Smooth cordgrass (*Spartina alterniflora*) dominates the tidal wetlands of the reserve sites. Mixed deciduous hardwood forests, dominated by red oak (*Quercus rubra*), white oak (*Q. alba*), black cherry (*Prunus serotina*), American beech (*Fagus grandifolia*), American holly (*Ilex opaca*), sassafras (*Sassafras albidum*), tulip poplar (*Liriodendron tulipifera*), and hickories (*Carya* spp.), characterize the uplands. Agricultural cropland is the predominant land cover in the DNERR, comprising 48% of the St. Jones River watershed area and 39% of the Blackbird Creek watershed area. The principal anthropogenic impacts of concern are those associated with nonpoint source pollution from an encroaching human population.

The terrestrial and aquatic habitats in the reserve sites are teeming with life, as revealed by field surveys conducted in the system from 1993 through 1997. A wide diversity of amphibians, reptiles, mammals, and birds typifies the DNERR watersheds. In the aquatic habitats, more than 40 phytoplankton taxa and more than 80 zooplankton (microzooplankton and mesozooplankton) taxa were recorded at each reserve site during the aforementioned survey period. Fewer benthic macroinvertebrate forms are evident, with 21 taxa registered in the Upper Blackbird Creek and 33 taxa in the Lower St. Jones River. Finfish assemblages are representative of estuarine taxa found in the lower Delaware Estuary; the Atlantic menhaden (*Brevoortia tyrannus*), mummichog (*Fundulus heteroclitus*), spot (*Leiostomus xanthurus*), and white perch (*Morone americana*) are most abundant in the Upper Blackbird Creek, and the Atlantic silverside (*Menidia menidia*), bay anchovy (*Anchovy mitchilli*), mummichog (*Fundulus heteroclitus*), and sheepshead minnow (*Cyprinodon variegatus*) are most abundant in the Lower St. Jones River. Aquatic mammals commonly observed in wetland habitats of the DNERR include the beaver (*Castor canadensis*), muskrat (*Ondatra zibethicus*), mink (*Mustela vison*), river otter (*Lutra canadensis*), and rice rat (*Oryzomys palustris*). The bottlenose dolphin (*Tursiops truncata*) may also occasionally occur in DNERR waters.

Various organisms in the DNERR and nearby Delaware River estuary are of commercial and recreational importance. Among these organisms are various furbearers (e.g., muskrat, *Ondatra zibenthicus*; mink, *Mustela vison*; and river otter, *Lutra canadensis*), waterfowl (e.g., the black duck, *Anas rubripes*; gadwall, *Anas strepera*; pintail, *A. acuta*; blue-winged teal, *A. discors*; green-winged teal, *A. crecca*;

wigeon, *A. americana*; wood duck, *Aix sponsa*; red-breasted merganser, *Mergus serrator*; bufflehead, *Bucephala albeola*; canvasback, *Aythya valisineria*; redhead, *Aythya americana*; lesser scaup, *Aythya affinis*; and greater scaup, *Aythya marila*), terrapins (*Malaclemys terrapin terrapin*), and snapping turtles (*Chelydra serpentina*). However, fisheries of the Delaware Estuary and its subestuaries are of overriding significance. Species of greatest importance in the recreational fishery are the bluefish (*Pomatomus saltatrix*), weakfish (*Cynoscion regalis*), striped bass (*Morone saxatilis*), and summer flounder (*Paralichthys dentatus*). The most valuable species landed in the commercial fishery are the American eel (*Anguilla rostrata*), American shad (*Alosa sapidissima*), Atlantic menhaden (*Brevoortia tyrannus*), bluefish (*Pomatomus saltatrix*), and weakfish (*Cynoscion regalis*). Shellfish species of principal economic importance in the system include the eastern oyster (*Crassostrea virginica*), blue crab (*Callinectes sapidus*), and horseshoe crab (*Limulus polyphemus*). Declining abundance of the eastern oyster and horseshoe crab remains a cause of concern for recreational and commercial fishermen, government regulatory agencies, and the general public.

REFERENCES

Clark, K.E. 1988. Delaware Bay Shorebird Project. Final Report, Endangered and Nongame Species Program, Division of Fish, Game, and Wildlife, New Jersey Department of Environmental Protection, Trenton, NJ.

Delaware Department of Natural Resources and Environmental Control. 1993. Comprehensive Conservation and Management Plan for Delaware's Tidal Wetlands. Delaware Department of Natural Resources and Environmental Control, Dover, DE.

Delaware Department of Natural Resources and Environmental Control. 1994. Superfund: A Year in Review. Superfund Annual Report, Delaware Department of Natural Resources and Environmental Control, Dover, DE.

Delaware Department of Natural Resources and Environmental Control. 1995. Delaware Nonpoint Source Pollution Program. Technical Report, Delaware Department of Natural Resources and Environmental Control, Dover, DE.

Delaware National Estuarine Research Reserve. 1999. Delaware National Estuarine Research Reserve: Estuarine Profiles. Technical Report, Delaware National Estuarine Research Reserve, Delaware Department of Natural Resources and Environmental Control, Dover, DE.

Dove, L.E. and R.M. Nyman (Eds.). 1995. Living Resources of the Delaware Estuary. Technical Report, The Delaware Estuary Program, U.S. Environmental Protection Agency, New York.

Frithsen, J.B., D.E. Strebel, and T. Schawitsch. 1995. Estimates of contaminant inputs to the Delaware Estuary. Technical Report, Versar Inc., Columbia, MD.

Herman, S.S. 1988. Zooplankton. In: Bryant, T.L. and J.R. Pennock (Eds.). *The Delaware Estuary: Rediscovering a Forgotten Resource*. Technical Report, University of Delaware Sea Grant College Program, Newark, DE, pp. 60–67.

Herman, S.S., B.R. Hargreaves, R.A. Lutz, L.W. Fritz, and C.E. Epifanio. 1983. Zooplankton and parabenthos. In: *The Delaware Estuary: Research as Background for Estuarine Management and Development*. Technical Report, Delaware River and Bay Authority, University of Delaware Sea Grant College Program, Newark, DE.

Jenkins, D. and L.A. Gelvin-Innvaer. 1995. Colonial wading birds. In: Dove, L.E. and R.M. Nyman (Eds.). *Living Resources of the Delaware Estuary.* Technical Report, The Delaware Estuary Program, U.S. Environmental Protection Agency, New York, pp. 335–345.

Kennish, M.J. (Ed.). 1997. *Practical Handbook of Estuarine and Marine Pollution.* CRC Press, Boca Raton, FL.

Kennish, M.J. (Ed.). 2000. *Estuary Restoration and Maintenance: The National Estuary Program.* CRC Press, Boca Raton, FL.

Kerlinger, P. and L. Widjeskog. 1995. Rails. In: Dove, L.E. and R.M. Nyman (Eds.). *Living Resources of the Delaware Estuary.* Technical Report, The Delaware Estuary Program, U.S. Environmental Protection Agency, New York, pp. 425–431.

Livingston, R.J. 2001. *Eutrophication Processes in Coastal Systems.* CRC Press, Boca Raton, FL.

Livingston, R.J. 2003. *Trophic Organizations in Coastal Systems.* CRC Press, Boca Raton, FL.

Marshall, H.G. 1992. *Assessment of Phytoplankton Species in the Delaware River Estuary.* Technical Report to the Delaware River Basin Commission by the U.S. Environmental Protection Agency, Philadelphia.

Marshall, H.G. 1995. Phytoplankton. In: Dove, L.E. and R.M. Nyman (Eds.). *Living Resources of the Delaware Estuary.* Technical Report, The Delaware Estuary Program, U.S. Environmental Protection Agency, New York, pp. 25–29.

Maurer, D., L. Watling, P. Kinner, W. Leathem, and C. Wethe. 1978. Benthic invertebrate assemblages of Delaware Bay. *Marine Biology* 45: 65–78.

McConnell, P.A. and J.L. Powers. 1995. Muskrat. In: Dove, L.E. and R.M. Nyman (Eds.). *Living Resources of the Delaware Estuary.* Technical Report, The Delaware Estuary Program, U.S. Environmental Protection Agency, New York, pp. 507–513.

Niles, L.J. and C. Sutton. 1995. Migratory raptors. In: Dove, L.E. and R.M. Nyman (Eds.). *Living Resources of the Delaware Estuary.* Technical Report, The Delaware Estuary Program, U.S. Environmental Protection Agency, New York, pp. 433–440.

O'Herron, J.C., III, T. Lloyd, and K. Laidig. 1994. A survey of fish in the Delaware Estuary from the area of the Chesapeake and Delaware Canal to Trenton. Technical Report, U.S. Environmental Protection Agency, New York.

Pennock, J.R. and J. Sharp. 1986. Phytoplankton production in the Delaware Estuary: temporal and spatial variability. *Marine Ecology Progress Series* 34: 143–155.

Price, K.S. and R.A. Beck. 1988. Finfish. In: Bryant, T.L. and J.R. Price (Eds.). *The Delaware Estuary: Research as Background for Estuarine Management and Development.* Technical Report, University of Delaware Sea Grant College Program, Newark, DE.

Sharp, J.H., L.A. Cifuentes, R.B. Coffin, M.E. Lebo, and J.R. Pennock. 1994. Eutrophication: Are Excessive Nutrient Inputs a Problem for the Delaware Estuary? Technical Report, University of Delaware Sea Grant College Program, Newark, DE.

Stearns, D.E. 1995. Copepods. In: Dove, L.E. and R.M. Nyman (Eds.). *Living Resources of the Delaware Estuary.* Technical Report, The Delaware Estuary Program, U.S. Environmental Protection Agency, New York, pp. 33–42.

Steimle, F. 1995. Soft (mud/sand) bottom pohyhaline communities. In: Dove, L.E. and R.M. Nyman (Eds.). *Living Resources of the Delaware Estuary.* Technical Report, The Delaware Estuary Program, U.S. Environmental Protection Agency, New York, pp. 119–124.

Sutton, C.C., J.C. O'Herron III, and R.T. Zappolorti. 1996. The Scientific Characterization of the Delaware Estuary. Technical Report, The Delaware Estuary Program (DRBC Project No. 321; HA File No. 93.21), U.S. Environmental Protection Agency, New York.

Watling, L., D. Bottom, A. Pembroke, and D. Maurer. 1979. Seasonal variations in Delaware Bay phytoplankton community structure. *Marine Biology* 52: 207–215.

Wetlands Research Associates, Inc. and Environmental Consulting Services, Inc. 1995. DNERR Comprehensive Site Description. Phase II — Characterization of Finfish and Aquatic Macroinvertebrate Communities and Waterbird Populations. Technical Report, Delaware Natural Estuarine Research Reserve, Dover, DE.

Case Study 4

5 Ashepoo–Combahee–Edisto (ACE) Basin National Estuarine Research Reserve

INTRODUCTION

The Ashepoo–Combahee–Edisto (ACE) Basin of South Carolina was designated as a National Estuarine Research Reserve (NERR) program site in 1992. The ACE Basin NERR encompasses an area of more than 56,000 ha in the lower coastal plain of South Carolina, located in parts of Colleton, Charleston, Beaufort, and Hampton counties. The reserve site is bounded along the southeast margin by the Atlantic Ocean, and it trends northwestward (inland) for ~35 km (Figure 5.1). The Ashepoo, Combahee, and Edisto rivers are the principal river systems in the basin, although numerous smaller streams also flow through the region. They drain into St. Helena Sound and the Atlantic Ocean. A wide range of habitats, supporting numerous biotic communities, exists in the ACE Basin, such as upland pine–mixed hardwoods, maritime forests, freshwater marshes, brackish marshes, salt marshes, tidal flats, tidal creeks and channels, open estuarine waters, and barrier islands. Salt marshes, tidelands, and open channels/water habitat cover the greatest area (SCDNR and NOAA, 2001).

Characteristic of coastal plain environments, the topography of the ACE Basin NERR consists primarily of flat, low-lying terrain (Colquhoun, 1969; Soller and Mills, 1991). Drainage is to the southeast, with stream gradients varying from ~20 to 375 cm/km (Bloxham, 1979, 1981). The Edisto River is the longest river in the region, originating in the middle coastal plain of South Carolina by the confluence of the North and South Fork Edisto Rivers, and diverging into the North Edisto River and South Edisto River near the coast. In contrast, the Ashepoo and Combahee Rivers both originate in swamps of the lower coastal plain and therefore are much less extensive. These rivers and their tributaries transport sediments seaward from the Piedmont, and the sediments are deposited downstream in the lower coastal plain as well as along the coast. Marine-derived sediments accumulate along the coastal margin of the basin in the Edisto Beach area and alongshore (Mathews et al., 1980; McIntyre, 1991; Soller and Mills, 1991).

Surface water flow in the basin is substantial. For example, Cooney et al. (1998) report that the mean annual streamflow of the Edisto River near Givhans is 74 m^3/sec. Much lower flow rates are observed elsewhere in the Edisto River sub-basin (Table 5.1). In addition to surface waters, six aquifer systems exist in the ACE Basin (Cape Fear, Middendorf, Black Creek, Tertiary Sand, Floridian Aquifer, and Shallow Aquifer systems). Three of these systems (Tertiary Sand, Floridian Aquifer, and

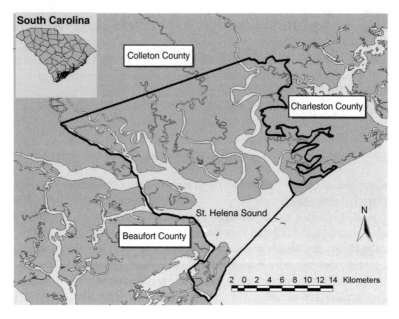

FIGURE 5.1 Map showing the ACE Basin National Estuarine Research Reserve and surrounding watershed areas. (From the South Carolina Department of Natural Resources and National Oceanic and Atmospheric Administration. 2001. *Characterization of the Ashepoo–Combahee–Edisto (ACE) Basin, South Carolina.* Special Scientific Report 17, South Carolina Marine Resources Center, Charleston, SC.)

TABLE 5.1
Streamflow Characteristics in the Edisto River Sub-Basin

Location	Drainage Area (km²)	Period	Daily Average Flow (m³/sec)
McTier Creek, near Monetta	39.6	1995–97	0.67
Dean Swamp Creek, near Salley	80.8	1980–Present	0.7
South Fork Edisto, near Denmark	1865	1931–Present	21.7
South Fork Edisto, near Cope	2090	1991–Present	23.4
South Fork Edisto, near Bamberg	683	1991–Present	32.3
North Fork Edisto at Orangeburg	1769	1938–Present	22.2
Cow Castle Creek, near Bowman	60.6	1971–Present	0.57
Edisto River, near Givhans	2730	1939–Present	74.0

Source: South Carolina Department of Natural Resources and National Oceanic and Atmospheric Administration. 2001. *Characterization of the Ashepoo–Combahee–Edisto (ACE) Basin, South Carolina.* Special Scientific Report 17, South Carolina Marine Resources Center, Charleston, SC.

Shallow Aquifer) serve as the primary sources of water for domestic, public, and commercial uses. All are subject to local saltwater intrusion (Park, 2001).

Human activities affect watershed and estuarine environments in the ACE Basin. Among the most significant effects are coastal development and associated construction, aquaculture, silviculture, logging operations, dredging and filling of wetlands, and ditching activities. Point source pollution and nonpoint source pollution both influence water quality conditions in the ACE Basin NERR. Principal point source discharges in the reserve include the CCX Fiberglass Products Plant in Walterboro (Ashepoo River), the wastewater treatment facility of the City of Walterboro (Ashepoo River), the SCE&G Canadys Power Station (Edisto River), and the Yemassee Wastewater Treatment Facility (Combahee River). A growing concern exists with regard to nonpoint source pollution associated with accelerated development, agriculture, forestry, and other anthropogenic activities (Wenner et al., 2001a).

In aquatic habitats of the ACE Basin, communities of phytoplankton, zooplankton, benthic flora and fauna, and fish are well represented. From wetland to upland habitats, numerous species of amphibians, reptiles, mammals, insects, and birds proliferate. Several endangered and threatened species inhabit the ACE Basin area; examples are the loggerhead sea turtle (*Caretta caretta*), shortnose sturgeon (*Acipenser brevirostrum*), wood stork (*Mycteria americana*), and American alligator (*Alligator mississippiensis*) (Riekerk et al., 2001).

A key initiative of the ACE Basin NERR is the protection of biotic and other natural resources. Land protection remains an important component of this initiative. In the ACE Basin, ~15% of the land area is now classified as protected, and much of it (~40%) consists of public land. Most of the private land in the basin (~60%) is subject to conservation easements, which protect wildlife habitat by preserving the natural value of the land. Conservation easements play a critical role in environmental protection of the ACE Basin NERR (Wenner, 2001a).

WATERSHED

Plant Communities

Wetland and upland communities comprise most of the area of the reserve, covering more than 32,000 ha. Coastal marshes are quite extensive. Smooth cordgrass (*Spartina alterniflora*) is the dominant salt marsh plant, although rushes (*Juncus roemerianus*) and saltworts (*Salicornia* spp.) are also abundant at higher elevations. In bottomland and upland–forested habitats, oaks (*Quercus* spp.), pines (*Pinus* spp.), red cedar (*Juniperus virginiana*), wax myrtle (*Myrica cerifera*), and palmettos (e.g., *Sabal palmetto*) predominate. The bald cypress (*Taxodium distichum*) is also relatively abundant. More than 5000 ha of wetlands and uplands constitute the core area of the reserve, with more than 50,000 ha of land and water habitat forming a broad buffer zone. Seven barrier islands occur in the core area; these are the Ashe, Beet, Big, Boulder, Otter, South Williman, and Warren islands. A variety of habitats and plant communities can be found on the barrier islands, such as dunes, salt marshes, fresh and brackish ponds, maritime estuarine and palustrine areas, and maritime forests (Upchurch, 2001).

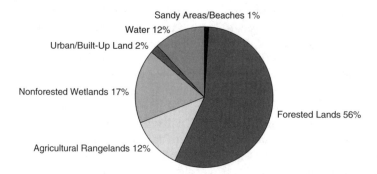

FIGURE 5.2 Percent land use cover in the ACE Basin. Note large percentage of forested lands. (From the South Carolina Department of Natural Resources and National Oceanic and Atmospheric Administration. 2001. *Characterization of the Ashepoo–Combahee–Edisto (ACE) Basin, South Carolina.* Special Scientific Report 17, South Carolina Marine Resources Center, Charleston, SC.)

While estuarine and maritime plant communities are well developed in the system (TNC, 1993), freshwater wetlands are spatially restricted. Despite their limited areal extent, freshwater marshes provide important habitat for numerous species of birds, reptiles, mammals, and other organisms. The Snuggedy Swamp, encompassing more than 900 ha near the South Edisto River, is a representative site of freshwater marsh and swamp communities. It is South Carolina's largest grass-sedge-marsh/loblolly-bay complex (SCDNR and NOAA, 2001).

Forested habitat is substantial. In Colleton County, for example, resource surveys reveal that forests account for 56% (more than 180,000 ha) of the land use cover (Figure 5.2) (Conner, 1993). Among the principal components are deciduous upland forest, mixed upland forest, evergreen upland forest, and upland pine and forested wetlands. A significant element of the upland forested vegetation is planted pine, which has increased pine and oak–pine stands by more than 10% (Koontz and Sheffield, 1993). Planted pine is the product of forestry efforts to grow pine for timber in place of low-quality hardwood stands. Timberland in Colleton County consists of the following stands in decreasing order of areal coverage:

1. Loblolly–shortleaf pine
2. Oak–gum–cypress trees
3. Longleaf–slash pines

Loblolly pine (*Pinus taeda*), shortleaf pine (*P. echinata*), oaks (*Quercus* spp.), sweet gum (*Liquidambar styraciflua*), bald cypress (*Taxodium distichum*), and longleaf pine (*P. palustris*) are important species. The predominant hardwood species include white oak (*Q. alba*), southern red oak (*Q. falcata*), and sweet gum. A thick understory is observed in many areas. Bracken fern (*Pteridium aquilinum*) blankets some pine forest floors. Wax myrtle (*Myrica cerifera*) is a common constituent of upland as well as maritime communities (Wenner and Zimmerman, 2001).

Some 30 plant communities have been identified in the ACE Basin area. Most of these communities (N = 16) occur in palustrine wetlands. Upland habitats contain seven plant community types, and estuarine wetlands support four plant communities. Only three plant communities are found on the barrier islands (Upchurch, 2001).

The ACE Basin watershed is divided into two sub-basins:

1. The Edisto River sub-basin
2. The Combahee–Coosawhatchie River sub-basin (Badr and Zimmerman, 2001)

The Edisto River, South Fork Edisto River, North Edisto River, and Four Hole Swamp are the principal tributaries draining the Edisto River sub-basin. The South Fork Edisto River and North Fork Edisto River, which flow through the upper coastal plain, merge in the middle coastal plain to form the Edisto River as noted above. Farther downstream in the lower coastal plain, the Edisto River diverges into the North Edisto River and South Edisto River, two subestuaries that drain coastal marshes in the watershed (SCWRC, 1983). River flows in this sub-basin supply freshwater for domestic, industrial, and commercial users, as well as for agricultural irrigation.

The Combahee–Coosawhatchie River sub-basin lies to the south of the Edisto River sub-basin. The Combahee River and its tributaries, the Salkehatchie and Little Salkehatchie rivers, deliver freshwater to St. Helena Sound. Flow of the Salkehatchie River is less variable than that of the Combahee River. Cooney et al. (1998) showed that at Miley the mean annual streamflow of the Salkehatchie River amounts to 9.8 m^3/sec. The Ashepoo River also discharges to St. Helena Sound. However, the Coosawhatchie flows into Port Royal Sound. The most extensive estuarine waters in South Carolina occur in the coastal areas of the Combahee–Coosawhatchie River sub-basin (SCWRC, 1983).

Wetland habitats are dominant features of the ACE Basin NERR. Salt marshes are particularly extensive, but freshwater and brackish marshes may be no less important. Meandering tidal creeks that flow through the coastal marshes provide habitat for many estuarine organisms.

In poorly drained wetland habitats of the ACE Basin, elliptical-shaped depressions (i.e., Carolina Bays) are relatively common features (Riekerk, 2001). These depressions are often filled with water during periods of high precipitation, but they frequently dry out during droughts or seasons of low rainfall. Bennett and Nelson (1991) recorded about 20 Carolina Bays larger than 0.8 ha in the coastal plain of Colleton County. These unique habitats are typified by temporally and spatially variable biotic communities subjected to extremes of environmental conditions.

Animal Communities

The following discussion of organisms in the ACE Basin watershed focuses on four faunal groups:

1. Amphibians and reptiles (i.e., herpetofauna)
2. Mammals
3. Birds
4. Insects

It is largely based on observations of the South Carolina Department of Natural Resources (SCDNR and NOAA, 2001). While the distribution of many herpetofaunal populations depends on the occurrence of standing water (e.g., ponds and lakes), other species are less restricted and can tolerate a broad range of environmental conditions. They may be seen near water bodies as well as in relatively dry habitats. The southern toad (*Bufo terrestris*) is an example (Riekerk and Rhodes, 2001).

Nearly 50 mammalian species inhabit the ACE Basin region. Among the commonly observed mammals are bats, raccoons, rabbits, deer, minks, foxes, beavers, and opossums. Some of the most conspicuous mammals (e.g., dolphins and manatees) do not reside in the watershed but occupy nearby coastal waters. Mammals can be found in nearly all ACE Basin watershed habitats, from marshes and lowland maritime forests to meadows and upland mixed forests (Zimmerman, 2001a).

Almost 300 species of birds have been documented in the ACE Basin. Many of these species are migratory forms that travel great distances from northern regions and overwinter or rest in the reserve area prior to resuming flight to the southern latitudes. An array of avifaunal groups utilizes ACE Basin habitats (e.g., shorebirds, waders, waterfowl, songbirds, and raptors). They include granivores, insectivores, omnivores, and carnivores, as well as a few scavengers (Zimmerman, 2001b).

Insects are the principal herbivorous component of the watershed. They are numerous and highly diverse. As a group, insects also play a major role as decomposers in the breakdown of plant matter and hence are important to the recycling of nutrients and other chemical constituents in the ecosystem. In addition, insects constitute a primary staple food for amphibians, reptiles, birds, fish, and other organisms (Thompson, 1984). Although many insect species are widely distributed across multiple habitats of the ACE Basin, others appear to be restricted to certain habitats (e.g., swamps, marshes, and moist woods) (Scholtens, 2001).

Amphibians and Reptiles

Herpetological surveys conducted in the ACE Basin indicate that 110 species of amphibians and reptiles either inhabit the area or are expected to occur there. They are differentiated into the following taxonomic groups in the order of decreasing species richness: 36 snake species, 20 frog species, 19 salamander species, 18 turtle species, 12 lizard species, 4 toad species, and 1 alligator species (Riekerk and Rhodes, 2001) (Table 5.2). Based on the work of Sandifer et al. (1980) and Conant and Collins (1998), most of these species occupy lacustrine littoral habitats (37 species), upland hardwood forests (35), palustrine freshwater wetlands (34), and maritime forests (32) (Table 5.3).

The South Carolina Department of Natural Resources (SCDNR) has examined amphibian and reptilian communities in specific habitats of the ACE Basin (Riekerk and Rhodes, 2001), and this work is reviewed here. Upland forested areas characterized primarily by dry conditions are inhabited by a variety of lizards and snakes that can tolerate low moisture levels. For example, in upland pine flatwoods, the dominant species of lizards include the eastern fence lizard (*Sceloporus undulatus*), slender glass lizard (*Ophisaurus attenuatus*), mimic glass lizard (*O. mimicus*), ground skink (*Scincella lateralis*), broadhead skink (*Eumeces laticeps*), and six-lined racerunner

TABLE 5.2
Reptiles and Amphibians That Occur or Potentially Occur in the ACE Basin

Common Name	Scientific Name
Snakes	
Copperhead	*Agkistrodon contortrix*
Cottonmouth	*Agkistrodon piscivorus*
Worm snake	*Carphophis amoenus*
Scarlet snake	*Cemophora coccinea*
Northern black racer	*Coluber constrictor constrictor*
Eastern diamondback rattlesnake	*Crotalus adamanteus*
Timber rattlesnake	*Crotalus horridus horridus*
Ringneck snake	*Diadophis punctatus*
Corn snake	*Elaphe guttata guttata*
Rat snake	*Elaphe obsoleta*
Mud snake	*Farancia abacura*
Rainbow snake	*Farancia erytrogramma*
Eastern hognose snake	*Heterodon platirhinos*
Southern hognose snake	*Heterodon simus*
Mole king snake	*Lampropeltis calligaster*
Eastern king snake	*Lampropeltis getula getula*
Scarlet king snake	*Lampropeltis triangulum*
Eastern coachwhip	*Masticophis flagellum*
Eastern coral snake	*Micrurus fulvius*
Redbelly water snake	*Nerodia erythrogaster*
Banded water snake	*Nerodia fasciata*
Florida green water snake	*Nerodia floridana*
Brown water snake	*Nerodia taxispilota*
Rough green snake	*Opheodrys aestivus*
Northern pine snake	*Pituophis melanoleucus melanoleucus*
Glossy crayfish snake	*Regina rigida*
Pine woods snake	*Rhadinea flavilata*
Black swamp snake	*Seminatrix pygaea*
Pigmy rattlesnake	*Sistrurus miliarius*
Brown snake	*Storeria dekayi*
Redbelly snake	*Storeria occipitomaculata*
Southeastern crowned snake	*Tantilla coronata*
Eastern ribbon snake	*Thamnophis sauritus sauritus*
Eastern garter snake	*Thamnophis sirtalis*
Rough earth snake	*Virginia striatula*
Smooth earth snake	*Virginia valeriae*
Alligator	
American alligator	*Alligator mississippiensis*

(continued)

TABLE 5.2 (CONTINUED)
Reptiles and Amphibians That Occur or Potentially Occur in the ACE Basin

Common Name	Scientific Name
Lizards	
Carolina anole	*Anolis carolinensis*
Six-lined racerunner	*Cnemidophorus sexlineatus*
Five-lined skink	*Eumeces faciatus*
Southeastern five-lined skink	*Eumeces inexpectatus*
Broadhead skink	*Eumeces laticeps*
Slender glass lizard	*Ophisaurus attenuatus*
Island glass lizard	*Ophisaurus compressus*
Mimic glass lizard	*Ophisaurus mimicus*
Eastern glass lizard	*Ophisaurus ventralis*
Texas horned lizard	*Phrynosoma cornutum*
Eastern fence lizard	*Sceloporus undulatus*
Ground skink	*Scincella lateralis*
Salamanders	
Flatwoods salamander	*Ambystoma cingulatum*
Mabee's salamander	*Ambystoma mabeei*
Spotted salamander	*Ambystoma maculatum*
Marbled salamander	*Ambystoma opacum*
Mole salamander	*Ambystoma talpoideum*
Tiger salamander	*Ambystoma tigrinum tigrinum*
Two-toed amphiuma	*Amphiuma means*
Southern dusky salamander	*Desmognathus auriculatus*
Southern two-lined salamander	*Eurycea cirrigera*
Three-lined salamander	*Eurycea longicauda*
Dwarf salamander	*Eurycea quadridigitata*
Dwarf waterdog	*Necturus punctatus*
Central newt	*Notophthalmus viridescens*
South Carolina slimy salamander	*Plethodon variolatus*
Dwarf siren	*Pseudobranchus striatus*
Eastern mud salamander	*Pseudotriton montanus montanus*
Lesser siren	*Siren intermedia*
Greater siren	*Siren lacertina*
Many lined salamander	*Stereochilus marginatus*
Frogs	
Southern cricket frog	*Acris gryllus*
Cope's gray treefrog	*Hyla chrysoscelis*
Green treefrog	*Hyla cinerea*
Pinewoods treefrog	*Hyla femoralis*
Barking treefrog	*Hyla gratiosa*
Squirrel treefrog	*Hyla squirella*
Northern spring peeper	*Pseudacris crucifer crucifer*

TABLE 5.2 (CONTINUED)
Reptiles and Amphibians That Occur or Potentially Occur in the ACE Basin

Common Name	Scientific Name
Little grass frog	*Pseudacris ocularis*
Brimley's chorus frog	*Pseudacris brimleyi*
Southern chorus frog	*Pseudacris nigrita*
Ornate chorus frog	*Pseudacris ornata*
Upland chorus frog	*Pseudacris triseriata*
Gopher frog	*Rana capito*
Bullfrog	*Rana catesbeiana*
Bronze frog	*Rana clamitans clamitans*
Pig frog	*Rana grylio*
River frog	*Rana heckscheri*
Pickerel frog	*Rana palustris*
Southern leopard frog	*Rana utricularia*
Carpenter frog	*Rana virgatipes*
Toads	
Oak toad	*Bufo quercicus*
Southern toad	*Bufo terrestris*
Eastern narrowmouth toad	*Gastrophryne carolinensis*
Eastern spadefoot toad	*Scaphiopus holbrooki*
Turtles	
Florida softshell	*Apalone ferox*
Spiny softshell	*Apalone spinifera*
Loggerhead	*Caretta caretta*
Green sea turtle	*Chelonia mydas*
Snapping turtle	*Chelydra serpentina*
Spotted turtle	*Clemmys guttata*
Chicken turtle	*Deirochelys reticularia*
Leatherback	*Dermochelys coriacea*
Hawksbill	*Eretmochelys imbricata*
Striped mud turtle	*Kinosternon baurii*
Eastern mud turtle	*Kinosternon subrubrum subrubrum*
Kemp's Ridley sea turtle	*Lepidochelys kempii*
Diamondback terrapin	*Malaclemys terrapin terrapin*
River cooter	*Pseudemys concinna*
Florida cooter	*Pseudemys floridana*
Stinkpot	*Sternotherus odoratus*
Eastern box turtle	*Terrapene carolina*
Yellowbelly slider	*Trachemys scripta*

Source: South Carolina Department of Natural Resources and National Oceanic and Atmospheric Administration. 2001. Characterization of the Ashepoo–Combahee–Edisto (ACE) Basin, South Carolina. Special Scientific Report 17, South Carolina Marine Resources Center, Charleston, SC.

TABLE 5.3
Total Number of Amphibian and Reptilian Species by Habitat Type in the ACE Basin[a]

Habitat	Species	Water Regime	Structure[b]	Salt Exposure
Maritime — coastal	5	Open water	Low	High
Maritime — dune	9	Dry to irregularly flooded	Intermediate	High
Maritime — forest	32	Dry to permanently flooded	High	Low
Estuarine — impoundment	1	Shallow water	Intermediate	Medium
Riverine — open water	13	Open water	Low	None
Lacustrine — limnetic	14	Open water	Low	None
Lacustrine — littoral	37	Shallow water	Intermediate	None
Palustrine — freshwater impoundment	30	Shallow water	Intermediate	Low
Palustrine — tidal emergent freshwater wetland	34	Shallow water	High	None
Palustrine — tidal forested freshwater wetland	33	Shallow water	High	None
Palustrine — nontidal forested freshwater wetland	31	Shallow water	High	None
Palustrine — inland wetland	30	Intermittent flooding	High	None
Upland — hardwood forest	35	Sporadic flooding	High	None
Upland — pine flatwoods, open field	31	Dry to seasonally wet	High	None

[a] Diversity of amphibians and reptiles increases with increasing amounts of emergent vegetation, trees, logs, and forest litter.

[b] Structure refers to the spatial complexity of the habitat.

Sources: Sandifer, P.A., J.V. Miglarese, D.R. Calder, J.J. Manzi, and L.A. Barclay. 1980. Ecological Characteristics of the Sea Island Coastal Region of South Carolina and Georgia, Vol. 3: Biological Features of the Characterization Area. FWS/OBS-79/42, U.S. Fish and Wildlife Service, Office of Biological Services, Washington, D.C.; South Carolina Department of Natural Resources and National Oceanic and Atmospheric Administration. 2001. Characterization of the Ashepoo–Combahee–Edisto (ACE) Basin, South Carolina. Special Scientific Report 17, South Carolina Marine Resources Center, Charleston, SC.

(*Cnemidophorus sexlineatus*). Commonly encountered species of snakes are the pine snake (*Pituophis melanoleucus melanoleucus*), corn snake (*Elaphe guttata guttata*), eastern garter snake (*Thamnophis sirtalis*), and eastern diamondback rattlesnake (*Crotalus adamanteus*). A conspicuous feature of the herpetofauna of this environment is the paucity of species typically found in moist settings, particularly salamanders. An exception is seen at isolated wetland habitats where the herpetofaunal communities are more diverse than elsewhere in the pine flatwoods.

Upland habitats associated with sporadic flooding, and thus typified by greater amounts of moisture than the upland pine flatwoods, harbor more species of salamanders. In the upland hardwood forests, for instance, the marbled salamander (*Ambystoma opacum*), spotted salamander (*A. maculatum*), and mole salamander (*A. talpoideum*) are often observed. The upland hardwood forests also support several species of snakes not commonly observed in the upland pine flatwoods, such as the timber rattlesnake (*Crotalus horridus horridus*), smooth earth snake (*Virginia valeriae*), black racer (*Coluber constrictor constrictor*), cottonmouth (*Agkistrodon piscivorus*), and copperhead (*A. contortrix*). Toads (eastern spadefoot toad, *Scaphiopus holbrooki*; southern toad, *Bufo terrestris*; and eastern narrow-mouth toad, *Gastrophryne carolinensis*) and treefrogs (squirrel treefrog, *Hyla squirella*; barking treefrog, *H. gratiosa*; Cope's gray treefrog, *H. chrysoscelis*; and green treefrog, *H. cinerea*) are well represented.

Perched wetlands (e.g., Carolina Bays and pocosins) in upland forested areas are sites of isolated still standing water where wetland-dependent species predominate. Because the moist conditions associated with these wetlands often disappear during extended dry periods, the herpetofaunal community can shift rather abruptly to those dominant forms capable of withstanding dry conditions. Isolated wetlands are typically ephemeral habitats, and despite the thick growth of trees, shrubs, and other vegetation found here during wet years, environmental conditions are subject to marked seasonal changes linked closely to the level and frequency of precipitation. Amphibian populations can increase substantially in numbers during wet years, while being absent during protracted dry periods; the hydroperiod is a critically important factor in regulating the occurrence of amphibians in the ACE Basin (Pechmann et al., 1991; Blaustein et al., 1994). Isolated intermittent ponds and pools in upland areas, therefore, are sites of large fluxes in abundance and composition of herpetologic fauna.

Moving away from the tidal portions of the major river systems, nontidal forested wetlands are a commonly observed feature. Nontidal forested wetland habitat generally exhibits drier conditions during the year than does tidal forested habitat, and thus some herpetofaunal compositional differences are evident. For example, species requiring permanent water sources are most conspicuous in tidal forested wetlands, and the more semiaquatic forms usually occur in the semipermanently flooded habitat characteristic of the nontidal forested wetlands. Among the more frequently occurring herpetofaunal species in the nontidal forested wetlands are the timber rattlesnake (*Crotalus horridus horridus*), black swamp snake (*Seminatrix pygaea*), southeastern crowned snake (*Tantilla coronata*), eastern glass lizard (*Ophisaurus ventralis*), slender glass lizard (*O. attenuatus*), three-lined salamander (*Eurycea longicauda*), many-lined salamander (*Stereochilus marginatus*), oak toad (*Bufo quercicus*), carpenter frog (*Rana virgatipes*), and little grass frog (*Pseudacris ocularis*).

Frogs, underrepresented in the drier nontidal forested wetlands, may be the most numerous herpetofauna in some tidal forested wetland habitats. Southern cricket frogs (*Acris gryllus*), upland chorus frogs (*P. triseriata*), southern leopard frogs (*Rana utricularia*), and river frogs (*R. heckscheri*) provide examples. Toads are much less abundant in the permanently flooded habitats of the tidal forested wetlands, since they prefer drier conditions. However, salamanders are well represented. Commonly observed species of salamanders include the spotted salamander (*Ambystoma maculatum*), Mabee's salamander (*A. mabeei*), dwarf salamander (*Eurycea quadridigitata*), South Carolina slimy salamander (*Plethodon variolatus*), two-toed amphiuma (*Amphiuma means*), and greater siren (*Siren lacertina*).

Turtles likewise frequent tidal forested wetlands. Among the species of significance are the Florida cooter (*Pseudemys floridana*), Florida softshell (*Apalone ferox*), eastern mud turtle (*Kinosternon subrubrum subrubrum*), yellowbelly slider (*Trachemys scripta*), and snapping turtle (*Chelydra serpentina*). Basking turtles, such as Florida scooters, are at times the most prominent members of the herpetofaunal community, often observed perched on rocks, logs, and other objects.

Some herpetofauna occur in both tidal and nontidal forested wetlands. Alligators (*Alligator mississippiensis*), for example, occupy both habitats. Several species of snakes also live in both wetland types; common species in this regard are brown water snakes (*Nerodia taxispilota*), redbelly water snakes (*N. erythrogastor*), banded water snakes (*N. fasciata*), rough green snakes (*Opheodrys aestivus*), cottonmouths (*Agkistrodon piscivorus*), and copperheads (*A. contortrix*).

The herpetofaunal community is particularly rich and diverse in freshwater tidal emergent wetlands and impoundments. Some of the aforementioned species inhabiting tidal and nontidal forested wetlands also occur in these habitats. In addition, various species that do not live in the forested habitats are found in the emergent wetlands. Snake species that often occupy areas of emergent wetland vegetation and impoundments include the brown water snake, banded water snake, black swamp snake, cottonmouth, and glossy crayfish snake (*Regina regida*). The Florida cooter, Florida softshell, yellowbelly slider, snapping turtle, stinkpot (*Sternotherus odoratus*), and chicken turtle (*Deirochelys reticularia*) are commonly observed species of turtles. Frogs are numerous, particularly ranids such as the bullfrog (*Rana catesbeiana*), pig frog (*R. grylio*), pickerel frog (*R. palustris*), and southern leopard frog. Other frequent frog inhabitants are the green treefrog, squirrel treefrog, and southern cricket frog. Although lizards are generally less abundant than the aforementioned groups, some salamanders (e.g., two-toed amphiuma and greater siren) attain significant numbers. Both the two-toed amphiuma and greater siren are species characteristic of more open water but have likewise been documented in other areas as well.

Alligators, turtles, and snakes are the most common herpetofauna of freshwater, shoreline, and open riverine habitats. Species of turtles observed here include the snapping turtle, yellowbelly slider, Florida cooter, river cooter (*Pseudemys concinna*), and spiny softshell (*Apalone spinifera*). Many of the snakes in creeks and mainstem rivers are also found in the previously described wetland habitats. For example, the brown water snake, redbelly water snake, banded water snake, and cottonmouth are reported along freshwater shoreline and open riverine habitats as well as in many wetland areas. Most of the species recorded in the open water of creeks and rivers also inhabit open water lacustrine habitats in the ACE Basin.

The species composition of herpetofaunal communities on barrier islands (e.g., Edisto Beach, Hunting Island, Otter Island, and Pine Island) changes significantly from the drier habitats of the coastal dunes and maritime dry grasslands to the moist habitats of the maritime forests. For example, in the coastal dunes as well as the maritime dry grassland and dune shrub thickets, herpetofauna capable of tolerating drier conditions predominate. These include the southern toad, eastern spadefoot toad, eastern diamondback rattlesnake, eastern coachwhip (*Masticophis flagellum*), and several lizard species (e.g., eastern glass lizard, six-lined racerunner, and island glass lizard, *Ophisaurus compressus*) (Gibbons and Harrison, 1981).

Species diversity of herpetofauna is relatively low in the aforementioned dune and maritime shrub thicket communities because of adverse environmental conditions associated with excessive heat, dessication, salt spray, and other harsh factors. As the amount of forested vegetation increases on the barrier islands, so does the diversity of herpetofauna. An array of frogs (southern leopard frog, squirrel treefrog, and green treefrog), snakes (cottonmouth snake, rough green snake, and southeastern crowned snake), lizards (ground skink, broadhead skink, and Carolina anole, *Anolis carolinensis*), and turtles (yellowbelly slider and chicken turtle) likely inhabits the barrier island maritime forests (Gibbons, 1978; Gibbons and Coker, 1978; Sandifer et al., 1980; Gibbons and Harrison, 1981). Although the diversity of herpetofauna is significantly higher in the maritime forests than in other habitats on the barrier islands, it remains lower than in nearby mainland (watershed) areas (Gibbons and Coker, 1978).

Mammals

Zimmerman (2001a) has examined in detail the mammalian communities of the ACE Basin region. Nearly 50 species of mammals are likely to occur in the ACE Basin (Table 5.4). Of these species, most reside in palustrine habitats (N = 46), upland habitats (42), and maritime forested habitats (34). Far fewer species occupy estuarine habitats (13), dune habitats (10), and coastal water habitats (2). Habitat diversity in the watershed is a key factor influencing mammalian species diversity. Therefore, the highest diversity of mammalian species is evident in palustrine environments, which have the greatest habitat diversity (hardwood forests, pine forests, mixed forests, meadows, swamps, marshes, and freshwater rivers) in the basin.

Mammalian populations are well represented in the agricultural fields and woodland habitats that dominate the uplands of the ACE Basin. Species commonly encountered here are the coyote (*Canis latrans*), red fox (*Vulpes vulpes*), white-tailed deer (*Odocoileus virginianus*), opossum (*Didelphis marsupialis*), striped skunk (*Mephitis mephitis*), raccoon (*Procyon lotor*), gray squirrel (*Sciurus carolinensis*), Norway rat (*Rattus norvegicus*), hispid cotton rat (*Sigmodon hispidus*), evening bat (*Nycticeius humeralis*), eastern harvest mouse (*Reithrodontomys humulis*), old-field mouse (*Peromyscus polionotus*), and eastern cottontail rabbit (*Sylvilagus floridanus*) (Webster et al., 1985). While some of the mammalian species (e.g., eastern cottontail rabbit) forage on vegetation in the open fields and forests, the larger forms (e.g., coyote, red fox, and raccoon) prey on other mammals in the region.

TABLE 5.4
List of Principal Mammalian Species That Occur or Are Expected to Occur in the ACE Basin

Common Name	Scientific Name
Bats	
Southeastern myotis	*Myotis austroriparius*
Silver-haired bat	*Lasionycteris noctivagans*
Eastern pipistrelle	*Pipistrellus subflavus*
Big brown bat	*Eptesicus fuscus*
Red bat	*Lasiurus borealis*
Seminole bat	*Lasiurus seminolus*
Hoary bat	*Lasiurus cinereus*
Northern yellow bat	*Lasiurus intermedius*
Evening bat	*Nycticeius humeralis*
Big-eared bat	*Coryrochinus rafinesquii*
Free-tailed bat	*Tadarida brasiliensis*
Rabbits, Hares, Pikas	
Eastern cottontail	*Sylvilagus floridanus*
Marsh rabbit	*Sylvilagus palustris*
Marsupials	
Opossum	*Didelphis marsupialis*
Rodents	
Gray squirrel	*Sciurus carolinensis*
Fox squirrel	*Sciurus niger*
Southern flying squirrel	*Glaucomys volans*
Beaver	*Castor canadensis*
Rice rat	*Oryzomys palustris*
Eastern harvest mouse	*Reithrodontomys humulis*
Old-field mouse	*Peromyscus polionotus*
Cotton mouse	*Peromyscus gossypinus*
Golden mouse	*Ochrotomys nuttalli*
Hispid cotton rat	*Sigmodon hispidus*
Eastern woodrat	*Neotoma floridana*
Meadow vole	*Microtus pennsylvanicus*
Pine vole	*Microtus pinetorum*
Roof rat	*Rattus rattus alexandrinus*
Black rat	*Rattus rattus rattus*
Norway rat	*Rattus norvegicus*
House mouse	*Mus musculus*
Carnivores	
Coyote	*Canis latrans*
Red fox	*Vulpes vulpes*
Gray fox	*Urocyon cinereoargenteus*
Black bear	*Ursus americanus*

TABLE 5.4 (CONTINUED)
List of Principal Mammalian Species That Occur or Are Expected to Occur in the ACE Basin

Common Name	Scientific Name
Raccoon	*Procyon lotor*
Long-tailed weasel	*Mustela frenata olivacea*
Mink	*Mustela vison*
Striped skunk	*Mephitis mephitis*
River otter	*Lutra canadensis*
Bobcat	*Lynx rufus*
Odontoceti	
Bottlenose dolphin	*Tursiops turncatus*
Sirenia	
West Indian manatee	*Trichechus manatus*
Artiodactyls	
White-tailed deer	*Odocoileus virginianus*
Feral swine	*Sus scrofa*

Sources: Webster, W.D., J.F. Parnell, and W.C. Biggs, Jr. 1985. *Mammals of the Carolinas, Virginia, and Maryland.* University of North Carolina Press, Chapel Hill, NC; South Carolina Department of Natural Resources and National Oceanic and Atmospheric Administration. 2001. *Characterization of the Ashepoo–Combahee–Edisto (ACE) Basin, South Carolina.* Special Scientific Report 17, South Carolina Marine Resources Center, Charleston, SC.

Some of the most suitable habitats for mammals in the ACE Basin exist in palustrine environments. Many mammalian species have relatively broad habitat preferences in palustrine areas. Examples are the white-tailed deer, raccoon, opossum, long-tailed weasel (*Mustela frenata olivacea*), and golden mouse (*Ochrotomys nuttalli*). Other species such as the mink (*M. vison*), beaver (*Castor canadensis*), and gray fox (*Urocyon cinereoargenteus*) have narrower habitat preferences. The mink, for example, inhabits swamps and freshwater marshes, preying heavily on various species of fish and other organisms (Baker and Carmichael, 1996). The cotton mouse (*Peromyscus gossypinus*) and marsh rabbit (*Sylvilagus palustris*) also live in swamp and freshwater marsh habitats. Similarly, beavers prefer swamps, as well as creeks and ponds. The gray fox occupies wetland and upland forests as well as open field habitats (Baker and Carmichael, 1996). Palustrine forests are favored habitats of the gray squirrel, fox squirrel (*Sciurus niger*), eastern woodrat (*Neotoma floridana*), short-tailed shrew (*Blarina brevicauda*), big brown bat (*Eptesicus fuscus*), and free-tailed bat (*Tadarida brasiliensis*) (Weakley, 1981; Webster et al., 1985; Mengak et al., 1987).

The marsh rabbit is an inhabitant of brackish marshes. The river otter (*Lutra canadensis*) and rice rat (*Oryzomys palustris*), in turn, prefer salt marsh habitats (Andre, 1981; Baker and Carmichael, 1996). Foxes and other predators often visit these habitats in search of prey such as the marsh rabbit. Overall, mammalian species diversity is lower in these wetland habitats than in upland and palustrine environments.

The number of mammalian species declines in the harsher environments of the coastal dunes, maritime dry grasslands, and maritime dune shrub thickets along the coast. Smaller mammalian species reported in dune habitats include the rice rat, marsh rabbit, eastern cottontail rabbit, eastern mole (*Scalopus aquaticus*), and house mouse (*Mus musculus*). Larger mammalian species observed on sand dunes are the raccoon, opossum, and white-tailed deer. These wide-ranging species are found in all major habitats of the ACE Basin — dunes, estuaries, maritime forests, palustrine areas, and uplands (Sandifer et al., 1980; Zimmerman, 2001a).

Mammals are more prominent in the maritime forests. Bats (evening bat, brown bat, and red bat), moles (star-nosed mole, *Condylura cristata*; and eastern mole), and shrews (least shrew, *Cryptotis parva*; southeastern shrew, *Sorex longirostris*; and short-tailed shrew) often predominate. These smaller mammalian species are important insectivores, consuming large numbers of insects in the forested habitat. The fox squirrel and gray squirrel are also important members of the maritime forest community (Webster et al., 1985; Whitney, 1998).

Larger members of this community include more ubiquitous forms such as the white-tailed deer, opossum, and raccoon. The carnivorous bobcat (*Lynx rufus*) is also relatively abundant. It may play a significant role in population control of smaller mammalian forms in the woodlands (Sandifer et al., 1980; Webster et al., 1985; Whitney, 1998).

Birds

Zimmerman (2001b) has provided a comprehensive description of the avifaunal community in the ACE Basin. The community is highly diverse, consisting of about 280 species in 17 orders (SCDNR and NOAA, 2001). While many of these species remain as year-round residents, others are migratory and transient. As in the case of the herpetofaunal and mammalian communities, the avifauna exhibits higher species diversity in wetland habitats than in upland farmfields and woodlands, reflecting differences in habitat diversity in these areas.

More than 50 species of birds inhabit the pine forests in upland areas. Insectivores (e.g., pine warbler, *Dendroica pinus*; common yellowthroat, *Geothlypis trichas*; red-bellied woodpecker, *Melanerpes carolinus*; and California wren, *Thryothorus ludovicianus*), granivores (mourning dove, *Zenaida macroura*; chipping sparrow, *Spizella passerina*; and brown-headed nuthatch, *Sitta pusilla*), and omnivores (American crow, *Corvus brachyrhynchos*; common grackle, *Quiscalus quiscula*; and northern bobwhite, *Colinus virginianus*) are well represented. In addition, raptors (red-shouldered hawk, *Buteo lineatus*; and red-tailed hawk, *Buteo jamaicensis*) also occur in the pine forests. At least three species of owls (i.e., great horned owl, *Bubo virginianus*; barn owl, *Tyto alba*; and eastern screech owl, *Otus asio*) have been reported as well.

Mixed upland forests support more species of birds than do the pine forests, with nearly 90 species estimated to occur here. Sandifer et al. (1980) attribute the greater species diversity in pine–hardwood forests to the well-developed subcanopy and understory vegetation that provides additional habitat. Insectivores are common, comprising several species of woodpeckers (i.e., hairy woodpecker, *Picoides villosus*; downy woodpecker, *P. pubescens*; pileated woodpecker, *Dryocopus pileatus*; red-headed woodpecker, *Melanerpes erythrocephalus*; and red-bellied woodpecker), warblers (yellow-throated warbler, *Dendroica dominica*; black-throated warbler, *D. caerulescens*; Cape May warbler, *D. tigrina*; magnolia warbler, *D. magnolia*; hooded warbler, *Wilsonia citrina*; and black-and-white warbler, *Mniotilta varia*), and flycatchers (Acadian flycatcher, *Empidonax virescens*; and great crested flycatcher, *Myiarchus crinitus*), as well as vireos (white-eyed vireo, *Vireo griseus*; and solitary vireo, *V. solitarius*) and Carolina wrens (*Thryothorus ludovicianus*). Granivores of significance include sparrows (house sparrow, *Passer domesticus*; song sparrow, *Melospiza melodia*; white-throated sparrow, *Zonotrichia albicollis*; fox sparrow, *Passerella iliaca*; lark sparrow, *Chondestes grammacus*; and chipping sparrow, *Spizella passerina*), northern cardinals (*Cardinalis cardinalis*), American goldfinches (*Carduelis tristis*), and mourning doves (*Zenaida macroura*).

Omnivorous forms of note are American crows and fish crows (*Corvus ossifragus*). Black vultures (*Coragyps atratus*) and turkey vultures (*Cathartes aura*) serve as the primary scavengers. Among the top predators are eagles (golden eagle, *Aquila chrysaetos*; and bald eagle, *Haliaeetus leucocephalus*), hawks (red-tailed hawk; red-shouldered hawk; Cooper's hawk, *Accipiter cooperii*; sharp-shinned hawk, *A. striatus*; and broad-winged hawk, *Buteo platypterus*), and owls (eastern screech-owl; great horned owl; and barred owl, *Strix varia*) (Potter et al., 1980; Sandifer et al., 1980; SCDNR and NOAA, 2001).

Nearly 75 species of birds comprise the old-field avifaunal communities. The farmlands, open fields, and grasslands form highly accessible feeding grounds for many different types of birds. Numerous species also nest in these habitats. Insectivores are abundant, including swallows (barn swallow, *Hirundo rustica*; northern rough-winged swallow, *Stelgidopteryx serripennis*; and tree swallow, *Tachycineta bicolor*), wrens (short-billed marsh wren, *Cistothorus platensis*; water wren, *Troglodytes troglodytes*; and Carolina wren), warblers (prairie warbler, *Dendroica discolor*; yellow-rumped warbler, *Dendroica coronata*; and orange-crowned warbler, *Vermivora celata*), brown thrashers (*Toxostoma rufum*), eastern meadowlarks (*Sturnella magna*), and common yellowthroats (*Geothlypis trichas*). Granivores of importance are sparrows (field sparrow, *Spizella pusilla*; house sparrow, *Passer domesticus*; and song sparrow), northern mockingbirds (*Mimus polyglottos*), indigo buntings (*Passerina cyanea*), and northern cardinals (*Cardinalis cardinalis*). Several omnivorous species congregate here as well, notably boat-tailed grackles (*Quiscalus major*), common grackles, red-winged blackbirds (*Agelaius phoeniceus*), northern bobwhites, American crows, and fish crows. Representative raptors are northern harriers (*Circus cyaneus*), merlin (*Falco columbarius*), sharp-shinned hawks, Cooper's hawks, and red-tailed hawks.

Wetlands of the ACE Basin support the most species of birds in the watershed. The estimated number of bird species in forested wetlands (N = 132) far exceeds that in nonforested wetlands (92) and estuarine-emergent vegetation (87) of the basin (Potter et al., 1980; Sandifer et al., 1980). The insectivores found in the forested wetlands are similar to those observed in upland areas and include mourning doves, Carolina wrens, winter wrens (*Troglodytes troglodytes*), white-eyed vireos, red-eyed vireos (*Vireo olivaceus*), common yellowthroats (*Geothlypis trichas*), yellow warblers (*Dendroica petechia*), black-and-white warblers (*Mniotilta varia*), woodpeckers (downy woodpeckers, pileated woodpeckers, and hairy woodpeckers), and other forms. Some of the granivores reported in the forested wetlands are sparrows (swamp sparrow, *Melospiza georgiana*; white-throated sparrow; and song sparrow), painted buntings (*Passerina ciris*), American goldfinches, and northern cardinals. Omnivores of significance are the common grackles, American crows, and red-winged blackbirds. Several raptors continually search for prey in the forested wetlands, frequently consuming small mammals, amphibians, reptiles, fish and other birds. Examples are osprey (*Pandion haliaetus*), red-tailed hawks, broad-winged hawks, red-shouldered hawks, sharp-shinned hawks, Cooper's hawks, golden eagles, bald eagles, long-eared owls (*Asio otus*), and barn owls (*Tyto alba*). Wading birds, such as egrets (e.g., great egret, *Casmerodius albus*), herons (e.g., little blue heron, *Egretta caerulea*), and ibises (e.g., white ibis, *Eudocimus albus*), use forested wetlands as feeding and nesting grounds.

Although fewer species of birds inhabit nonforested wetlands and estuarine-emergent wetlands than the forested wetlands of the ACE Basin, similar species groups are evident (Sandifer et al., 1980). For example, swallows, sparrows, and wrens occur in all three environments; in addition, raptors (osprey, hawks, bald eagles, northern harriers, owls, merlin, and peregrine falcons), waders (egrets, herons, and ibises), rails (Virginia rail, *Rallus limicola*; king rail, *R. elegans*; clapper rail, *R. longirostris*; and sora, *Porzana carolina*), and waterfowl (mallard, *Anas platyrhynchos*; blue-winged teal, *A. discors*; northern pintail, *A. acuta*; ruddy duck, *Oxyura jamaicensis*; and red-breasted merganser, *Mergus serrator*) feed and nest in many habitats in these wetland environments.

Nearly 90 species of birds inhabit salt marsh habitats, where they feed, breed, and nest (Potter et al., 1980; Sandifer et al., 1980). The list of resident and seasonal forms includes shorebirds (gulls, terns, plovers, and sandpipers), wading birds (egrets, herons, and ibises), rails (clapper rails, Virginia rails, and soras), and raptors (northern harriers, sharp-shinned hawks, peregrine falcons, and merlin). Bird populations are abundant in these habitats because of the rich food supply and favorable nesting sites.

Beaches, dunes, maritime dry grasslands, and maritime shrub communities are inhabited by fewer species of birds than are observed in maritime forests. The estimated number of bird species recorded in the beach, dune, and maritime shrub environments amounts to 44, 34, and 26, respectively (SCDNR and NOAA, 2001). These coastal habitats are harsh, thereby limiting the number of year-round residents. Some of the species of birds most commonly found on beaches in the ACE Basin include the piping plover (*Charadrius melodus*), Wilson's plover (*C. wilsonia*), sanderling (*Calidris alba*), black skimmer (*Rynchops niger*), least tern (*Sterna antillarum*), American oystercatcher (*Haematopus palliatus*), laughing gull (*Larus atricilla*), and herring gull (*L. argentatus*).

Maritime dunes are a favored habitat for various groups of shorebird species (plovers, sandpipers, and terns) that feed, nest, rest, and breed there. Insectivorous and granivorous birds often occur within the dune habitat. Granivorous birds (sparrows, doves, cardinals, and blackbirds) are common inhabitants because the supply of seeds and other grains is substantial. Among the insectivores, swallows (tree swallow, *Tachycineta bicolor*; barn swallow; and northern rough-winged swallow) and warblers (palm warbler, *Dendroica palmarum*; and yellow-throated warbler) are particularly abundant. Omnivores are likewise common constituents of the avifaunal community. Many birds observed in the maritime shrubs visit these habitats in search of food; they are typically residents of nearby environments such as the maritime forests (Zimmerman, 2001b).

Bay and inlet islands, also known as bird keys, provide valuable habitat in the ACE Basin for colonial nesting waterbirds. Predominant species occupying bird keys are the black skimmer (*Rynchops niger*), laughing gull (*Larus atricilla*), royal tern (*Sterna maxima*), and brown pelican (*Pelecanus occidentalis*) (Sandifer et al., 1980). Colonial nesting birds on these islands are sensitive to pollution and other anthropogenic impacts, and their numbers can vary considerably when human activities create stressful conditions.

Maritime forests contain more diverse habitats than the beach, dune, and maritime shrub environments and thus support more species of birds (N = 87). Passerines, mainly comprised of insectivores and granivores, are numerous in the maritime forests, and are represented by warblers (prairie warbler, pine warbler, yellow-throated warbler, hooded warbler, and black-and-white warbler), sparrows (field sparrow, song sparrow, swamp sparrow, and fox sparrow), grackles (boat-tailed grackle and common grackle), wrens (winter wren, Carolina wren, and house wren, *Troglodytes aedon*), swallows (tree swallow, barn swallow, and northern rough-winged swallow), vireos (white-eyed vireo, red-eyed vireo, and solitary vireo), buntings (indigo bunting and painted bunting), and other forms. Woodpeckers (red-bellied woodpecker, red-headed woodpecker, hairy woodpecker, and downy woodpecker) are abundant as well. The principal raptors are red-tailed hawks and red-shouldered hawks. Other predatory species of significance include the barred owl, eastern screech-owl, and great horned owl (Potter et al., 1980; Sandifer et al., 1980; SCDNR and NOAA, 2001).

Insects

Insects play a major role in energy flow of terrestrial ecosystems; they are the principal herbivorous component, consuming as much as 80% of the total plant matter (Price, 1997). They also constitute a main staple in the diets of many terrestrial and aquatic organisms (Thompson, 1984). Insects represent one of the most successful groups of organisms on Earth; more than 800,000 species have been described (Solomon et al., 1999).

Many gaps exist in the data associated with the insects of the ACE Basin (Scholtens, 2001). More information has been collected on insects in salt marshes than in any other habitat of the ACE Basin. Grasshoppers, aphids, thrips, moth larvae, and other insects feed on salt marsh plants. In addition to these herbivores, various salt marsh species are predators, detritivores, parasitoids, and parasites. Davis and Gray (1966), Vernberg and Sansbury (1972), and Davis (1978) have compiled comprehensive lists of salt marsh insects in the region. Davis and Gray (1966) and Davis (1978) listed more than 350 species of insects in this habitat, with most belonging to eight orders:

1. Coleoptera (beetles)
2. Diptera (true flies)
3. Hemiptera (true bugs)
4. Homoptera (hoppers)
5. Hymenoptera (ants, bees, and wasps)
6. Lepidoptera (butterflies and moths)
7. Odonata (damselflies and dragonflies)
8. Orthoptera (crickets, katydids, and grasshoppers)

The Diptera and Homoptera are most abundant (SCDNR and NOAA, 2001). Despite the high absolute abundance of salt marsh insects, species diversity is not great in the salt marshes.

Among the herbivorous forms, grasshoppers (*Orchelimum fidicinium*) and plant hoppers (*Prokelisia marginata*) are significant. *Orchelimum fidicinium* chews on *Spartina alterniflora* leaves; it consumes ~1% of the net aerial primary production of a cordgrass marsh in the region (Smalley, 1980). *Prokelisia marginata*, a sap-sucking species, obtains nutrition by extracting substances translocated through the vascular vessels of cordgrass (Pfeiffer and Wiegert, 1981). *Chaetopsis* spp., which are flies, also obtain nutrition from salt marsh plant fluids. Teal (1962) and Kraeuter and Wolf (1974) indicate that salt marsh herbivores may consume as much as 5–10% of the net annual *Spartina* primary production.

Dragonflies (*Erythemis* sp. and *Pachydiplax* sp.) are examples of predatory insects in the ACE Basin salt marshes. Species belonging to the Braconidae and Chalcidoidae families within the Hymenoptera are representative parasitoids. Parasitic insects include such major groups as the mosquitoes (Culicidae) and green-headed flies (Tabanidae).

Some insect groups (e.g., Lepidoptera) in the ACE Basin have been studied in detail. An estimated 125 species of butterflies and moths occur in this system (Gatrelle, 1975; Wallace, 1987; Opler and Malikul, 1998). However, much less is known about the other insect taxa, with the possible exception of some of the Diptera, Homoptera, and Orthoptera. Because the database on most insect groups in different ACE Basin habitats is not extensive, it has not been possible to precisely determine the total number of species inhabiting the region. The best estimate of insect species richness in the ACE Basin is approximately 8,000 to 10,000 species (Scholtens, 2001).

ESTUARY

Physical-Chemical Characteristics

St. Helena Sound is a drowned river valley. Based on hydrologic characteristics and stratification, it is also classified as a partially mixed estuary (Mathews et al., 1980;

TABLE 5.5
Physical and Hydrologic Features of St. Helena Sound

Estuarine drainage area (100 km^2)	37.5 km^2
Total drainage area (100 km^2)	120.0 km^2
Mean daily freshwater inflow (100 m^3/sec)	1.31 m^3/sec
Wetlands area	1747.2 km^2
Surface area	212.5 km^2
Mean depth	4.3 m
Volume (billion m^3)	9.93 m^3

Source: NOAA. 1990. Estuaries of the United States: Vital Statistics of a National Resource Base. NOAA 20th Annual Report, National Oceanic and Atmospheric Administration, Office of Ocean Resources Conservation and Assessment, Rockville, MD.

Orlando et al., 1994). Table 5.5 provides physical and hydrological data on the sound. Water depths are generally less than 20 m, and the bottom topography exhibits irregular contours. Because of strong tidal currents relative to freshwater inflow, as well as variable bottom topography, current-induced turbulent mixing creates a moderately mixed condition. When winds are strong, however, the water column is often well mixed (Orlando et al., 1994). Some areas of the estuary, therefore, experience complete mixing contingent on the intensity of tidal currents, rate of freshwater discharge, amount of turbulent eddies and vertical mixing, and wind velocity.

St. Helena Sound and contiguous waters are subject to semidiurnal tides. Tidal amplitude at the coast is ~2 m. Oceanic tides influence a large area of the ACE Basin, with tidal flux being observed far upstream in tributary systems. For example, Eidson (1993) documented tidal influences more than 60 km upstream in estuarine tributaries. Measurable salinity was recorded more than 30 km upstream in these systems.

The most extensive water quality database on the ACE Basin NERR exists for two System-wide Monitoring Program (SWMP) sites (one in Big Bay Creek and the other in St. Pierre Creek). The Big Bay Creek monitoring site is located at 32°09'37"N, 80°19'26"W and the St. Pierre Creek monitoring site at 32°01'43"N, 80°2'34"W. Both Big Bay Creek and St. Pierre Creek are tributaries of the South Edisto River, which discharges into St. Helena Sound. Six water quality parameters are measured semicontinuously (every 30 min) year-round at these two sites using automated data sondes (YSI 6000® or YSI 6600®) left unattended in the field. As at SWMP sites nationwide, these parameters include temperature, salinity, dissolved oxygen, pH, turbidity, and depth (Wenner et al., 2001b).

Wenner et al. (2001b) discussed the findings of water quality monitoring in the ACE Basin over the 3-year period from 1996 through 1998. During this period, the mean water temperature at the two aforementioned SWMP sites ranged from ~10 to 12°C in winter and ~27 to 29°C in summer. Salinity was higher at the Big Bay Creek site than at the St. Pierre Creek site. For example, mean salinity at the Big Bay Creek site during the 1996–1997 period ranged from ~25 to 27‰ in winter/spring to ~30‰ in summer, with extreme values of 0.3‰ and 41.7‰. At the St. Pierre Creek site, in turn, mean salinity in winter/spring ranged from ~22 to 24‰, and in summer it ranged from ~28 to 30‰, with extreme values of 0‰ and 41.7‰.

Mean dissolved oxygen levels recorded by the ACE Basin NERR at the two SWMP sites from March 1995 through December 1997 revealed generally well-oxygenated conditions (Wenner et al., 2001a, b). At the Big Bay Creek site, for example, dissolved oxygen levels averaged 6.6 mg/l and 83.6% saturation compared to 6.3 mg/l and 78.5% saturation at the St. Pierre site. Conditions of both hypoxia (<4 mg/l or <28% saturation) and supersaturation (>120% saturation) occurred at each site. Supersaturation was evident during all seasons. Hypoxic events were most frequently observed in summer; however, they were not common events. Dissolved oxygen levels were typically lowest in July and August and highest during the winter months. In St. Pierre Creek, hypoxia was documented more than 10% of the time during July and August. In Big Creek, it was registered 17% of the time during September.

The most consistent parameter at the two monitoring sites was pH. For example, pH values at the Big Bay Creek site ranged from 5.4 to 8.3 during the period from March 1995 through December 1997. Similarly, they ranged from 5.3 to 8.4 at the St. Pierre site during this interval (Wenner et al., 2001a).

Turbidity levels were comparable in both Big Bay Creek and St. Pierre Creek. Mean monthly turbidity levels (Nephelometry Turbidity Units, NTU) were generally less than 50 NTU at both monitoring sites. In sampling conducted during 1996 and 1997, highest values were found at both sites during 1997. Turbidity was more variable in St. Pierre Creek than in Big Bay Creek. Water depth was also similar at the two monitoring sites, ranging from less than 1 to more than 3 m (Wenner et al., 2001a).

The ACE Basin is a relatively pristine area. Nutrient levels exhibit similarities as well as differences when compared with those of other South Carolina estuaries. For example, nitrate concentrations in the Edisto River compare favorably with those reported in North Inlet, but are much lower (by a factor of 10) than those documented in Winyah Bay. However, ammonium levels in the Edisto River exceed those in the other two systems. Mean values of nitrate–nitrite, ammonium, and orthophosphate in the Edisto River amount to ~1.3, ~15.3, and 0.6 μmol/l, respectively (Table 5.6). Nitrate–nitrite levels recorded in the Ashepoo and Combahee Rivers over the period from 1986 through 1995 ranged from <0.05 to ~0.25 mg/l (SCDNR and NOAA, 2001). Extensive wetlands may serve as a source as well as a sink for nutrients in lower riverine and open estuarine areas.

Biotic Communities

Phytoplankton

Phytoplankton communities consist of free-floating, unicellular, filamentous, or chain-forming microscopic plants that inhabit surface waters (the photic zone) of marine, estuarine, and freshwater environments. These microscopic plants form the

TABLE 5.6
Mean Nutrient Concentrations (µmol/l) in the Edisto River Relative to Other South Carolina Estuarine Systems

Estuasry	Nitrate–Nitrite	Ammonium	Orthophosphate
Edisto River	1.26 ± 1.03	15.26 ± 27.14	0.63 ± 0.47
North Inlet	0.55 ± 0.79	1.73 ± 2.0	0.03 ± 0.01
Winyah Bay	16.57	14.07	0.55 ± 0.32
Cooper River	3.65 ± 1.18	5.24 ± 0.32	0.68 ± 0.04

Source: South Carolina Department of Natural Resources and National Oceanic and Atmospheric Administration. 2001. *Characterization of the Ashepoo–Combahee–Edisto (ACE) Basin, South Carolina*. Special Scientific Report 17, South Carolina Marine Resources Center, Charleston, SC.

base of the food chain in open water habitats of the ACE Basin, accounting for a large fraction of the total primary production. They may be grouped on the basis of size into four classes:

1. Picoplankton (<5 µm)
2. Nanoplankton (5–20 µm)
3. Microphytoplankton (20–100 µm)
4. Macrophytoplankton (>100 µm)

The picoplankton and nanoplankton pass through plankton nets (~30 to 64 µm apertures), which retain the larger plant cells.

The net plankton collected in estuarine waters generally consists of large numbers of diatoms (Bacillariophyceae) and dinoflagellates (Dinophyceae). However, several other classes of phytoplankton may be represented in these open water habitats, namely the Chlorophyceae, Chrysophyceae, Cryptophyceae, Euglenophyceae, Haptophyceae, and Raphidophyceae (Dawes, 1998). While little is known regarding the picoplankton and nanoplankton of many estuaries, they have been shown to be a numerically dominant component of some systems. For example, most of the newly listed phytoplankton species of the Barnegat Bay–Little Egg Harbor Estuary in New Jersey are phytoflagellate forms, including the numerically dominant members of the phytoplankton community (Olsen and Mahoney, 2001).

Few investigations have been conducted on phytoplankton in the ACE Basin (Zimmerman, 2001c). As part of an estuarine eutrophication survey of Georgia and South Carolina coastal waters (see Verity, 1998), NOAA (1996) examined the phytoplankton community of the St. Helena Sound system and the Stono/North Edisto River system in the ACE Basin. This survey showed that diatoms dominate the community of both systems. Chlorophyll *a* levels were less than 5 µg/l. Diatoms have also been shown to be major constituents of other estuarine waters of the South Carolina region. For example, Davis and Van Dolah (1992) reported that diatoms, notably *Skeletonema costatum*, dominated the phytoplankton community of Charleston Harbor, South Carolina, during the spring and fall. Flagellates, together with cyanobacteria, predominated during the summer and winter. In North Inlet, South Carolina, Lewitus et al. (1998) likewise revealed the numerical importance of diatoms in the phytoplankton community, with *Cylindrotheca closterium*, *Nitzschia* spp., and *Thalassiosira* spp. the most abundant species. Phytoflagellates were numerically dominant year-round in North Inlet; both picoplankton and nanoplankton populations attained high abundances.

Zooplankton

Zooplankton comprise the principal herbivorous component of estuarine ecosystems, serving as an essential link in the food web by converting plant biomass to animal matter. Although many zooplankton species in estuaries are herbivores that graze on phytoplankton populations, omnivorous and carnivorous species also occur in the community. Zooplankton grazing plays a significant role in regulating the standing crop of phytoplankton in estuarine systems (Omori and Ikeda, 1984; Kennish, 2001).

Based on size, zooplankton are categorized as microzooplankton ($<\sim 200$ µm), mesozooplankton (~ 200 to 500 µm), and macrozooplankton ($>\sim 500$ µm). They may also be classified as holoplankton, meroplankton, and tychoplankton based on the duration of planktonic life. While holoplankton live in the plankton their entire lives, meroplankton largely consist of early life history (larval) stages of benthic invertebrates that inhabit the water column. Tychoplankton are demersal populations periodically inoculated into the plankton from bottom habitats by various processes (e.g., waves, currents, and bioturbation). Among the predominant members of estuarine zooplankton communities are copepods, cladocerans, tintinnids, rotifers, and benthic invertebrate larvae (e.g., bivalve, gastropod, polychaete, and crustacean larvae), as well as the larger forms — ichthyoplankton and the jellyfish group (hydromedusae, comb jellies, and true jellyfishes) (Kennish, 2001).

Knott (1980, 2001) investigated the zooplankton community in tidal waters of the North Edisto River at Bluff Point, conducting weekly sampling over an annual cycle. Analysis of surface samples disclosed that the calanoid copepod, *Acartia tonsa*, numerically dominated the mesozooplankton at this location, followed by the harpacticoid copepod, *Euterpina acutifrons*. Sixty-three copepod species accounted for 78% of the total mesozooplankton abundance at the river site. Secondary dominant species included two calanoid copepods, *Parvocalanus crassirostris* and *Pseudodiaptomus coronatus*. Other dominants were barnacle larvae (cirripedes), copepod nauplii, and rotifers (Table 5.7). Meroplankton larvae comprised 3 to 21% of the total zooplankton abundance; aside from barnacle larvae, gastropod larvae also comprised a major fraction of the meroplankton.

Total zooplankton abundance exceeded 6000 individuals/m^3 throughout the year. Peak numbers occurred in spring. Highest densities (>23,000 individuals/m^3) were found in April, and densities exceeded 10,000 individuals/m^3 from April through June. The annual mean density of zooplankton in the North River at Bluff Point amounted to 10,148 individuals/m^3 (Knott, 1980, 2001). Both the zooplankton

TABLE 5.7
Taxonomic Composition and Relative Abundance of Mesozooplankton at Bluff Point in the North Edisto River over an Annual Cycle

Taxa	Abundance (Percent of Total)
Acartia tonsa	41.6
Euterpina acutifrons	10.4
Rotifera	8.4
Parvocalanus crassirostris	7.6
Copopod nauplii	6.3
Pseudodiaptomus coronatus	6.0
Cirripede larvae	5.5
62 other taxa	14.7

Source: South Carolina Department of Natural Resources and National Oceanic and Atmospheric Administration. 2001. *Characterization of the Ashepoo–Combahee–Edisto (ACE) Basin, South Carolina.* Special Scientific Report 17, South Carolina Marine Resources Center, Charleston, SC.

composition and seasonal variation in abundance were similar to those observed in other South Carolina estuarine waters (Knott, 2001).

Benthic Invertebrates

Based on mode of life, benthic fauna are broadly grouped into epifauna (those forms living on the estuarine bottom or attached to a firm substrate) and the infauna (those forms living in bottom sediments below the sediment–water interface). Some species, however, are more appropriately classified as interstitial, boring, swimming, and commensal–mutualistic forms. Benthic fauna can also be divided into five groups based on feeding habits: suspension feeders, deposit feeders, herbivores, carnivores, and scavengers (Levinton, 1982, 1995).

Based on size, benthic fauna are differentiated into four groups: microfauna, meiofauna, macrofauna, and megafauna. The microfauna are organisms less than 0.1 mm in size. This group consists largely of ciliates and foraminifera. Benthic fauna ranging from 0.1 to 0.5 mm in size constitute the meiofauna; important representatives of the meiofauna are nematodes, ostracods, harpacticoid copepods, rotifers, gastrotrichs, kinorhynchs, archiannelids, halacarines, tardigrades, and mystacocarids. The meiofauna also include the juvenile stages of polychaetes, oligochaetes, turbellarians, and other organisms. The macrofauna, in turn, are those benthic animals larger than 0.5 mm and smaller than 20 mm. Individuals exceeding 20 mm in size comprise the megafauna.

No data have been collected on the benthic microfauna and benthic meiofauna of the ACE Basin system (Van Dolah, 2001). However, Bell et al. (1978), Bell (1979), Coull and Bell (1979), Coull et al. (1979), and Coull and Dudley (1985) have investigated the meiofauna of intertidal and subtidal sediments of other South Carolina estuarine systems. These studies, as well as others (see Kennish, 1986, 2001), reveal the high abundances that these organisms attain in estuarine bottom habitats. For example, in South Carolina intertidal mudflat environments, the density of meiofauna can exceed 2.5×10^7 individuals/m^2, and the biomass may range from 1 to 2 g/m^2 (Bell, 1979). Bell (1979) found that nematodes are particularly abundant, comprising nearly 75% of the total meiofauna in a South Carolina estuarine benthic community. In some estuaries, annual meiofaunal production is greater than 20 g C/m^2/yr (Kennish, 1986, 2001).

Meiofauna concentrate in the uppermost portion of the sediment column. For example, Coull and Bell (1979) found that more than 95% of the meiofauna in their estuarine benthic samples occurred within the top 7 cm of the sediment column. In addition, 60 to 70% of the meiofauna inhabited the upper 2 cm of the bottom sediments. Because of high abundances near the sediment–water interface, the meiofauna provide a rich food source for benthic macrofauna and demersal fish (Bell and Coull, 1978; Kennish, 1986; Coull, 1990).

Coull and Bell (1979), Coull et al. (1979), and Coull and Dudley (1985) examined long-term patterns of meiofaunal abundance, distribution, and composition at North Inlet, South Carolina. Meiofauna in this system exhibited considerable temporal variation in abundance, evident over both seasonal and annual periods. Coull et al. (1979) observed conspicuous spatial distribution patterns of meiofaunal assemblages along an intertidal–subtidal gradient. Within the intertidal zone of salt marsh flats, *Nitocra lacustris* and *Schizopera knabeni* occupied the upper flats. Other species (i.e., *Diarthrodes aegideus, Nannopuus palustris, Pseudostenhelia wellsi,* and *Robertsonia propinqua*) were restricted to the lower marsh flats. A number of meiofauna populations, such as *Halectinosoma winonae* and *Pseudobradya pulchella,* were limited to subtidal habitats. Several species (e.g., *Enhydrosoma propinquuum, Halicyclops coulli,* and *Microarthridium littorale*) ranged across the entire intertidal–subtidal gradient.

Much more is known about the benthic macrofauna of the ACE Basin (Van Dolah, 2001). Since the mid-1970s, several studies of benthic macrofauna have been conducted within or near the ACE Basin NERR, including those of Calder and Boothe (1977a, 1977b), Calder et al. (1977), Van Dolah et al. (1979, 1984, 1991), and Hyland et al. (1996, 1998). These investigators sampled benthic organisms at a total of 26 stations. Oyster dredges and bottom grabs were employed to sample benthic epifauna and infauna, respectively. Results of these studies indicate that the epifauna and infauna of the benthic macroinvertebrate community of the ACE Basin are both abundant and diverse. Among the most commonly sampled epifaunal species were barnacles (*Balanus improvisus*), crabs (*Callinectes sapidus*), shrimp (*Penaeus setiferus*), chordates (*Molgula manhattensis*), bryozoans (*Aeverrillia setigera, Amathia distans, Anguinell palmata, Bowerbankia gracilis,* and *Electra monostachys*), and cnidarians (*Clytia kincaidi, Ectopleura dumortieri,* and *Obelia bidentata*). Highest epifaunal species diversity was recorded in the North Edisto River and at Rock Creek.

Subtidal estuarine benthic samples collected at 20 of the 26 field stations recovered more than 10 infaunal species per grab. The Shannon–Weaver (H′) diversity index of the benthic infaunal assemblages averaged 2.8 and ranged from 0.8 to 4.4, reflecting a generally undegraded condition. Most of the grab samples also contained more than 500 individuals/m^2. The mean infaunal density amounted to 2430 individuals/m^2. Species collected at the most sampling stations were the amphipods *Ampelisca vadorum, Batea catharinensis, Melita nitida,* and *Paracaprella tenuis*; the polychaetes *Heteromastus filiformis, Nereis succinea, Paraprionospio pinnata, Sabellaria vulgaris,* and *Streblospio benedicti*; and the bivalve *Mulina lateralis*. At nontidal, freshwater riverine habitats, the dominant macroinvertebrate taxa observed were amphipods (mainly *Crangonyx* sp., *Gammarus* sp., and *Hyallela azteca*), isopods (*Asellus* sp.), oligochaetes, crayfish (Cambaridae undet.), and insects (Van Dolah, 2001).

Several species of bivalves, gastropods, and crustaceans proliferate in tidal (*Spartina alterniflora*) salt marshes of the ACE Basin. The ribbed mussel (*Geukensia demissa*) has a highly clustered distribution, commonly found clumped in masses on the marsh surface. Important gastropod species include periwinkles (*Littorina irrorata*) and mud snails (*Ilyanassa obsoleta*). Fiddler crabs (*Uca* spp.) are also abundant constituents of the salt marsh benthic macroinvertebrate community of the ACE Basin (Wenner, 2001b).

The eastern oyster (*Crassostrea virginica*), an epibenthic suspension-feeding bivalve of considerable commercial and recreational importance, represents an important habitat former in intertidal estuarine areas of the ACE Basin (Wenner et al., 1996). Oyster reefs provide habitat for numerous benthic organisms (e.g., polychaete worms), decapod crustaceans (*Eurypanopeus depressus, Panopeus herbstii,* and *Menippe mercenaria*), and fish populations (e.g., bay anchovy, *Anchoa mitchilli*; naked goby, *Gobiosoma bosci*; sheepshead, *Archosargus probatocephaluss*; and red drum, *Sciaenops ocellatus*). They occur at the mouths of some tidal creeks, reducing water flow and facilitating the deposition of suspended sediments and particulate organic matter. Organic matter accumulating on the reef surface serves as a food source for some of the reef inhabitants. Birds and mammals often visit the reef habitat to forage. Oyster beds also exist in other shallow subtidal areas of the estuary and can be seen lining banks in the system (Wenner, 2001b).

Several recreationally and commercially important shellfish species, in addition to *Crassostrea virginica,* also inhabit oyster reefs in the ACE Basin. Of particular note are blue crabs (*Callinectes sapidus*) and penaeid shrimp (i.e., brown shrimp, *Penaeus aztecus*; pink shrimp, *P. duorarum*; and white shrimp, *P. setiferus*). These decapod crustaceans are seasonally abundant in ACE Basin waters, often attaining high numbers in estuarine subtidal rivers and tidal creeks (Wenner et al., 1991). Penaeid shrimp, grass shrimp (*Palaemonetes* spp.), and blue crabs rank among the most abundant decapod crustaceans in the ACE Basin (Table 5.8). They are abundant in natural habitats as well as in impoundments (Wenner, 1986, 2001b).

Fish

Wenner (2001c) has examined the fish assemblages of the ACE Basin. Numerous species of fish utilize waters of the basin as feeding, spawning, and nursery areas (Wenner et al., 1991; Beasley et al., 1996). These species consist of residents, seasonal

TABLE 5.8
Abundance and Biomass of Decapod Crustaceans Collected by Trawl Net Sampling in the ACE Basin from August 1993 to December 1997

Scientific Name	Total Number	Weight (kg)
Penaeus setiferus	35,982	180.002
Penaeus aztecus	1,779	13.034
Callinectes sapidus	1,641	122.310
Palaemonetes vulgaris	1,416	0.666
Callinectes similis	920	10.304
Palaemonetes pugio	511	0.193
Trachypenaeus constrictus	336	0.386
Rhithropanopeus harrisii	289	0.107
Penaeus duorarum	99	0.315
Xiphopenaeus kroyeri	75	0.122
Pagurus longicarpus	56	0.007
Alpheus heterochaelis	48	0.122
Libinia dubia	28	0.653
Neopanope sayi	26	0.013
Panopeus herbstii	24	0.029
Clibanarius vittatus	19	0.018
Ovalipes ocellatus	18	0.165
Menippe mercenaria	12	0.216
Portunus spinicarpus	8	0.076
Portunus spinimanus	8	0.097
Libinia sp.	6	0.273
Xanthidae	5	0.004
Pagurus pollicaris	4	0.101
Cancer irroratus	4	0.064
Portunus gibbesi	2	0.003
Callinectes ornatus	1	0.005
Panopeus occidentalis	1	0.001
Portunus sp.	1	0.001

Source: South Carolina Department of Natural Resources and National Oceanic and Atmospheric Administration. 2001. *Characterization of the Ashepoo–Combahee–Edisto (ACE) Basin,* South Carolina. Special Scientific Report 17, South Carolina Marine Resources Center, Charleston, SC.

migrants, anadromous forms, and strays. Fish abundance and community structure vary substantially over seasonal as well as annual periodicities due to species migrations, differences in reproductive and juvenile recruitment success, and variable responses to changes in environmental conditions (Wenner and Sedberry, 1989). Species assemblages are characterized by the numerical dominance of relatively few species.

Fish surveys have been periodically conducted in ACE Basin estuarine waters during the past three decades; the most recent effort was a trawl survey initiated along salinity gradients of the Ashepoo, Combahee, and Edisto rivers in 1993. Earlier trawl sampling in the North and South Edisto Rivers during 1973–1975 revealed that sciaenids dominated the fish faunas; of greatest significance were the star drum (*Stellifer lanceolatus*) — the most abundant species — and the Atlantic croaker (*Micropogonias undulatus*), spot (*Leiostomus xanthurus*), and weakfish (*Cynoscion regalis*). Species of secondary importance included the bay anchovy (*Anchoa mitchilli*), white catfish (*Ictalurus catus*), Atlantic bumper (*Chloroscombrus chrysurus*), Atlantic menhaden (*Brevoortia tyrannus*), hogchoker (*Trinectes maculatus*), and spotted hake (*Urophycis regia*). These 10 species accounted for more than 70% of the total fish biomass and more than 90% of the total fish abundance (Wenner et al., 1991; Wenner, 2001c).

Trawl sampling during the 1993–1997 period resulted in the collection of 67 species of fish from the Edisto River. In comparison, trawl sampling during the same period in the Ashepoo and Combahee Rivers yielded 68 and 49 species of fish, respectively. Species richness was greatest at sampling sites in higher salinity areas near the river mouths. In total, nearly 55,000 fish were collected during the sampling period in these three rivers. The star drum, Atlantic croaker, and bay anchovy dominated the collections, comprising more than 68% of the total number of individuals. While the star drum dominated the collections in the South Edisto and Combahee Rivers, the Atlantic croaker was numerically dominant in the Ashepoo River (Wenner, 2001c).

Trammel netting and rotenone sampling have been employed to sample fish assemblages in tidal creeks and other shallow water habitats of the ACE Basin. Salt marsh creeks of the basin are major finfish nursery areas heavily utilized by both larval and juvenile fishes (Shenker and Dean, 1979; Bozeman and Dean, 1980). The most recent shallow water sampling was conducted by the SCDNR Marine Resources Research Institute in the Ashepoo, Combahee, and South Edisto rivers as well as at sites in Two Sisters/Rock Creek, with trammel netting deployed monthly from 1994 to 1997. Nearly 11,000 fish belonging to 53 species were collected during the survey period. The most abundant species included the hardhead catfish (*Arius felis*), spotted seatrout (*Cynoscion nebulosus*), red drum (*Sciaenops ocellatus*), striped mullet (*Mugil cephalus*), and spot; they comprised more than 80% of all individuals collected. The hardhead catfish was the most abundant species during the spring and summer, whereas spot and spotted seatrout predominated during the fall and winter. Striped mullet dominated catches at Two Sisters/Rock Creek (Wenner, 2001c).

The SCDNR Marine Resources Research Institute also conducted monthly rotenone sampling at St. Pierre Creek in the South Edisto River during 1987–1988. Nearly 47,000 fish belonging to 39 taxa were collected during the two-year survey period, with Atlantic menhaden and spot accounting for more than 75% of all individuals. The seasonally dominant species were as follows (Wenner, 2001c):

1. Atlantic menhaden (*Brevoortia tyrannus*) and spot (*Leiostomus xanthurus*) in spring
2. Atlantic silverside (*Menidia menidia*) and mummichog (*Fundulus heteroclitus*) in summer
3. Atlantic silverside in fall
4. Spot in winter

The large number of individuals and taxa collected by the aforementioned rotenone and trammel net sampling surveys reflects the importance of tidal creeks and other shallow water habitats to fish communities in the ACE Basin.

It is important to note that the marsh surface also constitutes valuable fish habitat, especially for larval and juvenile stages. Among the most important species of fish inhabiting the marsh surface are gobies and killifishes (Weisberg et al., 1981; Kneib, 1986; Jackson, 1990). Other species encountered here include spotted seatrout, red drum, spot, striped mullet, southern flounder, and silver perch (*Bairdiella chrysoura*) (Hettler, 1989; Wenner, 2001c). Wenner et al. (1986) and McGovern and Wenner (1990) have shown that the striped mullet, inland silverside (*Menidia beryllina*), and ladyfish (*Elops saurus*) dominate impounded marshes in the ACE Basin.

A number of fish species found in ACE Basin waters are anadromous forms that pass through estuarine areas to riverine habitats during spawning migrations. Included here are the American shad (*Alosa sapidissima*), hickory shad (*A. mediocris*), blueback herring (*A. aestivalis*), Atlantic sturgeon (*Acipenser oxyrhynchus*), shortnose sturgeon (*A. brevirostrum*), and striped bass (*Morone saxatilis*). The blueback herring is most numerous. While adults spawn at upriver sites, juveniles occupy estuarine nursery grounds. Although early life history stages of these anadromous species occur in freshwater riverine and estuarine areas of the ACE Basin, most of their lives are spent in coastal marine waters (Rulifson et al., 1982).

Fish assemblages in blackwater areas of streams and rivers, above the influence of tides, have been surveyed mainly via electrofishing and rotenone sampling. Electrofishing has been conducted in streams, and rotenone sampling, in segments of major rivers. Most of the species collected belong to the Centrarchidae (sunfishes: e.g., redbreast sunfish, *Lepomis auritus*; bluegill, *L. macrochirus*; mud sunfish, *Acantharchus pomotis*; bluespotted sunfish, *Enneacanthus gloriosus*; and blackbanded sunfish, *E. chaetodon*), Ictaluridae (bullheads and madtoms: e.g., brown bullhead, *Ameiurus nebulosus*; yellow bullhead, *A. natalis*; flat bullhead, *A. platycephalus*; tadpole madtom, *Noturus gyrinus*; and margined madtom, *Noturus insignis*); Cyprinidae (minnows and carp: e.g., eastern mudminnow, *Umbra pygmaea*; and common carp, *Cyprinus carpio*); and Percidae (darters: e.g., tessellated darter, *Etheostoma olmstedi*). In freshwater areas of the Edisto River Basin, 87 species of fish belonging to 25 families have been collected since the mid-1960s. Sunfishes, minnows, and carps have dominated electrofishing collections. Sunfishes, bullheads, and suckers (e.g., spotted sucker, *Minytrema melanops*), in turn, have dominated the rotenone collections. The American eel (*Anguilla rostrata*), a catadromous species, has been documented in both tidal and nontidal freshwater habitats of the ACE Basin (Thomason et al., 1993; Marcy and O'Brien-White, 1995; Wenner, 2001c).

COASTAL MARINE WATERS

Animal Communities

Fish

Many coastal marine fish migrate into estuaries to spawn, forage, or use nursery areas. The coastal marine fish community consists of year-round residents as well

TABLE 5.9
Species of Fish Collected in the Nearshore Coastal Zone by the Southeast Area Monitoring and Assessment Program

Species	Total Number	Weight (kg)
Leiostomus xanthurus	707,212	44,870.22
Micropogonias undulatus	579,578	34,519.59
Chloroscombrus chrysurus	159,514	3,229.46
Stenotomus sp.	154,304	7,830.07
Anchoa hepsetus	148,610	1,029.87
Lagodon rhomboides	103,047	5,869.37
Stellifer lanceolatus	82,835	1,298.21
Larimus fasciatus	78,910	4,992.59
Cynoscion nothus	8,203	3,310.77
Menticirrhus americanus	73,320	7,596.82

Source: South Carolina Department of Natural Resources and National Oceanic and Atmospheric Administration. 2001. *Characterization of the Ashepoo–Combahee–Edisto (ACE) Basin, South Carolina*. Special Scientific Report 17, South Carolina Marine Resources Center, Charleston, SC.

as various migrant species. In a study of coastal trawl collections from Cape Fear, North Carolina to St. John's River, Florida, Wenner and Sedberry (1989) reported that drums (sciaenids) comprised most of the total abundance (56%) and biomass (66%) of the fish samples, reflecting migration and juvenile recruitment patterns. The Atlantic croaker and spot dominated the collections during all seasons. In winter and spring, the Atlantic menhaden was also numerically abundant. The southern kingfish (*Menticirrhus americanus*), Atlantic croaker, and spot contributed much of the total biomass. The Southeast Area Monitoring and Assessment Program reported that the spot and the Atlantic croaker numerically dominated fish collections in the nearshore coastal zone during the 1990s (Table 5.9) (Beatty and Boylan, 1997). The striped anchovy (*Anchoa hepsetus*), pinfish (*Lagodon rhomboides*), scup (*Stenotomus* spp.), Atlantic bumper, and star drum were other abundant species (Wenner, 2001c). Along sandy beach nearshore habitats, the pompano (*Trachinotus carolinus*), Gulf whiting (*Menticirrhus littoralis*), white mullet (*Mugil curema*), bay anchovy, and Atlantic silverside attained greatest abundance (Anderson et al., 1977; Delancey, 1984).

Reptiles

Several sea turtles inhabit coastal marine waters seaward of the ACE Basin. These include the loggerhead (*Carretta caretta*), leatherback (*Dermochelys coriacea*), green sea turtle (*Chelonia mydas*), Kemp's Ridley sea turtle (*Lepidochelys kempii*), and hawksbill turtle (*Eretmochelys imbricata*). Of these species, only the endangered loggerhead turtle enters ACE Basin estuarine waters on a consistent basis. However, ongoing development and other human activities pose a threat to this species by degrading nesting habitat and increasing mortality via commercial trawling (Riekerk and Rhodes, 2001).

Mammals

The bottlenose dolphin (*Tursiops turncatus*) and West Indian manatee (*Trichechus manatus*) are two resident species of marine mammals found in coastal waters offshore of the ACE Basin. There are also several species of whales (e.g., killer whale, *Orcinus orca*; pygmy whale, *Foresa attenuata*; and sperm whale, *Physeter macrocephalus*) and other dolphins (e.g., Atlantic spotted dolphin, *Stenella frontalis*; and striped dolphin, *S. coeruleoalba*) occurring in these waters. Bottlenose dolphins (*Tursiops* spp.) are carnivores preying on shellfish (crabs and shrimp), fish, and squid. West Indian manatees, in contrast, are herbivores, feeding on *Spartina alterniflora* and other plants (Zimmerman, 2001a).

Birds

The avifauna most frequently observed in coastal marine habitats are piscivorous forms that prey on fish, although other feeding types are also common (Zimmerman, 2001b). Examples include cormorants (double-crested cormorant, *Phalacrocorax auritus*), loons (common loon, *Gavia immer*; and red-throated loon, *G. stellata*), gulls (herring gull, *Larus argentatus*; laughing gull, *L. atricella*; ring-billed gull, *L. delawarensis*; great black-backed gull, *L. marinus*; and Bonaparte's gull, *L. philadelphia*), skimmers (black skimmer, *Rynchops niger*), terns (common tern, *Sterna hirundo*; Forster's tern, *S. forsteri*; least tern, *S. antillarum*; royal tern, *S. maxima*; sandwich tern, *S. sandvicensis*; Caspian tern, *S. caspia*; and black tern, *Chlidonias niger*), gannets (northern gannet, *Morus bassanus*), and pelicans (eastern brown pelican, *Pelecanus occidentalis*). Grebes (horned grebe, *Podiceps auritus*), petrels (Wilson's petrel, *Oceanites oceanicus*), scoters (black scoter, *Melanitta nigra*; and surf scoter, *M. perspicillata*), and various waterfowl (ruddy duck, *Oxyura jamaicensis*; canvasback, *Aythya valisineria*; and lesser scaup, *A. affinis*) also utilize this habitat (Potter et al., 1980; Sandifer et al., 1980).

ENDANGERED AND THREATENED SPECIES

Several plant and animal species inhabiting the ACE Basin are classified as endangered or threatened (Riekerk et al., 2001). Among the federally endangered species are Canby's dropwort (*Oxypolyis canbyi*), American peregrine falcon (*Falco peregrinus anatum*), wood stork (*Mycteria americana*), red-cockaded woodpecker (*Picoides borealis*), shortnose sturgeon (*Acipenser brevirostrum*), and West Indian manatee (*Trichechus manatus*). The least tern (*Sterna antillarum*) is a state endangered species. The federally threatened species include the American alligator (*Alligator mississippiensis*), loggerhead sea turtle (*Caretta caretta*), and bald eagle

(*Haliaeetus leuocephalus*) (Riekerk et al., 2001). Endangered and threatened species have been defined by SCDNR and NOAA (2001) as follows:

- *Endangered Species* — "Any species that is in danger of extinction through all or a significant portion of its range; a species of native fish, wildlife, or plants threatened with extinction because its habitat is threatened with destruction, drastic modification, or severe curtailment, or because of over-exploitation, disease, predation, or other factors affecting its survival."
- *Threatened Species* — "Any species which is likely to become an endangered species within the foreseeable future throughout all or a significant portion of its range and which has been designated in the Federal Registry by the Secretary of Interior as a threatened species."

It is critically important to provide protection for these species against anthropogenic impacts. However, it must also be noted that threats to some of these species derive from natural events as well. Hence, efforts are ongoing to improve environmental conditions in the ACE Basin, thereby making habitats in the system more favorable for the long-term success of these valuable species.

ANTHROPOGENIC IMPACTS

Most estuaries in the U.S. and many abroad are affected in some way by anthropogenic activities either in adjoining coastal watersheds or on the water bodies themselves. The impacts are numerous and varied, including point and nonpoint source pollution, pathogen inputs, water quality degradation, habitat loss and alteration, nutrient overenrichment, organic loading, chemical contaminant accumulation, overfishing, freshwater diversions, introduced species, coastal subsidence (associated with groundwater, oil, and gas withdrawal), watercraft effects, electric generating station effects (thermal discharges as well as entrainment and impingement of organisms), and litter (Kennish, 1992, 1997, 2000, 2001). These impacts can alter biotic communities by directly disrupting constituent populations or by altering estuarine habitats that support them. Due to multiple environmental threats, the long-term outlook for these vital coastal ecosystems is tenuous (Kennish, 2002).

The principal sources of land-based impacts on the ACE Basin system are residential and commercial development, agriculture, and silviculture (SCDNR and NOAA, 2001). Land cover studies indicate that 56% of the ACE Basin study area consists of timberland (Figure 5.2) (Connor, 1993). Another 12% of the land cover is farmland, and urban development accounts for ~2% of the area (USEPA, 1997). According to Mathews and Sanger (2001), only 13 National Pollution Discharge Elimination System (NPDES) permits have been issued for point source discharges in the ACE Basin (e.g., the Yamassee wastewater treatment facility, the Walterboro City Wastewater Treatment Facility, and the SCE&G Canadys Power Station). Of even greater concern is nonpoint source pollution associated with agricultural production and runoff as well as clearcutting of hardwood timber.

There are several potentially significant impacts attributable to agricultural production in the ACE Basin. First, heavy irrigation can deplete surface water flows leading to lower water levels in area streams, shifts in salinity levels in tidal reaches, and the loss of habitat for aquatic organisms. The use of pesticides and other chemical treatments to enhance agricultural production may result in unintentional impacts on nontarget organisms in receiving water bodies (Scott et al., 1994). Other contaminants can also enter these waters from fertilizer applications on lawns and golf courses, as well as runoff from impervious surfaces in developed areas. Inputs of nutrient elements, particularly nitrogen, and animal waste from farmlands raise eutrophication concerns.

Soil cultivation and construction often hasten erosion and sediment load in surface waters (Wenner, 2001d). The potential also exists for accelerated erosion and sediment loading to surface waters due to hardwood clearing and logging. Associated impacts include habitat loss and water quality degradation. In Colleton County, however, sustainable forestry management practices are carefully followed; for example, the management of forest plantations involves rotation cycles that promote long-term maintenance of the resource. The focus is on conservation of the resource and protection of the environment using best management practices (Hunter, 1990; Meffe and Carroll, 1994; Allen et al., 1996).

Scott et al. (1998) reported on sediment chemical contamination in the ACE Basin NERR during the 1994 to 1996 period. Sediment samples taken at eight sites in each of the major rivers (i.e., Ashepoo, Combahee, and Edisto rivers), three sites in St. Helena Sound, and seven sites in small tidal creeks and the Intracoastal Waterway were analyzed for trace metals and organic contaminants (PCBs, organochlorine pesticides, and polycyclic aromatic hydrocarbons [PAHs]). The concentrations of both trace metals and organic contaminants were shown to be low. Of the trace metals, only arsenic was present at levels that could possibly cause adverse biological effects. The input of PCBs, organochlorine pesticides, and PAHs appears to be minimal; the concentrations of all of these contaminants are well below the thresholds where adverse biological effects would be observed based on effects range–low and effects range–median criteria (Long et al., 1995). However, in another study, Marcus and Mathews (1987) determined that sediments in Campbell Creek near a chemical plant had PCB concentrations greater than 24,000 ppb. The extremely high concentrations of PCBs at this "hot-spot" location indicate that adverse biological effects are probable in organisms exposed to the sediment. Despite this impacted site, the overall results of the two studies show that the ACE Basin remains a generally pristine system with respect to chemical contaminants, reflecting in part the relatively limited amount of development in neighboring watershed areas (Mathews and Sanger, 2001).

Other anthropogenic activities that have potentially significant environmental impacts are ditching and impounding of wetlands. These land changes can cause the loss or alteration of substantial habitat area in the system. However, impoundments in the ACE Basin also support rich and diverse communities of organisms (Wenner, 2001b, c).

Introduced species have also had an impact on some organisms in the ACE Basin (Wenner, 2001c). For example, Allen (1997) revealed that introduction of the nonendemic flathead catfish (*Ictalurus furcatus*) to the Edisto River has caused depletion of the redbreast sunfish population. Abundance of native bullhead catfish (*Ameiurus* spp.) has also declined in areas where the flathead catfish population is flourishing. Such changes in the composition of the fish community can contribute to shifts in the food web structure and other alterations of biotic communities in the river, and thus must be carefully monitored.

SUMMARY AND CONCLUSIONS

The ACE Basin NERR consists of more than 50,000 ha of diverse coastal plain habitats, including pine-mixed hardwoods, forested wetlands, maritime forests, brackish marshes, salt marshes, marsh islands, tidal creeks, tidal flats, barrier beaches and dunes, and open riverine and estuarine waters. Although residential and urban land use in the ACE Basin area has increased in recent years, the reserve habitats remain relatively pristine, and management programs are in place under direction of the SCDNR to protect biotic communities and ensure sustainability of resources. The ACE Basin NERR encompasses much of the Ashepoo, Combahee, and Edisto river systems, which support a wide array of aquatic and terrestrial organisms. An abundance of clams, oysters, shrimps, crabs, and finfishes in estuarine waters provides bountiful catches for commercial and recreational fishermen and shellfishermen. A number of endangered and threatened species, such as the American alligator (*Alligator mississippiensis*), bald eagle (*Haliaeetus leucocephalus*), loggerhead sea turtle (*Caretta caretta*), and shortnose sturgeon (*Acipensier brevirostrum*), occur in the basin.

Most of the land cover in the ACE Basin area consists of forested habitat (56% of the total area), with nonforested wetlands (17%), agricultural rangelands (12%), and open water (12%) accounting for much of the remaining land use cover (SCDNR and NOAA, 2001). The harvesting of timber in the ACE Basin NERR is a major industry of economic importance for South Carolina. Sustainable forest management practices have been implemented to promote conservation of the resource.

Numerous species of amphibians, reptiles, mammals, birds, and insects reside in watershed habitats. As many as 110 species of amphibians and reptiles, nearly 50 species of mammals, and almost 300 species of birds are likely to inhabit the ACE Basin region. Up to 10,000 species of insects are estimated to occur there.

Estuarine waters harbor robust communities of phytoplankton, zooplankton, benthic invertebrates, and fish. Tidal creeks are important nursery areas for many species of fish. Several anadromous forms (American shad, *Alosa sapidissima*; hickory shad, *A. mediocris*; blueback herring, *A. aestivalis*; shortnose sturgeon, *Acipenser brevirostrum*; and striped bass, *Morone saxatilis*) utilize estuarine waters as a pathway during spawning runs to freshwater sites upriver.

Human activities are a potential threat to biotic communities in watershed and estuarine habitats. The primary anthropogenic impacts of concern are associated with residential and commercial development, agriculture, and silviculture. Ditching and impounding of wetlands, the introduction of nonendemic species, and the removal of surface waters for heavy irrigation of crops can adversely affect various components of the ACE Basin system. Chemical contaminant accumulation in the basin is generally low, and hence much of the estuarine habitat remains relatively pristine with respect to toxic substances.

REFERENCES

Allen, A.W., Y.K. Bernal, and R.J. Moulton. 1996. Pine plantations and wildlife in the southeastern United States: An Assessment of Impacts and Opportunities. Information and Technology Report No. 3, U.S. Department of the Interior, National Biological Service, Washington, D.C.

Allen, D.E. 1997. Flathead Catfish Investigations in the Edisto River: Fisheries Investigations in Lakes and Streams. Project No. F-63-2-6, Freshwater Fisheries District VI, Columbia, SC.

Anderson, W.D., J.K. Dias, D.M. Cupka, and N.A. Chamberlain. 1977. The Macrofauna of the Surf Zone off Folly Beach, South Carolina. Technical Report NMFS SSRF-704, National Oceanic and Atmospheric Administration, Washington, D.C.

Andre, J.B. 1981. Habitat use and relative abundance of the small mammals of a South Carolina barrier island. *Brimleyana* 5: 129–134.

Badr, B. and L. Zimmerman. 2001. Surface water. In: South Carolina Department of Natural Resources and National Oceanic and Atmospheric Administration. *Characterization of the Ashepoo-Combahee-Edisto (ACE) Basin, South Carolina.* Special Scientific Report 17, South Carolina Marine Resources Center, Charleston, SC.

Baker, O.E., III and D.B. Carmichael, Jr. 1996. South Carolina's Furbearers. Technical Report, Department of Natural Resources, Division of Wildlife and Freshwater Fisheries, Charleston, SC.

Beasley, B.R., W.D. Marshall, A.H. Miglarese, J.D. Scurry, and C. Van den Houten. 1996. Managing Resources for a Sustainable Future: The Edisto River Basin Project Report. Report 12, South Carolina Department of Natural Resources, Water Resources Division, Charleston, SC.

Beatty, H.R. and J.M. Boylan. 1997. Results of the Southeast Area Monitoring and Assessment Program — South Atlantic (SEAMAP-SA) Nearshore, Day/Night Trawl Sampling in the Coastal Habitat of the South Atlantic Bight during 1987 and 1988. Special Report 65, Atlantic States Marine Fisheries Commission, Washington, D.C.

Bell, S.S. 1979. Short- and long-term variation in a high marsh meiofauna community. *Estuarine and Coastal Marine Science* 9: 331–342.

Bell, S.S. and B.C. Coull. 1978. Field evidence that shrimp predation regulates meiofauna. *Oecologia* 35: 141–148.

Bell, S.S., M.C. Watzin, and B.C. Coull. 1978. Biogenic structure and its effect on the spatial heterogeneity of meiofauna in a salt marsh. *Journal of Experimental Marine Biology and Ecology* 35: 99–107.

Bennett, S.H. and J.B. Nelson. 1991. Distribution and Status of Carolina Bays in South Carolina. Publication No. 1, South Carolina Wildlife and Marine Resources Department, Columbia, SC.

Blaustein, A.R., D.B. Wake, and W.P. Sousa. 1994. Amphibian declines: judging stability, persistence, and susceptibility of populations to local and global extinctions. *Conservation Biology* 8: 60–71.

Bloxham, W.M. 1979. Low-flow Frequency and Flow-duration of South Carolina Streams. Report No. 11, South Carolina Water Resources Commission, Columbia, SC.

Bloxham, W.M. 1981. Low-flow Characterization of Ungaged Streams in the Piedmont and Lower Coastal Plain in South Carolina. Report No. 14, South Carolina Water Resources Commission, Columbia, SC.

Bozeman, E.L., Jr. and J.M. Dean. 1980. The abundance of estuarine larval and juvenile fish in a South Carolina intertidal creek. *Estuaries* 3: 89–97.

Calder, D.R. and B.B. Boothe, Jr. 1977a. Some Subtidal Epifaunal Assemblages in South Carolina Estuaries. Data Report 4, South Carolina Marine Resources Center, South Carolina Wildlife and Marine Resources Department, Charleston, SC.

Calder, D.R. and B.B. Boothe, Jr. 1977b. Data from Some Subtidal Quantitative Benthic Samples Taken in Estuaries of South Carolina. Data Report 3, South Carolina Marine Resources Center, South Carolina Wildlife and Marine Resources Department, Charleston, SC.

Calder, D.R., B.B. Boothe, Jr., and M.S. Maclin. 1977. A Preliminary Report on Estuarine Macrobenthos of the Edisto and Santee River Systems, South Carolina. Technical Report 22, South Carolina Marine Resources Center, South Carolina Wildlife and Marine Resources Department, Charleston, SC.

Colquhoun, D.W. 1969. Geomorphology of the Lower Coastal Plain of South Carolina. Technical Report, Division of the Geology State Development Board, Columbia, SC.

Conant, R. and J.T. Collins. 1998. *A Field Guide to Reptiles and Amphibians of Eastern and Central North America.* 3rd ed. Houghton Mifflin, New York.

Conner, R.C. 1993. Forest Statistics for South Carolina, 1993. Resource Bulletin SE-141, U.S. Department of Agriculture, Forest Service, Southeastern Forest Experiment Station, Asheville, NC.

Cooney, T.W., K.H. Jones, P.A. Drewes, J.W. Gissendanner, and B.W. Church. 1998. Water Resources Data, South Carolina, Water Year 1997. Technical Report, U.S. Geological Survey, Columbia, SC.

Coull, B.C. 1990. Are members of the meiofauna food for higher trophic levels? *Transactions of the American Microscopical Society* 109: 233–246.

Coull, B.C. and S.S. Bell. 1979. Perspectives of marine meiofaunal ecology. In: Livingston, R.J. (Ed.). *Ecological Processes in Coastal Marine Systems.* Plenum Press, New York, pp. 189–207.

Coull, B.C., S.S. Bell, A.M. Savory, and B.W. Dudley. 1979. Zonation of meiobenthic copepods in a southeastern United States marsh. *Estuarine and Coastal Marine Science* 9: 181–188.

Coull, B.C. and B.W. Dudley. 1985. Dynamics of meiobenthic copepod populations: a long-term study (1973–1983). *Marine Ecology Progress Series* 23: 219–229.

Davis, L.V. 1978. Class Insecta. In: Zingmark, R.G. (Ed.). *An Annotated Checklist of the Biota of the Coastal Zone of South Carolina.* University of South Carolina Press, Columbia, SC, pp. 186–220.

Davis, L.V. and I.E. Gray. 1966. Zonal and seasonal distribution of insects in North Carolina salt marshes. *Ecological Monographs* 36: 275–295.

Davis, R.B. and R.F. Van Dolah (Eds.). 1992. Characterization of the Physical, Chemical, and Biological Conditions and Trends in Three South Carolina Estuaries 1970–1985. Volume 1: Charleston Harbor Estuary. Technical Report, South Carolina Sea Grant Consortium, Charleston, SC.

Dawes, C.J. 1998. *Marine Botany,* 2nd ed. John Wiley & Sons, New York.

Delancey, L.B. 1984. An Ecological Study of the Surf Zone at Folly Beach, South Carolina. M.S. thesis, College of Charleston, Charleston, SC.

Eidson, J.P. 1993. Water quality. In: *Assessing Change in the Edisto River Basin: An Ecological Characterization.* Report No. 177, South Carolina Water Resources Commission, Columbia, SC.

Gatrelle, R.R. 1975. The Hesperioidea of the south coastal area of South Carolina. *Journal of the Lepidopterists' Society* 29: 56–59.

Gibbons, J.W. 1978. Reptiles. In: Zingmark, R.G. (Ed.). *An Annotated Checklist of the Biota of the Coastal Zone of South Carolina.* University of South Carolina Press, Columbia, SC, pp. 270–276.

Gibbons, J.W. and J.W. Coker. 1978. Herpetofaunal colonization patterns of Atlantic coast barrier islands. *American Midland Naturalist* 99: 212–233.

Gibbons, J.W. and J.R. Harrison. 1981. Reptiles and amphibians of Kiawah and Capers Islands, South Carolina. *Brimleyana* 5: 145–162.

Hettler, W.F., Jr. 1989. Nekton use of regularly flooded saltmarsh cordgrass habitat in North Carolina, USA. *Marine Ecology Progress Series* 56: 111–118.

Hunter, M.L. 1990. *Wildlife, Forests, and Forestry: Principles of Managing Forests for Biodiversity.* Prentice-Hall, Englewood Cliffs, NJ.

Hyland, J.L., T.J. Herrlinger, T.R. Snoots, A.H. Ringwood, R.F. Van Dolah, C.T. Hackney, G.A. Nelson, J.S. Rosen, and S.A. Kokkinakis. 1996. Environmental Quality of Estuaries of the Carolinian Province: 1994. Annual Statistical Summary for the 1994 EMAP–Estuaries Demonstration Project in the Carolinian Province. NOAA Technical Memorandum NOS ORCA 97, NOAA/NOS, Office of Ocean Resources Conservation and Assessment, Silver Spring, MD.

Hyland, J.L., L. Balthis, C.T. Hackney, G. McRae, A.H. Ringwood, T.R. Snoots, R.F. Van Dolah, and T.L. Wade. 1998. Environmental Quality of Estuaries of the Carolinian Province: 1995. Annual Statistical Summary for the 1995 EMAP–Estuaries Demonstration Project in the Carolinian Province. NOAA Technical Memorandum NOS ORCA 123, NOAA/NOS, Office of Ocean Resources Conservation and Assessment, Silver Spring, MD.

Jackson, C.D. 1990. Recruitment of Fish Larvae to Shallow Estuarine Habitats along a Salinity Gradient. M.S. thesis, College of Charleston, Charleston, SC.

Kennish, M.J. 1986. *Ecology of Estuaries,* Vol. 2, *Biological Effects.* CRC Press, Boca Raton, FL.

Kennish, M.J. 1992. *Ecology of Estuaries: Anthropogenic Effects.* CRC Press, Boca Raton, FL.

Kennish, M.J. (Ed.). 1997. *Practical Handbook of Estuarine and Marine Pollution.* CRC Press, Boca Raton, FL.

Kennish, M.J. (Ed.). 2000. *Estuary Restoration and Maintenance: The National Estuary Program.* CRC Press, Boca Raton, FL.

Kennish, M.J. (Ed.) 2001. *Practical Handbook of Marine Science,* 3rd ed. CRC Press, Boca Raton, FL.

Kennish, M.J. 2002. Environmental threats and environmental future of estuaries. *Environmental Conservation* 29: 78–107.

Kneib, R.T. 1986. Size-specific patterns in the reproductive cycle of the killifish, *Fundulus heteroclitus*, (Pisces: Fundulidae) from two South Carolina, USA salt marshes. *Wetlands* 17: 65–81.

Knott, D.M. 1980. The Zooplankton of the North Edisto River and Two Artificial Saltwater Impoundments. M.S. thesis, College of Charleston, Charleston, SC.

Knott, D.M. 2001. Zooplankton. In: South Carolina Department of Natural Resources and National Oceanic and Atmospheric Administration, *Characterization of the Ashepoo–Combahee–Edisto (ACE) Basin, South Carolina*. Special Scientific Report 17, South Carolina Marine Resources Center, Charleston, SC.

Koontz, B.L. and R.M. Sheffield. 1993. Forest Statistics for the Southern Coastal Plain of South Carolina, 1993. Resource Bulletin SE-140, U.S. Department of Agriculture, Forest Service, Southeastern Forest Experiment Station, Asheville, NC.

Kraeuter, J.N. and P.L. Wolf. 1974. The relationships of marine macroinvertebrates to salt marsh plants. In: Reimhold, R.J. and W.H. Queen (Eds.). *Ecology of Halophytes*. Academic Press, New York, pp. 449–462.

Levinton, J.S. 1982. *Marine Ecology*. Prentice-Hall, Englewood Cliffs, NJ.

Levinton, J.S. 1995. *Marine Biology: Function, Biodiversity, Ecology*. Oxford University Press, New York.

Lewitus, A.J., E.T. Koepfler, and J.T. Morris. 1998. Seasonal variation in the regulation of phytoplankton by nitrogen and grazing in a salt marsh estuary. *Limnology and Oceanography* 43: 636–646.

Long, E.R., D.R. MacDonald, S.L. Smith, and F.D. Calder. 1995. Incidence of adverse biological effects within ranges of chemical concentrations in marine and estuarine sediments. *Environmental Management* 19: 81–97.

Marcus, J.M. and T.D. Mathews. 1987. Polychlorinated biphenyls in blue crabs from South Carolina. *Bulletin of Environmental Contamination and Toxicology* 39: 857–862.

Marcy, B.C., Jr. and S.K. O'Brien-White (Eds.). 1995. Fishes of the Edisto River Basin, South Carolina. Edisto River Basin Project Report 6, South Carolina Department of Natural Resources, Water Resources Division, Columbia, SC.

Mathews, T. and D. Sanger. 2001. Hydrochemistry and pollution. In: South Carolina Department of Natural Resources and National Oceanic and Atmospheric Administration, *Characterization of the Ashepoo–Combahee–Edisto (ACE) Basin, South Carolina*. Special Scientific Report 17, South Carolina Marine Resources Center, Charleston, SC.

Mathews, T.D., F.W. Stapor, Jr., C.R. Richter, J.V. Miglarese, M.D. McKenzie, and L.A. Barclay (Eds.). 1980. Ecological Characterization of the Sea Island Coastal Region of South Carolina and Georgia, Volume 1: Physical Features of the Characterization Area. FWS/OBS-79/40, U.S. Fish and Wildlife Service, Office of Biological Services, Washington, D.C.

McGovern, J.C. and C.A. Wenner. 1990. Seasonal recruitment of larval and juvenile fishes into impounded and non-impounded marshes. *Wetlands* 10: 203–221.

McIntyre, M.P., H.P. Eilers, and J.W. Mairs. 1991. *Physical Geography*. John Wiley & Sons, New York.

Meffe, G.K. and C.R. Carroll. 1994. *Principles of Conservation Biology*. Sinauer Associates, Sunderland, MA.

Mengak, M.T., D.C. Gwynn, Jr., J.K. Edwards, D.L. Sanders, and S.M. Miller. 1987. Abundance and distribution of shrews in western South Carolina. *Brimleyana* 13: 63–66.

National Oceanic and Atmospheric Administration. 1996. Estuarine Eutrophication Survey, Volume 1: South Atlantic Region. Technical Report, Office of Ocean Resources Conservation Assessment, Silver Spring, MD.

Olsen, P.S. and J.B. Mahoney. 2001. Phytoplankton in the Barnegat Bay–Little Egg Harbor estuarine system: species composition and picoplankton bloom development. *Journal of Coastal Research* Special Issue 32, pp. 115–143.

Omori, M. and T. Ikeda. 1984. *Methods in Marine Zooplankton Ecology*. John Wiley & Sons, New York.

Opler, P.A. and V. Malikul. 1998. *A Field Guide to Eastern Butterflies*. Houghton Mifflin, Boston.

Orlando, S.P., Jr., P.H. Wendt, C.J. Klein, M.E. Pattillo, K.C. Dennis, and G.H. Ward. 1994. Salinity Characteristics of South Atlantic Estuaries. Technical Report, Office of Ocean Resources Conservation and Assessment, National Oceanic and Atmospheric Administration, Silver Spring, MD.

Park, D. 2001. Ground water. In: South Carolina Department of Natural Resources and National Oceanic and Atmospheric Administration, *Characterization of the Ashepoo–Combahee–Edisto (ACE) Basin, South Carolina*. Special Scientific Report 17, South Carolina Marine Resources Center, Charleston, SC.

Pechmann, J.H.K., D.E. Scott, R.D. Semlitsch, J.P. Caldwell, L.J. Vitt, and J.W. Gibbons. 1991. Declining amphibian populations: the problem of separating human impacts from natural fluctuations. *Science* 253: 892–895.

Pfeiffer, W.J. and R.G. Wiegert. 1981. Grazers on *Spartina* and their predators. In: Pomeroy, L.R. and R.G. Wiegert (Eds.). *The Ecology of a Salt Marsh*. Springer-Verlag, New York, pp. 87–112.

Potter, E.F., J.F. Parnell, and R.P. Teulings. 1980. *Birds of the Carolinas*. University of North Carolina Press, Chapel Hill, NC.

Price, P.W. 1997. *Insect Ecology*, 3rd ed. John Wiley & Sons, New York.

Riekerk, G. 2001. Geomorphology. In: South Carolina Department of Natural Resources and National Oceanic and Atmospheric Administration, *Characterization of the Ashepoo–Combahee–Edisto (ACE) Basin, South Carolina*. Special Scientific Report 17, South Carolina Marine Resources Center, Charleston, SC.

Riekerk, G. and W. Rhodes. 2001. Herpetofauna. In: South Carolina Department of Natural Resources and National Oceanic and Atmospheric Administration, *Characterization of the Ashepoo–Combahee–Edisto (ACE) Basin, South Carolina*. Special Scientific Report 17, South Carolina Marine Resources Center, Charleston, SC.

Riekerk, G., S. Upchurch, M. Brouwer, A. Sievert, C. Layman, and T. Marcinko. 2001. Threatened and Endangered Species. In: South Carolina Department of Natural Resources and National Oceanic and Atmospheric Administration, *Characterization of the Ashepoo–Combahee–Edisto (ACE) Basin, South Carolina*. Special Scientific Report 17, South Carolina Marine Resources Center, Charleston, SC.

Rulifson, R.G., M.T. Huish, and R.W. Thoesen. 1982. Anadromous Fish in the Southeastern United States and Recommendations for Development of a Management Plan. Technical Report, U.S. Fish and Wildlife Service, Fishery Resources, Region 4, Atlanta, GA.

Sandifer, P.A., J.V. Miglarese, D.R. Calder, J.J. Manzi, and L.A. Barclay. 1980. Ecological Characteristics of the Sea Island Coastal Region of South Carolina and Georgia, Vol. 3: Biological Features of the Characterization Area. FWS/OBS-79/42, U.S. Fish and Wildlife Service, Office of Biological Services, Washington, D.C.

Scholtens, B. 2001. Insects. In: South Carolina Department of Natural Resources and National Oceanic and Atmospheric Administration, *Characterization of the Ashepoo—Combahee—Edisto (ACE) Basin, South Carolina*. Special Scientific Report 17, South Carolina Marine Resources Center, Charleston, SC.

Scott, G.I., M.H. Fulton, D.W. Moore, G.T. Chandler, P.B. Key, T.W. Hampton, J.M. Marcus, K.L. Jackson, D.S. Baughman, A.H. Trim, L. Williams, C.J. Louden, and E.R. Patterson. 1994. Agricultural Insecticide Runoff Effects on Estuarine Organisms: Correlating Laboratory and Field Toxicity Testing, Ecophysiology Assays, and Ecotoxicological Biomonitoring. Report EPA/600/R-94/004; Report PB94–160678, University of South Carolina, School of Public Health, Columbia, SC.

Scott, G.I., M.H. Fulton, D. Bearden, M. Sanders, A. Dias, L.A. Reed, S. Sivertsen, E.D. Strozier, P.B. Jenkins, J.W. Daugomah, J. De Vane, P.B. Key, and W. Ellenberg. 1998. Chemical Contaminant Levels in Estuarine Sediment of the Ashepoo–Combahee–Edisto River (ACE) Basin National Estuarine Research Reserve and Sanctuary Site. Technical Report, National Oceanic and Atmospheric Administration, Charleston, SC.

Shenker, J.M. and J.M. Dean. 1979. The utilization of an intertidal salt marsh creek by larval and juvenile fishes: abundance, diversity, and temporal variation. *Estuaries* 2: 154–163.

Smalley, A.E. 1980. Energy flow of a salt marsh grasshopper population. *Ecology* 41: 672–685.

Soller, D.R. and H.H. Mills. 1991. Surficial geology and geomorphology. In: Horton, J.W. and V.A. Zullo (Eds.). *The Geology of the Carolinas*. The 50th Anniversary Volume, Carolina Geological Society, University of Tennessee Press, Knoxville, TN.

Solomon, E.P., L.R. Berg, and D.W. Martin. 1999. *Biology,* 5th ed. Saunders College Publishing, Philadelphia.

South Carolina Department of Natural Resources and National Oceanic and Atmospheric Administration. 2001. *Characterization of the Ashepoo–Combahee–Edisto (ACE) Basin, South Carolina*. Special Scientific Report 17, South Carolina Marine Resources Center, Charleston, SC.

South Carolina Water Resources Commission. 1983. South Carolina State Water Assessment. Technical Report No. 140, South Carolina Water Resources Commission, Columbia, SC.

Teal, J.M. 1962. Energy flow in the salt marsh ecosystem of Georgia. *Ecology* 43: 614–624.

The Nature Conservancy. 1993. ACE Basin Biological Inventory Report, 1990–1992. Technical Report, The Nature Conservancy, Charleston, SC.

Thomason, C.S., D.E. Allen, and J.S. Crane. 1993. A Fisheries Study of the Edisto River. Wallop–Breaux Projects F-30 and F-32, South Carolina Wildlife and Marine Resources Department, Columbia, SC.

Thompson, J.N. 1984. Insect diversity and the trophic structure of communities. In: Huffaker, C.B. and R.L. Rabb (Eds.). *Ecological Entomology*. John Wiley & Sons, New York, pp. 591–606.

Upchurch, S. 2001. Plants. In: South Carolina Department of Natural Resources and National Oceanic and Atmospheric Administration, *Characterization of the Ashepoo–Combahee–Edisto (ACE) Basin, South Carolina*. Special Scientific Report 17, South Carolina Marine Resources Center, Charleston, SC.

U.S. Environmental Protection Agency. 1997. National Wetlands Inventory Classification. Technical Document, U.S. Environmental Protection Agency, Washington, D.C.

Van Dolah, R. 2001. Benthic invertebrates. In: South Carolina Department of Natural Resources and National Oceanic and Atmospheric Administration, *Characterization of the Ashepoo–Combahee–Edisto (ACE) Basin, South Carolina*. Special Scientific Report 17, South Carolina Marine Resources Center, Charleston, SC.

Van Dolah, R.F., D.R. Calder, F.W. Stapor, Jr., R.H. Dunlap, and C.R. Richter. 1979. Atlantic Intracoastal Waterway Environmental Studies at Sewee Bay and North Edisto River. Technical Report No. 39, South Carolina Marine Resources Center, South Carolina Wildlife and Marine Resources Department, Charleston, SC.

Van Dolah, R.F., D.R. Calder, and D.M. Knott. 1984. Effects of dredging and open water disposal on benthic macroinvertebrates in a South Carolina estuary. *Estuaries* 7: 28–37.

Van Dolah, R.F., P.H. Wendt, and M.V. Levisen. 1991. A study of the effects of shrimp trawling on benthic communities in two South Carolina sounds. *Fisheries Research* 12: 139–156.

Verity, P.G. 1998. Phytoplankton of the South Atlantic Bight. Technical Report, LUCES–South Carolina Sea Grant Consortium, Skidaway Institute of Oceanography, Skidaway, GA.

Verity, P.G. and G.A. Paffenhöfer. 1996. On assessment of prey ingestion by copepods. *Journal of Plankton Research* 18: 1767–1779.

Vernberg, F.J. and C.E. Sansbury. 1972. Studies on salt marsh invertebrates of Port Royal Sound. In: South Carolina Water Resources Commission (Ed.). *Port Royal Sound Environmental Study*. South Carolina Water Resources Commission, Columbia, SC.

Wallace, F.L. 1987. A Checklist of the Richard B. Dominick Moth and Butterfly Collection. International Center for Public Health Research, University of South Carolina, Columbia, SC.

Weakley, A.S. 1981. Natural Features Inventory and Management Recommendations for Mary's Island Plantation, Colleton County, South Carolina. Ph.D. thesis, Duke University, Durham, NC.

Webster, W.D., J.F. Parnell, and W.C. Biggs, Jr. 1985. *Mammals of the Carolinas, Virginia, and Maryland*. University of North Carolina Press, Chapel Hill, NC.

Weisberg, S.B., R. Whalen, and V.A. Lotrich. 1981. Tidal and diurnal influence on food consumption of a salt marsh killifish, *Fundulus heteroclitus*. *Marine Biology* 61: 243–246.

Wenner, E.L. 1986. Decapod crustacean community. In: De Voe, M.R. and D.S. Baughman (Eds.). *South Carolina Coastal Wetland Impoundments: Ecological Characterization, Management, Status, and Use*, Volume 2: *Technical Synthesis*. Publication No. SL-SG-TR-82-2, South Carolina Sea Grant Consortium, Charleston, SC.

Wenner, E.L. 2001a. Protected lands. In: South Carolina Department of Natural Resources and National Oceanic and Atmospheric Administration, *Characterization of the Ashepoo–Combahee–Edisto (ACE) Basin, South Carolina*. Special Scientific Report 17, South Carolina Marine Resources Center, Charleston, SC.

Wenner, E.L. 2001b. Decapod crustaceans. In: South Carolina Department of Natural Resources and National Oceanic and Atmospheric Administration, *Characterization of the Ashepoo–Combahee–Edisto (ACE) Basin, South Carolina*. Special Scientific Report 17, South Carolina Marine Resources Center, Charleston, SC.

Wenner, E.L. 2001c. Fish. In: South Carolina Department of Natural Resources and National Oceanic and Atmospheric Administration, *Characterization of the Ashepoo–Combahee–Edisto (ACE) Basin, South Carolina*. Special Scientific Report 17, South Carolina Marine Resources Center, Charleston, SC.

Wenner, E.L. 2001d. Agriculture. In: South Carolina Department of Natural Resources and National Oceanic and Atmospheric Administration, *Characterization of the Ashepoo–Combahee–Edisto (ACE) Basin, South Carolina*. Special Scientific Report 17, South Carolina Marine Resources Center, Charleston, SC.

Wenner, C.A. and G.R. Sedberry. 1989. Species Composition, Distribution, and Relative Abundance of Fishes in the Coastal Habitat off the Southeastern United States. Technical Report NMFS 79, National Oceanic and Atmospheric Administration, Prepared by South Carolina Wildlife and Marine Resources Department, Marine Resources Division, Charleston, SC.

Wenner, E.L. and L. Zimmerman. 2001. Forestry. In: South Carolina Department of Natural Resources and National Oceanic and Atmospheric Administration, *Characterization of the Ashepoo–Combahee–Edisto (ACE) Basin, South Carolina*. Special Scientific Report 17, South Carolina Marine Resources Center, Charleston, SC.

Wenner, C.A., J.C. McGovern, R. Martone, H.R. Beatty, and W.A. Roumillat. 1986. Ichthyofauna. In: De Voe, M.R. and D.S. Baughman (Eds.). *South Carolina Coastal Wetland Impoundments: Ecological Characterization, Management, Status, and Use,* Vol. III: *Technical Synthesis.* Publication No. SL-SG-TR-82–2, South Carolina Sea Grant Consortium, Charleston, SC, pp. 415–526.

Wenner, E.L., W.P. Coon, III, P.A. Sandifer, and M.H. Shealy, Jr. 1991. A Comparison of Species Composition and Abundance of Decapod Crustaceans and Fishes from the North and South Edisto Rivers in South Carolina. Technical Report No. 78, South Carolina Marine Resources Division, South Carolina Marine Resources Center, Charleston, SC.

Wenner, E., H.R. Beatty, and L. Coen. 1996. A method for quantitatively sampling nekton on intertidal oyster reefs. *Journal of Shellfish Research* 15: 769–775.

Wenner, E.L., M. Thompson, and D. Sanger. 2001a. Water quality. In: South Carolina Department of Natural Resources and National Oceanic and Atmospheric Administration, *Characterization of the Ashepoo–Combahee–Edisto (ACE) Basin, South Carolina.* Special Scientific Report 17, South Carolina Marine Resources Center, Charleston, SC.

Wenner, E.L., A.F. Holland, M.D. Arendt, Y. Chen, D. Edwards, C. Miller, M. Meece, and J. Caffrey. 2001b. A Synthesis of Water Quality Data from the National Estuarine Research Reserve's System-wide Monitoring Program. Technical Report, National Oceanic and Atmospheric Administration, Silver Spring, MD.

Whitney, M. 1998. 1996 Fox Squirrel Sighting Survey Report. Technical Report, Department of Natural Resources, Division of Wildlife and Freshwater Fisheries, Charleston, SC.

Zimmerman, L. 2001a. Mammals. In: South Carolina Department of Natural Resources and National Oceanic and Atmospheric Administration, *Characterization of the Ashepoo–Combahee–Edisto (ACE) Basin, South Carolina.* Special Scientific Report 17, South Carolina Marine Resources Center, Charleston, SC.

Zimmerman, L. 2001b. Birds. In: South Carolina Department of Natural Resources and National Oceanic and Atmospheric Administration, *Characterization of the Ashepoo–Combahee–Edisto (ACE) Basin, South Carolina.* Special Scientific Report 17, South Carolina Marine Resources Center, Charleston, SC.

Zimmerman, L. 2001c. Phytoplankton. In: South Carolina Department of Natural Resources and National Oceanic and Atmospheric Administration, *Characterization of the Ashepoo–Combahee–Edisto (ACE) Basin, South Carolina.* Special Scientific Report 17, South Carolina Marine Resources Center, Charleston, SC.

Case Study 5

6 Weeks Bay National Estuarine Research Reserve

INTRODUCTION

Weeks Bay was designated as a National Estuarine Research Reserve site in 1986. Covering an area of ~2400 ha, the Weeks Bay National Estuarine Research Reserve (Weeks Bay NERR) encompasses a variety of watershed and estuarine habitats, including upland forests, maritime and palustrine plant communities, swamps, freshwater marshes, salt marshes, tidal flats, and open estuarine waters and bay bottom. Uplands and tidelands cover nearly 80% of the reserve area.

The Weeks Bay NERR is one of three active NERR sites in the Gulf of Mexico region; the other two are Rookery Bay NERR near Naples, Florida (designated in 1978), and Apalachicola NERR at Apalachicola, Florida (designated in 1979). It is located in Baldwin County, Alabama, an area known during the past century for its agriculture and silviculture industries. Timber production remains an important industry, with several paper companies operating in the region. Forested habitat — pine-rich woodlands — represents a major land use category in the county. Farmland also constitutes a major land use category. Residential development accounts for a rather small percentage (<2%) of the total land area of Baldwin County, although a significant increase in the amount of developed land surrounding Weeks Bay is anticipated during the next decade (Arcenaux, 1996).

Weeks Bay is a small estuary and hence may be more susceptible to anthropogenic activities in adjoining watershed areas. However, despite considerable agriculture and silviculture in the watershed, no evidence exists of acute pollution or extensive habitat impacts in the bay (Lytle and Lytle, 1995; Lytle et al., 1995; Valentine and Lynn, 1996). Nevertheless, more data must be collected on the effects of anthropogenic activities on the biotic communities and habitats of the estuary. Currently, only a limited database has been compiled on this subject area, and more information must be obtained before definitive assessment of the system can be completed.

WEEKS BAY

Physical Description

Miller-Way et al. (1996) have conducted a detailed investigation of the physical–chemical and biological characteristics of Weeks Bay. With a surface area of only $\sim 7 \times 10^6$ m^2 and an average depth of less than 2 m (Crance, 1971; Schroeder et al., 1992), the bay is a tributary estuary of Mobile Bay (Schroeder, 1996). It is one of the smallest estuaries in the NERR system, measuring less than 4 km in length and width. Located along the eastern shore of Mobile Bay, Weeks Bay is oriented with its long axis trending

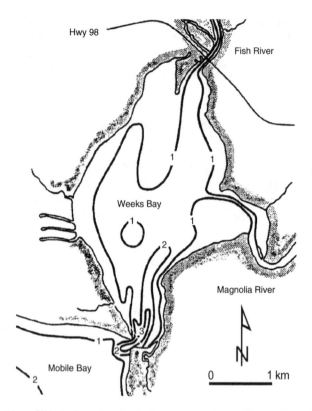

FIGURE 6.1 Map of Weeks Bay showing bathymetric contours. (From Schroeder, W.W., S.P. Dinnel, and W.J. Wiseman, Jr. 1992. Salinity structure of a shallow tributary estuary. In: D. Prandle (Ed.). *Dynamics and Exchanges in Estuaries and the Coastal Zone.* Vol. 40, Coastal and Estuarine Studies, American Geophysical Union, Washington, D.C., pp. 155–171.)

north–south such that hydrologic communication with Mobile Bay occurs through a narrow inlet at the mouth of the bay in the southern perimeter (Figure 6.1). Weeks Bay is a microtidal estuary characterized by diurnal tides with a range of 0.4 m. Currents at the mouth of the bay exceed 1 m/sec, but they decline appreciably within the bay to less than half of this value (Schroeder et al., 1990).

Most freshwater enters Weeks Bay via discharges from the Fish River and Magnolia River with a combined flow of ~9 m/sec. The Fish River, which flows into the northern bay, delivers nearly 75% of the total freshwater input. Much of this freshwater input flows southward along the bay's western perimeter. Water entering the bay at its mouth from Mobile Bay flows northward along the eastern margin, thereby creating essentially a counterclockwise circulation pattern. Freshwater discharge from the Magnolia River enters about midway along the eastern shore of Weeks Bay, and it mixes with the northward-flowing Mobile Bay water (Schroeder et al., 1990; Schroeder, 1996).

The salinity regime is highly variable in Weeks Bay because of the salinity flux of Mobile Bay water entering at its mouth, as well as changes in the volume of freshwater discharges from the Fish and Magnolia Rivers. In addition, variable wind and tidal conditions contribute to shifts in the temporal and spatial salinity structure of Weeks Bay. Hence, salinities in the bay generally range from near 0 to ~20‰, with horizontal salinity gradients varying from weak to strong depending on the aforementioned freshwater inputs and salinity of Mobile Bay water. The vertical salinity structure likewise is variable; both well-mixed and strongly stratified conditions have been documented in the bay (Schroeder et al., 1992).

Water depths are generally deeper in the lower bay (~2–3 m) than in the upper bay (1 m or less) as shown in Figure 6.1. The deepest areas (3–4 m) occur at the mouth of the bay and probably reflect the effects of tidal current scour. An even deeper bathymetric depression (~5–7 m) lies immediately upstream of the Fish River mouth (Schroeder, 1996). Sediments are actively accumulating in Weeks Bay, particularly along the western side (Hardin et al., 1976), and thus the long-term bathymetric condition appears to be one of shoaling. Most of the bottom sediments in the bay consist of a mixture of silts and clays (Figure 6.2). However, sand predominates at the mouth of the bay and in a relatively narrow band abutting the shoreline and surrounding much of the periphery of the bay. A tongue-like mass of sandy sediment also extends about 1 km into the bay from the western bank of the Fish River at its mouth. Sediment in the bay largely derives from the Fish and Magnolia Rivers. Some of the sediment in the area of the bay mouth originates from Mobile Bay (Haywick et al., 1994).

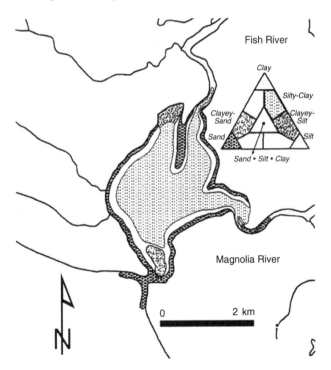

FIGURE 6.2 Sediment distribution and composition in Weeks Bay. (From Haywick, D.W., W.F. Geers, and M.D. Cooper. 1994. Preliminary Report of Grain Size Distribution in Weeks Bay, Baldwin County, Alabama. Technical Report, National Estuarine Research Reserve, Silver Spring, MD.)

WATERSHED

PLANT COMMUNITIES

Upland Habitats

Upland pine forests provide valuable habitat for herpetofauna, mammals, birds, and other animals in the Weeks Bay watershed. Loblolly pine (*Pinus taeda*), long leaf pine (*P. palustris*), and slash pine (*P. elliottii*) occur in this coastal plain habitat. The eastern red cedar (*Juniperus virginiana*), white oak (*Quercus alba*), laurel oak (*Q. laurifolia*), and live oak (*Q. virginiana*) are also found in these forests, along with other species of hardwood trees (Miller-Way et al., 1996).

Wetland Habitats

Stout (1987) showed that palustrine forested wetlands (bottomland hardwood swamps) are the dominant emergent habitat of the reserve, comprising nearly 90% of the mapped area. The canopy vegetation in this habitat consists primarily of pine trees (long leaf pine and slash pine) and various broadleaved deciduous trees (e.g., red maple, *Acer rubrum*; sweetbay, *Magnolia virginiana*; and swamp tupelo, *Nyssa sylvatica* var. *biflora*). The subcanopy includes robust species such as hollies (*Ilex* spp.) and Virginia willow (*Itea virginica*). Under this shrub subcanopy is a plush herbaceous ground cover layer consisting of poison ivy (*Toxicodendrom radicans*), ferns (royal fern, *Osmunda regalis*; and cinnamon fern, *O. cinnamomea*), and sundews (*Drosera* spp.).

Palustrine marshes are much less extensive than palustrine forested wetlands, covering less than 1% of the total Weeks Bay NERR habitat area. They typically concentrate in limited patches near the mouths of small streams. Among the species of plants growing in this habitat are the cattail (*Typha angustifolia*), common reed (*Phragmites australis*), saw grass (*Cladium jamaicense*), alligator weed (*Alternathera philoxeroides*), pickerel weed (*Ponderia cordata*), arrow arum (*Peltandra virginica*), and arrow leaf (*Sagittaria lancifolia*) (Stout, 1996).

The black needlerush (*Juncus roemerianus*) dominates salt marsh biotopes along the bay. Species of secondary abundance are the giant cordgrass (*Spartina cynosuroides*), which inhabits brackish areas near the head of the bay, and the smooth cordgrass (*S. alterniflora*), which concentrates near the mouth of the bay. Moving up the marsh, the salt meadow cordgrass (*S. patens*) and salt grass (*Distichlis spicata*) appear and eventually give way at higher elevations to sea myrtle (*Baccharus halmifolia*) and marsh elder (*Iva frutescens*) in irregularly flooded habitat (Stout and Lelong, 1981; Stout, 1996).

ANIMAL COMMUNITIES

Herpetofauna

Watershed habitats in the Weeks Bay NERR support nearly 50 species of amphibians and reptiles (Table 6.1). Marion and Dindo (1987, 1988) determined that the herpetofaunal community inhabiting the reserve is relatively rich, especially bordering the Fish and Magnolia Rivers. Amphibians are represented by an array of frogs, toads, salamanders, and amphiumas. Pine snakes, mud snakes, king snakes, and

TABLE 6.1
Herpetofaunal Species That Occur or Are Likely to Occur in the Weeks Bay National Estuarine Research Reserve

Common Name	Scientific Name
Amphibians	
One-toed amphiuma	*Amphiuma pholeter*
Two-toed amphiuma	*Amphiuma means*
Three-toed amphiuma	*Amphiuma tridactylum*
Bronze frog	*Rana clamitans clamitans*
Bullfrog	*Rana catesbeina*
Dusky gopher frog	*Rana aureolata sevosa*
Pig frog	*Rana grylio*
River frog	*Rana heckscheri*
Southern leopard frog	*Rana pipiens sphenocephala*
Southern chorus frog	*Pseudacris nigrita*
Southern cricket frog	*Acris gryllus gryllus*
Barking treefrog	*Hyla gratiosa*
Cope's gray treefrog	*Hyla chrysoscelis*
Green treefrog	*Hyla cinerea*
Pine woods treefrog	*Hyla femoralis*
Squirrel treefrog	*Hyla squirella*
Northern spring peeper	*Hyla crucifer crucifer*
Eastern lesser siren	*Siren intermedia intermedia*
Greater siren	*Siren lacertina*
Fowler's toad	*Bufo woodhousii fowleri*
Oak toad	*Bufo quercicus*
Southern toad	*Bufo terrestris*
Narrowmouth toad	*Gastrophryne carolinensis*
Flatwoods salamander	*Ambystoma cingulatum*
Mole salamander	*Ambystoma talpoideum*
Dwarf salamander	*Manculus quadridigitatus*
Gulf Coast mud salamander	*Pseudotriton montanus*
Slimy salamander	*Plethodon glutinosus*
Southern dusky salamander	*Desmognathus fuscus auriculatus*
Southern red salamander	*Pseudotriton ruber vioscai*
Two-lined salamander	*Eurycea bislineata*
Three-lined salamander	*Eurycea longicauda*
Red-spotted newt	*Notopthalmus viridescens*
Reptiles	
Scarlet king snake	*Lampropeltis triangulum*
Eastern king snake	*Lampropeltis getula getula*
Speckled king snake	*Lampropeltis getula holbrooki*
Pine woods snake	*Rhadinaea flavilata*
Black pine snake	*Pituophis melanoleucus lodingi*
Florida pine snake	*Pituophis melanoleucus mugitus*

(continued)

TABLE 6.1 (CONTINUED)
Herpetofaunal Species That Occur or Are Likely to Occur in the Weeks Bay National Estuarine Research Reserve

Common Name	Scientific Name
Florida green water snake	*Natrix cyclopion floridana*
Gulf salt marsh water snake	*Natrix fasciata clarki*
Banded water snake	*Nerodia fasciata*
Green water snake	*Nerodia cyclopion*
Yellow-bellied water snake	*Nerodia erythrogaster flavigaster*
Water moccasin	*Agkistrodon piscivorus*
Northern black racer	*Coluber constrictor constrictor*
Coral snake	*Micrurus fulvius*
Corn snake	*Elaphe guttata guttata*
Eastern diamondback	*Crotalus adamanteus*
Eastern garter snake	*Thamnophis sirtalis*
Eastern ribbon snake	*Thamnophis sauritus sauritus*
Eastern indigo snake	*Drymarchon corais couperi*
Eastern mud snake	*Farancia abacura*
Rainbow snake	*Farancia erytrogramma*
Gray rat snake	*Elaphe obsoleta spiloides*
Ringneck snake	*Diadophis punctatus*
Rough green snake	*Opheodrys aestivus*
Eastern glass lizard	*Ophisaurus ventralis*
Green anole	*Anolis carolinensis*
Broadheaded skink	*Eumeces laticeps*
Five-lined skink	*Eumeces fasciatus*
Ground skink	*Scincella lateralis*
Six-lined racerunner	*Cnemidophorus sexlineatus*
Snapping turtle	*Chelydra serpentina*
Florida softshell turtle	*Trionyx ferox*
Gulf Coast box turtle	*Terrapene carolina major*
Atlantic Ridley turtle	*Lepidochelys kempii*
Loggerhead musk turtle	*Sternotherus minor*
Stinkpot musk turtle	*Sternotherus odoratus*
Gopher tortoise	*Gopherus polyphemus*
Yellow-bellied pond slider	*Pseudemys scripta*
River cooter	*Pseudemys concinna*
Florida cooter	*Pseudemys floridana*
Alabama red-bellied turtle	*Pseudemys alabamensis*
Alligator snapping turtle	*Macroclemys temminckii*
Mississippi diamondback terrapin	*Malaclemys terrapin pileata*
American alligator	*Alligator mississippiensis*

Source: Miller-Way, T., M. Dardeau, and G. Crozier (Eds.). 1996. Weeks Bay National Estuarine Research Reserve: An Estuarine Profile and Bibliography. Dauphin Island Sea Lab Technical Report 96–01, Dauphin Island, AL.

skinks are also common. While some turtles are seasonally abundant (e.g., Gulf Coast box turtle, *Terrapene carolina major*), others (e.g., Mississippi diamondback terrapin, *Malaclemys terrapin pileata*) rarely appear.

Mammals

The list of mammals recorded in the Weeks Bay NERR is not extensive (<40 species) (Table 6.2). Marion and Dindo (1987, 1988) characterized the mammalian species diversity of the reserve as somewhat limited. Dardeau (1996) reported that marsh rabbits (*Sylvilagus palustris*) and raccoons (*Procyon lotor*) dominate the marsh and shoreline habitats of the reserve. Other common inhabitants include bats (e.g., evening bat, *Nycticeius humeralis*), squirrels (e.g., eastern gray squirrel, *Sciurus carolinensis*), opossums (*Didelphis marsupialis*), and foxes (e.g., gray fox, *Urocyon cinereoargenteus*).

Birds

Gulls, cormorants, terns, coots, grebes, kingfishers, waders, flycatchers, warblers, grackles, sparrows, goldfinches, wrens, doves, plovers, sandpipers, vireos, owls, and hawks frequent Weeks Bay NERR habitats. All major feeding groups are represented (i.e., granivores, insectivores, omnivores, herbivores, piscivores, and carnivores). More than 300 species of birds either occur or are likely to occur in the reserve, reflecting the importance of its location within the migratory corridor. Marion and Dindo (1987, 1988), conducting shoreline surveys in the reserve, noted that only six species of birds were common during all seasons of the year; these included the laughing gull (*Larus atricilla*), common tern (*Sterna hirundo*), least tern (*S. antillarum*), royal tern (*S. maxima*), great blue heron (*Ardea herodias*), and belted kingfisher (*Ceryle alcyon*). While coots, cormorants, gulls, grebes, terns, and long-legged waders were observed in the Weeks Bay area either seasonally or year-round, other species were rarely (if at all) seen. For example, small wading birds, marsh ducks, and black skimmers (*Rynchops niger*) were not registered by these investigators. Their absence is probably due to either the limited extent of suitable habitat or insufficient food sources for these birds in the reserve (Dardeau, 1996).

ESTUARY

PLANT COMMUNITIES

Phytoplankton and Microphytobenthos

Schreiber (1994), Schreiber and Pennock (1995), and Pennock (1996) have investigated the nutrient dynamics and microalgal production of Weeks Bay. They noted that Weeks Bay is generally nutrient-rich and productive for several reasons, most importantly:

1. Nutrient inputs from the Fish and Magnolia Rivers as well as Mobile Bay
2. Nutrient enrichment from anthropogenic activities in the watershed
3. Shallow water depths enabling light transmission through the water column to the bay bottom, particularly during the productive summer months when turbidity is generally low

TABLE 6.2
Mammalian Species That Occur or Are Likely to Occur in the Weeks Bay National Estuarine Research Reserve

Common Name	Scientific Name
Armadillo	*Dasypus novemcinctus*
Atlantic bottlenose dolphin	*Tursiops truncatus*
Big brown bat	*Eptesicus fuscus*
Bobcat	*Felis rufus*
Cotton mouse	*Peromyscus gossypinus*
Eastern cottontail	*Sylvilgus floridanus*
Eastern gray squirrel	*Sciurus carolinensis*
Eastern mole	*Scalopus aquaticus*
Eastern pipistrelle	*Pipistrellus subflavus*
Eastern woodrat	*Neotoma floridana*
Evening bat	*Nycticeius humeralis*
Florida black bear	*Ursus americanus floridanus*
Gray fox	*Urocyon cinereoargenteus*
Hispid cotton rat	*Sigmodon hispidus*
House mouse	*Mus musculus*
Marsh rabbit	*Sylvilagus palustris*
Mink	*Mustela vison*
Muskrat	*Ondatra zibethica*
Norway rat	*Rattus norvegicus*
Nutria	*Myocastor coypus*
Opossum	*Didelphis marsupialis*
Raccoon	*Procyon lotor*
Red bat	*Lasiurus borealis*
Red fox	*Vulpes vulpes*
Rice rat	*Oryzomys palustris*
River otter	*Lutra canadensis*
Seminole bat	*Lasiurus seminolus*
Southern flying squirrel	*Glaucomys volans*
Southern short-tailed shrew	*Blarina carolinensis*
Striped skunk	*Mephitis mephitis*
White-tailed deer	*Odocoileus virginianus*
West Indian manatee	*Trichechus manatus*

Source: Miller-Way, T., M. Dardeau, and G. Crozier (Eds.). 1996. Weeks Bay National Estuarine Research Reserve: An Estuarine Profile and Bibliography. Dauphin Island Sea Lab Technical Report 96–01, Dauphin Island, AL.

Over an annual cycle, the concentrations of ammonium, nitrate, phosphate, and silicate in the bay typically range from 1 to 10, 0 to >85, 0 to 8, and 20 to 140 μM, respectively (Pennock, 1996). Although nitrate is the predominate nitrogen form in the estuary and a major factor in microalgal growth, phosphate may be the principal limiting nutrient for phytoplankton growth because of its low concentrations in the bay relative to those of nitrate.

Pennock (1996) reported that the mean production of phytoplankton in Weeks Bay amounts to 348 g C/m^2/yr, which is about fivefold greater than microphytobenthos production. He also estimated that phytoplankton biomass per unit area ranges from 10 to 90 mg chl/m^2 over an annual cycle compared to microphytobenthos biomass values of 5 to 30 mg chl/m^2 over a seasonal cycle. Peak phytoplankton production occurs during the summer months, while highest phytoplankton biomass (up to 80 μg chl/l) takes place during the winter months when algal blooms generally develop.

Most of Weeks Bay contains unvegetated soft bottom, with submerged aquatic vegetation (SAV) contributing little, if any, production to the system (Stout, 1996). While Stout and Lelong (1981) documented small beds of SAV (i.e., *Myriophyllum spicatum, Potamogeton pectinatus*, and *Vallisneria americana*) near the mouth of the bay, these beds may no longer be present there. Thus, the contribution of primary production from the benthos is mainly attributed to the microphytobenthos.

Animal Communities

Zooplankton

Several studies have examined the zooplankton of Weeks Bay, the most detailed being those of Bain and Robinson (1990), Stearns et al. (1990), and Dardeau (1996). These studies indicate that rotifers and copepods are the most abundant groups, with rotifers numerically dominant. Maximum zooplankton numbers appear during the summer when the density of copepods (e.g., *Acartia tonsa, Halicyclops fosteri*, and *Oithona* spp.) is greatest, and minimum zooplankton numbers are evident during the winter. *Acartia tonsa* outnumbers all other species over an annual cycle; it overwhelmingly predominates during all seasons except summer, when other copepod species increase appreciably in abundance.

Stearns et al. (1990) discerned distinct spatial distribution patterns in the zooplankton community of Weeks Bay. For example, they showed that diel vertical migration is conspicuous among zooplankton in the water column despite the shallow depths of the bay. Cladocerans are mostly found in limnetic and oligohaline waters. Some copepod species (e.g., *Oithona colcarva* and *Saphirella* sp.) prefer mesohaline areas. Others (e.g., the calanoid copepod, *Eurytemora* sp.; and the harpacticoid copepod, *Leptocaris kunzi*) concentrate in vegetated habitats, such as marsh tidal creeks bordered by *Spartina alterniflora* and *Juncus roemerianus*. However, most of the zooplankton species are widely distributed in the bay, where they exert significant grazing pressure on phytoplankton populations in unvegetated open water areas.

Benthic Fauna

The benthic community of Weeks Bay has not been well characterized. Only two studies, Bault (1970) and Bain and Robinson (1990), have focused on the benthic fauna of the bay. Dardeau (1996) has reviewed this work. Sampling in the mid-bay, Bault (1970) identified three species of polychaetes (*Eteone* sp.,

Hobsonia florida, and *Laeonereis culveri*) and several rhynchocoels, as well as insect larvae. In a more comprehensive investigation, Bain and Robinson (1990) recorded the predominance of polychaetes, which accounted for more than 80% of the infauna collected in samples along the peripheral areas of the bay. Amphipods, mysids, bivalves, gastropods, oligochaetes, and rhynchocoels, which were also collected in these samples, together comprised less than the 20% of the infauna. Abundance of the benthic populations varied, in part, with the sediment type and season of sampling.

Fish

Weeks Bay serves as an important nursery for an array of fish species such as the striped mullet (*Mugil cephalus*), Gulf menhaden (*Brevoortia patronus*), and Atlantic croaker (*Micropogonias undulatus*) (Dardeau, 1996). Postlarval forms often numerically dominate the assemblages. They typically enter the bay in the spring (e.g., sand seatrout, *Cynoscion arenarius*) and winter (e.g., spot, *Leiostomus xanthurus*) and grow rapidly on the rich food supply in the embayment (Shipp, 1987).

Species composition varies considerably due to migrants that enter the bay seasonally and strays that appear sporadically and remain for different periods of time. Largely due to seasonal migration, species richness is greater in the spring and fall than in the summer and winter (Swingle and Bland, 1974; Dardeau, 1996). Similarly, total fish abundance varies seasonally, with the greatest number of individuals observed in the spring and fall and the least number of individuals in the summer and winter (Swingle, 1971; Swingle and Bland, 1974; Dardeau, 1996). These seasonal abundance patterns are largely ascribed to the occurrence of the bay anchovy (*Anchoa mitchilli*), Gulf menhaden, and striped mullet (Bain and Robinson, 1990; Dardeau, 1996).

Bain and Robinson (1990) emphasized that salinity and water clarity are also important factors influencing the spatial distribution and community structure of fish in the bay. They ascertained that the distribution of fish species occurs along a nearshore to offshore depth gradient or along a riverine to estuarine salinity gradient controlled by the amount of freshwater inflow. These environmental controls have also been shown to influence the trophic relationships of fish in the bay. Dardeau (1996) differentiated five feeding strategies of Weeks Bay fish:

1. Planktivores (e.g., anchovies, menhaden, and shad)
2. Detritivores (e.g., mullet)
3. Pelagic omnivores (e.g., tidewater silverside, *Menidia peninsulae*)
4. Benthic omnivores (e.g., Atlantic croaker, gobies, flatfish, spot, and black drum)
5. Epibenthic carnivores (e.g., seatrout and silver perch, *Bairdiella chrysoura*)

Seasonal migration of fish and responses of fish to shifts in the salinity regime of the bay greatly affect the species composition and hence the relative abundance of the different feeding groups in the estuary. The fish community of Weeks Bay, therefore, is one characterized by considerable variation in species composition and abundance year-round.

Several species of fish are commercially important in Weeks Bay. One of the most significant is the Gulf menhaden. Others include the striped mullet, sheepshead, seatrout, and flounder (*Paralichthys* spp.). Aside from fish, several other nektonic groups support commercial fisheries, notably penaeid shrimp (white shrimp, *Penaeus setiferus*; brown shrimp, *P. astecus*; and pink shrimp, *P. duorarum*), the eastern oyster (*Crassostrea virginica*), and the blue crab (*Callinectes sapidus*) (Dardeau, 1996).

ANTHROPOGENIC IMPACTS

Weeks Bay remains a relatively pristine estuarine system despite increasing development in surrounding watershed areas and greater human use of the embayment itself (Arcenaux, 1996). Habitat disturbance is linked to residential development, agriculture, silviculture, and the removal of naval stores and dirt in the watershed (Stout, 1996). Escalating residential and commercial development in Southern Baldwin County has contributed to the loss of some prime watershed habitat and soil disruption in upland buffer areas, as well as sediment and nutrient loading that has adversely affected aquatic habitats and biotic communities in the reserve. Land acquisition is part of a long-term management plan to minimize detrimental impacts in the watershed, Fish River, and Magnolia River. An array of lands — marshes, buffer habitat, and upland areas — is targeted for purchase. Another important component of the management plan involves the application of best management practices (BMPs) during construction and other invasive land use activities to mollify nonpoint source pollution, which continues to be a potential threat to the estuary (Adams et al., 1996).

Valentine and Lynn (1996) discussed water pollution in the Weeks Bay NERR. They state that agricultural development affects more than 80% of the land within Baldwin County, including extensive areas along the Fish and Magnolia Rivers. Aerial application of pesticides may pose a hazard to the water quality of these rivers as well as that of the estuary. Fertilizers and fungicides frequently used in agricultural operations also contain trace metals such as arsenic, cadmium, copper, lead, and mercury, which can be detrimental to estuarine and marine organisms (Kennish, 1992, 1997; Weber et al., 1992). The accumulation of polychlorinated biphenyls (PCBs) and polycyclic aromatic hydrocarbons (PAHs) in bottom sediments and biota must be more clearly delineated. Fertilizer use (for domestic purposes), municipal wastewater discharges, malfunctioning septic systems, and wildlife wastes augment these pollutant inputs and, if uncontrolled, can create nutrient enrichment problems that promote algal blooms. Such events may result in reduced dissolved oxygen concentrations. However, long-term dissolved oxygen measurements in the Weeks Bay system do not indicate anoxic or hypoxic problems (Scott Phipps, Weeks Bay NERR, personal communication, 2003).

Waters in the Weeks Bay system have been continually monitored for fecal coliform bacteria levels. A large percentage of the residences in the Weeks Bay watershed employ septic tanks for wastewater treatment, and faulty systems are a potential source of fecal coliforms for riverine and estuarine waters. Although fecal coliform concentrations have been consistently low (<50 per 100 ml) near the mouth of the bay, higher coliform densities near the head of the bay have exceeded thresholds for shellfish harvesting (Valentine and Lynn, 1996).

Ongoing housing and industrial construction and other human activities in the watershed have accelerated soil erosion and sediment inputs to surface waters in the Weeks Bay NERR. Estimates of annual sediment inputs to Weeks Bay approach 20,000 metric tons (Valentine and Lynn, 1996). Various shoreline modifications, particularly along the Fish and Magnolia Rivers, likely facilitate the transport of sediments, nutrients, and chemical contaminants to open waters of the system. It is vital to implement BMPs in the lower watershed as well as in the uplands to mitigate the inputs of sediments and pollutants to receiving waters of the reserve, thereby reducing the probability of future habitat and biotic community impacts.

SUMMARY AND CONCLUSIONS

Weeks Bay is a shallow tributary estuary of Mobile Bay. This small microtidal estuary is part of the Weeks Bay NERR, which encompasses upland pine forests, palustrine forested wetlands, and fringing salt marshes in the adjoining lower watershed of Baldwin County. The salinity regime in Weeks Bay indicates that oligohaline and mesohaline conditions predominate. Most freshwater enters the bay from the Fish and Magnolia Rivers. During periods of large freshwater inflow, limnetic conditions exist in the northern bay. Bottom sediments mainly consist of sands along the perimeter and silts and clays in the central bay areas.

Numerous species of plants and animals inhabit the Weeks Bay watershed. Loblolly pine, long leaf pine, and slash pine, together with red cedar and oaks, characterize upland pine forests. Long leaf pine, slash pine, and various broadleaved deciduous trees form a relatively thick canopy in bottomland hardwood swamps. Palustrine marshes comprised of cattails, common reed, pickerel weed, arrow arum, and several other plant species cover a very limited area (<1% of all habitat area). Black needlerush and cordgrass represent the dominant salt marsh plants bordering the bay; salt meadow cordgrass and salt grass occupy the higher marsh.

The Weeks Bay watershed is also replete with many species of amphibians, reptiles, mammals, birds, and other faunal groups. Nearly 50 species of herpetofauna have been recorded in the watershed. Fewer mammalian species (<40) occur there. Birds are much more diverse, with more than 300 species chronicled.

Weeks Bay is a relatively nutrient-rich estuary, and phytoplankton production (nearly 350 g $C/m^2/yr$) compares favorably with that of many other U.S. estuaries. Microphytobenthos production amounts to only ~20% of the phytoplankton production. The benthic habitat is essentially devoid of submerged aquatic vegetation. Zooplankton graze heavily on phytoplankton in the bay, with rotifers and copepods dominating the zooplankton community. Maximum zooplankton abundance takes place during the summer months.

Only limited benthic sampling has been conducted in the bay. However, polychaetes dominate the benthic infauna by far, accounting for more than 80% of the total faunal abundance. Amphipods, mysids, bivalves, gastropods, oligochaetes, and rhynchocoels are of secondary numerical abundance.

Nearly 200 species of fish have been identified in the Weeks Bay system. The estuary is an important nursery area for many of these species. Fish assemblages vary in composition year-round due to seasonal migration and the occurrence of marine and freshwater strays. Species richness and total abundance of fish are greater in the spring and fall than in the summer and winter. Among the most abundant fish are anchovies, silversides, gobies, menhaden, mullet, croaker, and spot. Some species (e.g., Gulf menhaden, striped mullet, sheepshead, and flounder) are of commercial importance.

Despite various anthropogenic impacts in the watershed and estuary, Weeks Bay remains a relatively pristine system. The levels of coliform bacteria, dissolved oxygen, nutrients, and chemical contaminants do not indicate that the system is significantly threatened by human activities. In addition, the amounts of habitat loss and alteration in the watershed and estuary do not reflect a heavily disturbed system. Nevertheless, management plans have been implemented to ensure the long-term viability and resource stability of this critically important estuary.

REFERENCES

Adams, L.G., T. Lynn, and R. McCormack. 1996. Management issues. In: Miller-Way, T. M. Dardeau, and G. Crozier (Eds.). *Weeks Bay National Estuarine Research Reserve: An Estuarine Profile and Bibliography.* Dauphin Island Sea Lab Technical Report 96–01, Dauphin Island, AL, pp. 75–79.

Arcenaux, C. 1996. Land use. In: Miller-Way, T., M. Dardeau, and G. Crozier (Eds.). *Weeks Bay National Estuarine Research Reserve: An Estuarine Profile and Bibliography.* Dauphin Island Sea Lab Technical Report 96–01, Dauphin Island, AL, pp. 53–61.

Bain, M.B. and C.L. Robinson. 1990. Abiotic and biotic factors influencing microhabitat use by fish and shrimp in Weeks Bay National Estuarine Research Reserve. Technical Report to NOAA, National Estuarine Research Reserves, Silver Spring, MD.

Bault, E.I. 1970. A Survey of the Benthic Organisms in Selected Coastal Streams and Brackish Waters of Alabama. Technical Report to the U.S. Department of Interior, Bureau of Commercial Fisheries, Washington, D.C.

Crance, J.H. 1971. Description of Alabama estuarine areas — cooperative Gulf of Mexico estuarine inventory. *Alabama Marine Resources Bulletin* 6: 1–85.

Dardeau, M.R. 1996. Estuarine consumers. In: Miller-Way, T., M. Dardeau, and G. Crozier (Eds.). *Weeks Bay National Estuarine Research Reserve: An Estuarine Profile and Bibliography.* Dauphin Island Sea Lab Technical Report 96–01, Dauphin Island, AL, pp. 37–51.

Hardin, J.D., C.D. Sapp, J.L. Emplaincourt, and K.E. Richter. 1976. Shoreline and Bathymetric Changes in the Coastal Area of Alabama: A Remote Sensing Approach. Geological Survey of Alabama, Information Series 50, Mobile, AL.

Haywick, D.W., W.G. Geers, and M.D. Cooper. 1994. Preliminary Report of Grain Size Distribution in Weeks Bay, Baldwin County, Alabama. Technical Report to NOAA, National Estuarine Research Reserves, Silver Spring, MD.

Kennish, M.J. 1992. *Ecology of Estuaries: Anthropogenic Effects.* CRC Press, Boca Raton, FL.

Kennish, M.J. (Ed.). 1997. *Practical Handbook of Estuarine and Marine Pollution.* CRC Press, Boca Raton, FL.

Lytle, J.S. and T.F. Lytle. 1995. Toxicity evaluation of sediments from a national estuary using emergent macrophytes. Abstract, SETAC National Meeting, Vancouver, Canada.

Lytle, J.S. T.F. Lytle, and H. Cui. 1995. Pesticide residues in rooted estuarine macrophytes. Abstract, SETAC National Meeting, Vancouver, Canada.

Marion, K.R. and J.J. Dindo. 1987. The Use of Indicator Species as a Means of Assessing the Environmental Condition of the Weeks Bay National Estuarine Research Reserve. Technical Report to NOAA, National Estuarine Research Reserves, Silver Spring, MD.

Marion, K.R. and J.J. Dindo. 1988. Enhancing Public Awareness of Estuaries: A Natural History Survey of the Weeks Bay National Estuarine Research Reserve. Technical Report to NOAA, National Estuarine Research Reserves, Silver Spring, MD.

Miller-Way, T., M. Dardeau, and G. Corzier (Eds.). 1996. Weeks Bay National Estuarine Research Reserve: An Estuarine Profile and Bibliography. Dauphin Island Sea Lab Technical Report 96–01, Dauphin Island, AL.

Pennock, J.R. 1996. Nutrients and aquatic primary production. In: Miller-Way, T., M. Dardeau, and G. Crozier (Eds.). *Weeks Bay National Estuarine Research Reserve: An Estuarine Profile and Bibliography.* Dauphin Island Sea Lab Technical Report 96–01, Dauphin Island, AL, pp. 30–36.

Schreiber, R.A. 1994. The Contribution of Benthic Microalgae to Primary Production in Weeks Bay, Alabama. M.S. thesis, University of Alabama, Tuscaloosa, AL.

Schreiber, R.A. and J.R. Pennock. 1995. The relative contribution of benthic microalgae to total microalgal production in a shallow subtidal estuarine environment. *Ophelia* 42: 335–352.

Schroeder, W.W. 1996. Environmental setting. In: Miller-Way, T., M. Dardeau, and G. Crozier (Eds.). *Weeks Bay National Estuarine Research Reserve: An Estuarine Profile and Bibliography.* Dauphin Island Sea Lab Technical Report 96–01, Dauphin Island, AL, pp. 15–25.

Schroeder, W.W., W.J. Wiseman, Jr., and S.P. Dinnel. 1990. Wind and river-induced fluctuations in a small, shallow tributary estuary. In: Cheng, R.T. (Ed.). *Residual Currents and Long-term Transport.* Vol. 38, Coastal and Estuarine Studies, Springer-Verlag, New York, pp. 481–493.

Schroeder, W.W., S.P. Dinnel, and W.J. Wiseman, Jr. 1992. Salinity structure of a shallow tributary estuary. In: Pringle, D. (Ed.). *Dynamics and Exchanges in Estuaries and the Coastal Zone.* Vol. 40, Coastal and Estuarine Studies, American Geophysical Union, Washington, D.C., pp. 155–171.

Shipp, R.L. 1987. Temporal distribution of finfish eggs and larvae around Mobile Bay. In: Lowrey, T. (Ed.). *Symposium on the Natural Resources of the Mobile Bay Estuary.* Mississippi–Alabama Sea Grant 87–007, Mobile, AL.

Stearns, D.E., M.R. Dardeau, and A.J. Planchart. 1990. Zooplankton Community Composition, Species Abundance, and Grazing Impact in Weeks Bay, Alabama: Tidal, Monthly, Seasonal, and Habitat Differences. Technical Report to NOAA, National Estuarine Research Reserves, Silver Spring, MD.

Stout, J.P. 1987. Delineation of Emergent Habitats of the Weeks Bay NERR. Dauphin Island Sea Lab Technical Report 870–004, Dauphin Island, AL.

Stout, J.P. 1996. Estuarine habitats. In: Miller-Way, T., M. Dardeau, and G. Crozier (Eds.). *Weeks Bay National Estuarine Research Reserve: An Estuarine Profile and Bibliography.* Dauphin Island Sea Lab Technical Report 96–01, Dauphin Island, AL, pp. 27–29.

Stout, J.P. and M.J. Lelong. 1981. Wetlands Habitats of the Alabama Coastal Area. Part II. An Inventory of Wetland Habitats South of the Battleship Parkway. Alabama Coastal Area Board Technical Publication CAB-81–01.

Swingle, H.A. 1971. Biology of Alabama Estuarine Areas — Cooperative Gulf of Mexico Estuarine Inventory. *Alabama Marine Resources Bulletin* 5: 1–123.

Swingle, H.A. and D.G. Bland. 1974. A study of fishes of the coastal watercourses of Alabama. *Alabama Marine Resources Bulletin* 10: 17–102.

Valentine, J.F. and T. Lynn. 1996. Pollution. In: Miller-Way, T., M. Dardeau, and G. Crozier (Eds.). *Weeks Bay National Estuarine Research Reserve: An Estuarine Profile and Bibliography.* Dauphin Island Sea Lab Technical Report 96–01, Dauphin Island, AL, pp. 62–74.

Weber, M., R.T. Townsend, and R. Bierce. 1992. *Environmental Quality in the Gulf of Mexico,* 2nd ed. Center for Marine Conservation, Washington, D.C.

Case Study 6

7 Tijuana River National Estuarine Research Reserve

INTRODUCTION

Tijuana River in Southern California was designated as a National Estuarine Research Reserve (NERR) site in 1982. The 1025-ha site, located in San Diego County at the U.S./Mexico border, encompasses a wide array of habitats, including uplands, coastal sage, upland–wetland transition, salt marshes, tidal creeks and channels, and mudflats and sandflats, as well as dunes and beaches. A significant fraction of the reserve area consists of salt marsh and riparian wetland habitats (Table 7.1). The intertidal salt marsh habitat is particularly well developed. The Tijuana River watershed is extensive, covering 1731 km^2 and bisecting the estuary into northern and southern regions. More than 75% of the watershed area lies within Mexico (Figure 7.1) (Nordby and Zedler, 1991; Zedler et al., 1992; Zedler, 2001).

The Tijuana Estuary is located at 32°34′N, 117°7′W. It is a coastal plain estuary consisting of a network of stream channels with no well-defined embayment. Hydrographic conditions in this intermittent estuary are highly variable; stream flow is greatly reduced during much of the year when drought conditions often exist, but significant precipitation during the winter generally results in a river-dominated system. Because of the large seasonal flux in stream flow, the estuary changes from a river-dominated system in winter to one that is partially mixed or vertically homogeneous during other seasons. Zedler and Beare (1986) reported a mean annual discharge of the Tijuana River amounting to ~20,820 MLD (million liters per day), although they noted the flow was extremely variable. For example, the coefficient of variability reported by these investigators was 325%. Flooding commonly occurs in years of very heavy precipitation, such as during the period from 1978 to 1980. Tidal flushing peaks at the channel mouths and declines upstream.

The Tijuana River NERR is an urbanized system, surrounded by the cities of San Diego, Imperial Beach, and Tijuana. The proximity of these cities to the Tijuana Estuary results in a number of anthropogenic impacts. For example, sewage-contaminated inflows from the city of Tijuana in past years have created water quality problems in the estuary (Seamans, 1988). In addition, urban runoff and irrigation runoff from farmlands in the U.S. have increased sedimentation downestuary, decreased salinity levels, and enabled exotic species to invade and establish viable populations in upper salt marsh habitats. Sediment influx during catastrophic flooding events has increased salt marsh elevations and reduced tidal inundation (Zedler, 2001). As a consequence, the Tijuana River NERR is challenged by several management problems, most notably wastewater inflows from Mexico, sediment influx

TABLE 7.1
List of Habitats and Their Areas in the Tijuana River National Estuarine Research Reserve

Habitat Type	Hectare Area	Percent Area
Transition	246.9	24.1
Salt marsh/salt panne	177.7	17.3
Disturbed	148.1	14.5
Riparian	100.4	9.8
Transition/disturbed	73.2	7.2
Salt marsh	71.2	7.0
Channels and ponds	70.0	6.8
Coastal sage	61.1	6.0
Dunes and beach	50.2	4.9
Mudflats	13.4	1.3
Brackish marsh	12.1	1.2

Source: Entrix Inc., PERL, and PWA, Ltd. 1991. Tijuana Estuary Tidal Restoration Program. Draft Environmental Impact Report/Environmental Impact Statement. California Coastal Conservancy and U.S. Fish and Wildlife Service, SCC, Oakland, CA, Volumes I–III.

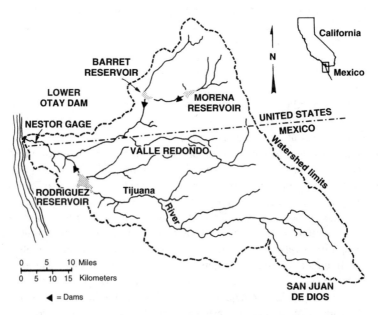

FIGURE 7.1 Map of the Tijuana Estuary showing surrounding watershed areas. (From Zedler, J.B., C.S. Nordby, and K.B. Kus. 1992. The Ecology of Tijuana Estuary, California: A National Estuarine Research Reserve. Technical Report, NOAA Office of Coastal Resource Management, Sanctuaries and Reserves Division, Washington, D.C.)

into the estuary from the watershed, and endangered species. A major goal of the reserve is habitat conservation and restoration.

Because human activities have significantly impacted the Tijuana Estuary, restoration projects are underway to revitalize the system (Vivian-Smith, 2001). These efforts have focused on salt marsh revitalization as well as sediment and flood control of upstream areas (Restore America's Estuaries, 2002). Most of the Tijuana River NERR is publicly owned; it is managed by the U.S. Fish and Wildlife Service, U.S. Navy, California Department of Parks and Recreation, County of San Diego, and City of San Diego. Public ownership of lands has significantly reduced development in the watershed and afforded considerable protection of endangered species and their habitat. However, sewage spills, dike construction, and gravel extraction have degraded substantial areas of the estuary and watershed. The Tijuana Estuary is a historically variable system characterized by large fluxes in rainfall, stream flow, periodic flooding, sedimentation, and channel morphology.

The Tijuana River NERR has established three long-term water quality monitoring sites in the Tijuana estuarine system. These include:

1. The Oneonta Slough site located ~1.5 km from the river mouth at 32°34'04.8"N, 117°07'52.3"W
2. The Tidal Linkage site located ~2.7 km from the river mouth at 32°34'27.9"N, 117°07'37.8"W
3. The Model Marsh site located ~1.6 km from the river mouth at 32°32'52.5"N, 117°07'37.7"W

Physical–chemical data recorded semicontinuously at these sites using YSI 6-Series data loggers include water temperature, salinity, dissolved oxygen, pH, turbidity, and water depth (NERRS, 2002).

WATERSHED

An array of habitats characterizes the Tijuana River NERR (Figure 7.2). Estuarine channels and tidal creeks cover only 70 ha or 6.8% of the total area of the reserve (Table 7.1). The remaining habitats comprising the adjoining Tijuana River watershed exhibit considerable spatial heterogeneity, and they support a high diversity of organisms, including more than 20 rare, threatened, or endangered species, such as the salt marsh bird's beak (*Cordylanthus maritimus* ssp. *maritimus*), California least tern (*Sterna antillarum browni*), light-footed clapper rail (*Rallus longirostris levipes*), least Bell's vireo (*Vireo belli pusilius*), California brown pelican (*Pelicanus occidentalis*), and American peregrine falcon (*Falco peregrinus anatum*) (Zedler et al., 1992). Among the most expansive habitats in the watershed, the upland–wetland transition zone covers the greatest area (>300 ha). Salt marshes dominate the wetland habitats. They have been the focus of a number of comprehensive research and restoration programs (Zedler, 1977, 2001; Zedler and Nordby, 1986; Zedler et al., 1992; Vivian-Smith, 2001). Riparian habitat is also quite extensive, encompassing 100 ha. Anthropogenic disturbances have altered many hectares of watershed habitat in the Tijuana River NERR (Zedler, 2001).

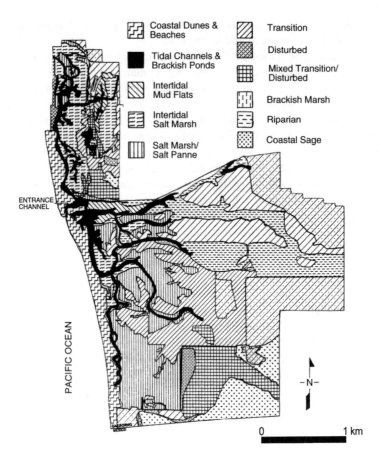

FIGURE 7.2 Map of habitat types at the Tijuana Estuary. (From Zedler, J.B., C.S. Nordby, and K.B. Kus. 1992. The Ecology of Tijuana Estuary, California: A National Estuarine Research Reserve. Technical Report, NOAA Office of Coastal Resource Management, Sanctuaries and Reserves Division, Washington, D.C.)

Habitat

Salt Marsh

Vegetation in the salt marsh habitat varies with elevation and thus can be differentiated into three assemblages, consisting of low marsh, mid-marsh plain, and high marsh (Desmond et al., 2001). The plant species are distributed in broadly overlapping bands, with peak occurrence of individual species typically observed at different elevation bands. Algal mats often grow as dense cover under the canopy of salt marsh plants. Zedler (1982) reported that 100 plant species have been identified in the edaphic algal mats of the estuary. The salt marsh habitat is a physically stressful environment where salinity levels range from 40 to 100‰, and the plants are subject to variable tidal flushing.

The Pacific cordgrass (*Spartina foliosa*) dominates the low marsh area. Zedler (1980) measured net primary productivity values of 340 g C/m^2/yr for *S. foliosa* and 276 g C/m^2/yr for benthic algae in salt marsh habitat of the estuary. Winfield (1980), in turn, recorded net aboveground productivity values of 224–307 g dry wt/m^2/yr for *S. foliosa* marsh. He also observed that cordgrass productivity accounted for less than 50% of the total marsh productivity. This vascular plant is a largely clonal species that provides year-round habitat for the endangered light-footed clapper rail. Other fauna inhabiting the low marsh include the fiddler crab (*Uca crenulata*), yellow shore crab (*Hemigrapsus oregonensis*), lined shore crab (*Pachygrapsus crassipes*), and California horn snail (*Cerithidea californica*), as well as two dipteran insect species (*Cricotopus* sp. and *Incertella* sp.) (Zedler et al., 1992).

The mid-marsh plain is a rather harsh environment characterized by periodic tidal inundation. The perennial pickleweed (*Salicornia virginica*) frequently predominates here and may form monotypic stands in some areas. However, this species is more commonly associated with other succulents in a mixed assemblage. Some of the co-inhabitants include the annual pickleweed (*Salicornia bigelovii*), arrow-grass (*Triglochin concinnum*), saltwort (*Batis maritima*), sea blite (*Suaeda esteroa*), fleshy jaumea (*Jaumea carnosa*), and sea lavender (*Limonium californicum*) (Desmond et al., 2001). Of these plant species, the saltwort may be most widespread; together with the annual pickleweed, it attains highest abundance along intertidal pools. Aside from the aforementioned succulents, numerous species of green and blue-green algae form extensive mats beneath the salt marsh canopy. These epibenthic algae contribute significant primary production to the system (Zedler, 2001).

The mid-marsh habitat supports many different species of animals. Gastropods are well represented; for example, the snails *Assiminea californica, Cerithidea californica,* and *Melampus olivaceus* commonly occur in the mid-marsh zone. Intertidal pools harbor substantial numbers of forage fishes such as the California killifish (*Fundulus parvipinnis*). Insects (e.g., water boatmen, *Trichocorixia* spp.; and flies, *Ephydra* sp.) are likewise abundant. The endangered Belding's Savannah sparrow (*Ammodramus sandwichensis beldingi*) inhabits areas vegetated by perennial pickleweed. Other species of birds frequently observed in the mid-marsh are the great egret (*Casmerodius albus*), great blue heron (*Ardea herodias*), willet (*Catoptrophorus semipalmatus*), marbled godwit (*Limosa fedoa*), and long-billed curlew (*Numenius americanus*) (Zedler et al., 1992).

Some of the most extreme environmental conditions occur in the high marsh, where extended periods of drought are punctuated by inundation. Flora and fauna of the high marsh must also adapt to frequent disturbance associated with burrowing mammals, human activities, and other factors. The mounded topography typifying this zone creates irregular terrain and contributes to greater complexity of biotic communities.

Predominant perennial plants that vegetate the higher marsh are the glasswort (*Salicornia subterminalis*), shoregrass (*Monanthocloe littoralis*), alkali heath (*Frankenia salina*), sea lavender (*Limonium californicum*), and Watson's salt bush (*Atriplex watsonii*). Salt grass (*Distichlis spicata*) may also be observed in this area. Annual plant assemblages likewise occupy the higher marsh habitat, often beneath perennial canopies. In addition, the high salt marsh supports sensitive plant species (e.g., Coulter goldfields, *Lasthenia glabrata coulteri*) as well as the endangered salt marsh bird's beak (*Cordylanthus maritimus* ssp. *maritimus*) (Desmond et al., 2001). Both of these species are annuals present during a limited growing season.

A number of spiders, mites, and other insects commonly occur in the higher marsh. Two prevalent species of spiders are *Pardosa ramulosa* and *Tetragnatha laboriosa*. The carabid beetle (*Tachys corax*) is widespread across the marsh surface. While insects are relatively abundant in the higher marsh, other invertebrates are not. Only a few snakes and lizards (e.g., *Phrynosoma coronatum blainvillei*) inhabit the higher marsh, but small mammals such as rabbits (*Lepus californicus* and *Sylvilagus audubonii sactidiegi*) and ground squirrels (*Spermophilus beechyii*) are quite abundant. The hummocky mound topography, which typifies the higher marsh, is attributed to burrowing activities of ground squirrels (Cox and Zedler, 1986).

Small mammals utilizing the high marsh provide prey for several species of raptors, notably the golden eagle (*Aquila chrysaetos*), American kestrel (*Falco sparverius*), and northern harrier (*Circus cyaneus*). Other species of birds forage on smaller organisms. Examples are the horned lark (*Eremophila alpestris*), loggerhead shrike (*Lanius ludovicianus*), song sparrow (*Melospiza melodia*), white-crowned sparrow (*Zonotrichia leucophrys*), and western meadowlark (*Sturnella neglecta*) (Zedler et al., 1992).

Salt Pannes

Among the most variable habitats in the Tijuana River watershed is the salt panne, an open salt flat area in the upper intertidal zone that lacks vegetation. Salt pannes are conspicuous during dry summer periods when evaporation causes soil salinities to increase up to 200‰, resulting in the formation of a salt crust. While the excessively high soil salinities are not favorable for plant growth, some animal populations may be observed on the dry panne surface. For example, the California least tern (*Sterna antilarrum browni*) and western snowy plover (*Charadrius alexandrinus nivosus*) nest here, and the Belding's Savannah sparrow (*Ammodramus sandwichensis beldingi*) feeds on insects. Rove beetles (*Bledius* spp.) and tiger beetles (*Cicindela* spp.) are two insect species that burrow in the dry sediments of the salt panne. Various reptilian and mammalian species also utilize the salt panne habitat (Zedler et al., 1992).

Increased precipitation augmented by saline water inundation during the early winter marks the onset of a wet phase when the salt pannes temporarily transform to shallow aquatic systems. The wet conditions enable algae and widgeon grass (*Ruppia maritima*) to flourish, albeit for only several months. Northern pintails (*Anas acuta*) and other waterfowl also use the habitat at this time, along with small waders (e.g., snowy plovers). Aquatic insects are relatively abundant. The winter wet season is of shorter duration (December–February) than the summer dry season when barren conditions predominate. Therefore, the barren habitat of the salt panne is the predominant condition (Desmond et al., 2001).

Brackish Marsh

Brackish marshes occur in the reserve in areas where salinities range from ~0.5 to 30‰. They are spatially restricted and are found in natural settings near freshwater springs and seepages and along braided channels. Human activities also create conditions that enable brackish marshes to proliferate. For example, impoundments where urban and agricultural runoff accumulates, as well as areas that receive wastewater discharges and sewage spills, are sites of brackish marsh development. Hydrologic and topographic modifications, therefore, appear to be significant factors in the formation of this habitat (Zedler et al., 1992). Intermittent freshwater inflows, variable precipitation, and evaporation result in fluctuating water levels and salinities that also influence development of this habitat. Alternating flood and drought conditions control the location and longevity of many brackish marshes. Such major shifts in environmental conditions are manifested by the ephemeral nature of brackish marsh habitats in some areas of the reserve.

Many organisms found in Tijuana River NERR brackish marshes also inhabit freshwater marsh habitat. Dominant emergent plant species include cattails (*Typha domingensis*) and bulrushes (*Scirpus californicus*). Spring rush (*Juncus acutus*) is likewise common. Widgeon grass (*Ruppia maritima*) is the dominant submerged aquatic vegetation. Brackish marshes are important habitat for insects (e.g., dragonflies, *Anax junius*). Birds are frequent inhabitants. For example, the red-winged blackbird (*Agelaius phoeniceus*) is often observed in the emergent vegetation. Other species wade and feed in marsh pools. Examples are the black-bellied plover (*Pluvialis squatarola*), killdeer (*Charadrius vociferus*), semipalmated plover (*C. semipalmatus*), least sandpiper (*Calidris minutilla*), black-necked stilt (*Himantopus mexicanus*), American avocet (*Recurvirostra americana*), willet (*Catoptrophorus semipalmatus*), dowitcher (*Limnodromus* sp.), and snowy egret (*Egretta thula*). These birds can be seen foraging on invertebrates and other organisms in the marshes (Zedler et al., 1992).

Riparian Habitat

Although riparian vegetation comprises less than 10% of the total Tijuana River NERR habitat area, it supports some of the most diverse avian, reptilian, mammalian, and finfish communities. Insect populations are abundant. Some mammals (e.g., bobcats and long-tailed weasels) occupy riparian woodlands.

Riparian habitats occur along moist perimeter areas of freshwater streams, rivers, and lakes. The phreatophytic vegetation of this habitat requires stream-, river-, or lake-influenced groundwater for growth and reproduction. Riparian zones may include wetland habitats, but they may lack hydric soil properties, thus preventing their designation as regulated wetlands (Tiner, 1999). Moving upstream along the Tijuana River, the riparian habitat consists of floodplain low shrub vegetation (i.e., mulefat scrub) such as salt bushes (*Baccharis glutinosa*) and sandbar willow (*Salix hindsiana*). Forests comprised of cottonwoods (*Populus fremontii*), willows (*Salix* spp.), and other species of larger plants replace the shrub vegetation, rising up to ~20 m in height (Zedler et al., 1992). As stated by Cox (1996, p. 158), "When the boundaries of riparian areas are based on their potential function as buffers, the distance from the stream will be highly variable depending on local soils and hydrological, topographical, and vegetation characteristics and should be determined for each unique stream reach." The riparian corridor continues to be threatened by anthropogenic activities (e.g., development, agriculture, sand and gravel mining, and flood control operations) in the Tijuana River NERR that are altering and destroying sensitive habitat at an alarming rate.

Wetland–Upland Transition

A transition community of wetland and upland vegetation exists in the Tijuana River NERR at elevations between ~1.4 and 2.1 m. Much of the transition community has been disturbed in the reserve, with only small remnants persisting, such as at the northern end of the Tijuana Estuary. The most impacted area in the transition zone occurs in peripheral uplands where habitat alteration has been considerable. Desmond et al. (2001) have described the plant species in this transition community based on observations in the Tijuana Estuary and other sites in southern California coastal wetlands. Salt marsh species characteristic of this community are the salt grass (*Distichlis spicata*), perennial glasswort (*Salicornia subterminalis*), sea lavender (*Limonium californicum*), alkali heath (*Frankenia salina*), and Watson's salt bush (*Atriplex watsonii*). Upland species characteristic of the transition zone include the box-thorn (*Lycium californicum*), California salt bush (*Atriplex californica*), yerba reuma (*Frankenia palmeri*), and pineapple weed (*Ambylopappus pusillus*). Ferren (1985) and Cox and Zedler (1986) reported several invasive species in this habitat as well, namely the Australian salt bush (*Atriplex semibaccata*) and crystal ice plant (*Mesembryanthemum crystallinum*). Desmond et al. (2001) also documented several exotic annual species, specifically the European sicklegrass (*Parapholis incurva*), rabbitfoot beardgrass (*Polypogon monspeliensis*), and little ice plant (*Mesembryanthemum nodilforum*).

Zedler et al. (1992) provided a list of the most abundant plant species comprising the wetland–upland transition community at the northernmost part of the Tijuana Estuary. Among these species are *Atriplex semibaccata, A. watsonii, Artemisia californica, Cressa truxillensis, Distichlis spicata, Eriogonum fasciculatum, Frankenia grandifolia, Haplopappas venetus, Limonium californicum, Lycium californicum, Monanthocloe littoralis, Salicornia subterminalis, S. virginica,* and *Rhus laurina*. Coastal sage scrub forms the upland community here; it is comprised of various shrub species, such as evergreens (e.g., jojoba, *Simmondsia chinensis*; laurel sumac, *Rhus laurina*; and lemonadeberry, *R. integrifolia*) and drought-deciduous forms (e.g., California sagebrush, *Artemisia californica*; and golden bush, *Haplopappus venetus*). Some exotic species (e.g., *Atriplex semibaccata*) have adapted to the upland zone. Zedler et al. (1992) noted that the steep topography in the area results in a narrow band of overlap between the upland and wetland plant assemblages. The occurrence of the salt grass (*Distichlis spicata*) high on the slope marks the upper margin of the wetland community. Other wetland plants (e.g., alkali heath, *Frankenia grandifolia;* and alkali weed, *Cressa truxillensis*) may also appear near this upper margin. Additional information on the plants in the wetland–upland transition can be obtained elsewhere (e.g., Zedler and Cox, 1985; James and Zedler, 2000).

An array of reptilian, mammalian, and avian species utilizes the wetland–upland transition habitat. Reptilian species of significance include the side-blotched lizard (*Uta stansburiana*), San Diego gopher snake (*Pituophis melanoleucus annectens*), and California king snake (*Lampropeltis getulus californiae*). These herpetofauna are relatively abundant in this habitat (Desmond et al., 2001).

Small mammals reported in the transition habitat are the brush mouse (*Peromyscus boyli*), cactus mouse (*P. eremicus*), deer mouse (*P. maniculatus*), San Diego pocket mouse (*P. fallax*), western harvest mouse (*Reithrodontomys megalotis*), agile kangaroo rat (*Dipodomys agilis*), and dusky-footed woodrat (*Neotoma fuscipes*). Squirrels (e.g., California ground squirrel, *Spermophilus beechyi*), rabbits (e.g., desert cottontail, *Sylvilagus audoboni*; and California jackrabbit, *Lepus californicus*), and opossum (e.g., *Didelphis marsupialis*) are also present. Larger carnivorous species of note include the long-tailed weasel (*Mustela frenata*), striped skunk (*Mephitis mephitis*), and coyote (*Canis latrans*) (Zedler et al., 1992; Zedler, 2001).

Birds are frequent visitors to the transition habitat. They rest and feed there. Several species (e.g., northern harrier, *Circus cyaneus*; short-eared owl, *Asio flammeus*; and black-shouldered kite, *Elanus caeruleus*) prey on some of the aforementioned small mammals (Zedler et al., 1992).

Dunes and Beach Habitat

Human activities and stochastic natural events (e.g., major storms) have altered the structure of the dunes and beach habitat and the associated native plant and animal communities. The dunes and beach habitat is highly dynamic and physically controlled. Strong wind and wave action, as well as human disturbance due to construction and recreational pursuits, have eliminated much of the natural vegetation in some areas. The trampling of dune vegetation and habitat by people, horses, and vehicles has caused destabilization of the dunes. The denudation of native dune plants has facilitated sediment erosion and enabled exotic plants to invade the habitat. The establishment of such exotic species has adversely affected native floral and faunal communities along the shoreline (Zedler et al., 1992).

Dune vegetation at the Tijuana Estuary consists of dune ragweed (*Ambrosia chamissonis*), dune primrose (*Camissonia cheiranthifolia* ssp. *suffruticosa*), sand verbena (*Abronia umbellata*), salt bush (*Atriplex leucophylla*), sea rocket (*Cakile maritima*), and hottentot-fig (*Carpobrotus edulis*). The salt bush, sea rocket, and hottentot-fig are invasive species that have adapted to the dune habitat subsequent to anthropogenic disturbance. The exotic dune plants have outcompeted some of the native plant species. Anthropogenic impacts appear to be responsible for the extirpation of the lemonadeberry (*Rhus integrifolia*) from the dunes (Zedler, 2001).

The native dune vegetation provides excellent habitat for various insects, particularly several species of burrowing beetles. Included here are the sand dune tiger beetle (*Coelus latesignata latesignata*), the sandy beach tiger beetle (*C. hirticollis gravida*), and the globose dune beetle (*C. globosus*). Both harvest and wood ants are abundant in the dunes. The coast horned lizard (*Phrynosoma coronatum blainvillei*) feeds on these ants. Two other species of lizards observed on the dunes are the side-blotched lizard (*Uta stansburiana*) and the silvery legless lizard (*Anniella pulchra pulchra*) (Entrix et al., 1991).

A number of bird species feed, nest, or rest on the dunes. For example the Belding's Savannah sparrow (*Ammodramus sandwichensis beldingi*) forages heavily on flies and other small insects on the dunes. Western snowy plovers (*Charadrius alexandrinus nivosus*) consume invertebrates along the shoreline. The California least tern (*Sterna antillarum*) is a colonial nesting species that prefers to lay its eggs in sand depressions on the dunes. It nests in areas nearly devoid of vegetation but is vulnerable to predators in this exposed setting. In addition to the snowy plover, several other shorebird species use the dunes and beach habitat, most conspicuously the black-bellied plover (*Pluvialis squatarola*), dowitcher (*Limnodromus* sp.), marbled godwit (*Limosa fedoa*), willet (*Catoptrophorus semipalmatus*), sanderling (*Calidris alba*), least sandpiper (*C. minutilla*), western sandpiper (*C. mauri*), and whimbrel (*Numenius phaeopus*) (Boland, 1981).

Intertidal Flats

Mudflats and sandflats rank among the most important habitats in the Tijuana estuarine system, supporting the most diverse invertebrate assemblages of the coastal wetland. Both of these habitats are often visited by shorebirds that feed on benthic invertebrates at low tide. Bivalves, crabs, gastropods, polychaetes, and amphipods are abundant members of the benthic communities of these habitats (Levin et al., 1998; Desmond et al., 2001).

Several invertebrate species are particularly abundant, including the yellow shore crab (*Hemigrapsus oregonensis*), lined shore crab (*Pachygrapsus crassipes*), fiddler crab (*Uca crenulata*), and California horn snail (*Cerithidea californica*). The California horn snail is an irruptive species, reaching peak densities greater than 1000 individuals/m^2. It can outnumber all other benthic species at some locations (Zedler et al., 1992; Zedler, 2001).

The intertidal flats of the Tijuana Estuary are favored sites for shorebirds, which use this habitat more than any other type. Waders are well represented. Among the small waders often observed feeding along the flats are the western snowy plover (*Charadrius alexandrinus nivosus*), semipalmated plover (*C. semipalmatus*), western sandpiper (*Calidris mauri*), least sandpiper (*C. minutilla*), sanderling (*C. alba*), red knot (*C. canutus*), and light-footed clapper rail (*Rallus longirostris levipes*). Large waders also commonly feed on the flats; for example, the marbled godwit (*Limosa fedoa*), dowitcher (*Limnodromus* sp.), long-billed curlew (*Numenius americanus*), whimbrel (*N. phaeopus*), American avocet (*Recurvirostra americana*), willet (*Catoptrophorus semipalmatus*), and yellowlegs (*Tringa* spp.) frequently probe the sediment surface in search of food (Boland, 1981).

ESTUARY

Aquatic Habitat: Tidal Creeks and Channels

The most important aquatic habitats in the Tijuana River NERR occur in a network of small tidal creeks and larger channels that support a diversity of phytoplankton, benthic algae, invertebrates, fish, birds, and wildlife. The tidal creeks and channels, which encompass more than a 60-ha area of the reserve, generally average ~1.4 and ~10 m in width, respectively. The tidal creeks are typified by lower tidal flows than the channels, which are distributed closer to marine habitats. Because the tidal creeks usually drain completely at low tide while the channels remain inundated during all tidal stages, the tidal creeks are only occasionally vegetated, while the channels support submerged aquatic vegetation (Desmond et al., 2001).

Plants

The phytoplankton community consists of an array of diatoms, dinoflagellates, and blue-green algae. Smaller forms, picoplankton (~1 to 2 μm in size) of uncertain taxonomy, are particularly abundant in the tidal creeks and channels. Phytoplankton blooms develop in the estuary from March to June and may be facilitated by reduced tidal flushing (Fong, 1986). Peak cell counts and biomass occur in spring.

Benthic macroalgae are common along the creek and channel bottoms. Chlorophytes (e.g., *Enteromorpha* sp. and *Ulva* sp.) predominate, and they often appear drifting along the bottom or floating to the water surface. Both macroalgae and phytoplankton attain highest biomass in small tidal creeks with low flow velocities. Diminished tidal flushing and reduced circulation promote the accumulation of algal biomass. A greater standing crop of the macroalgae also correlates positively with prevailing winds that move floating algal mats along the water surface (Rudnicki, 1986).

Benthic Invertebrates

Table 7.2 provides a species list of invertebrates, primarily benthic forms, found in the Tijuana Estuary. The abundance and distribution of benthic fauna are greatly affected by the sediment type and hydrologic conditions in the estuary. For example, deposit feeders, such as *Callianassa californiensis, Glycera dibranchiata, Pectinaria californiensis,* and *Scoloplos armiger*, inhabit areas with fine-grained sediments. In contrast, filter feeders, such as *Crassostrea* sp., *Cryptomya californica, Protothaca staminea,* and *Tagelus californianus,* prefer coarser, medium-sized sediments. Rapid increases in stream flow during periods of heavy precipitation in winter can markedly shift salinities as well as the sediment structure along the bottom of the tidal creeks and channels, thereby impacting the benthic populations (Zedler et al., 1984; Nordby and Zedler, 1991).

Since 1970, significant changes have occurred in the benthic community of the Tijuana Estuary, as evidenced by the investigations of Ford et al. (1971), Smith (1974), Peterson (1975), International Boundary and Water Commission (1976), Hosmer (1977), Rehse (1981), Zedler et al. (1992), and Zedler (2001). Bivalves dominated the benthic community between 1970 and 1977. Investigations by Peterson (1975) in the early 1970s, for example, revealed that the purple clam (*Sanquinolaria nuttalli*) and littleneck clam (*Protothaca staminea*) were the most abundant species in the sandy bottom community, attaining densities of 75 and 35 individuals/m^2, respectively. Other bivalves that were numerically abundant included the California jackknife clam (*Tagelus californianus*) and false mya (*Cryptomya californica*). Gastropods, decapod crustaceans, and polychaete worms were also relatively abundant. Heavy precipitation and flooding events in winter during the 1978–1980 period resulted in marked changes in the benthic community. Later hydrologic disturbances, notably reduced salinity associated with wastewater inflows, caused additional compositional change.

TABLE 7.2
Invertebrate Species Identified in the Tijuana Estuary

Bivalves
Chione californiensis
Chione fluctifraga
Chione undatella
Cistenides brevicoma
Cryptomya californica
Cooperella subdiaphana
Crassostrea sp.
Diplodonta orbella
Donax californicus
Leporimetis obesa
Laevicardium substriatum
Leptopecten latiauratus
Lucina nuttalli
Macoma nasuta
Macoma secta
Mactra californica
Mytilus edulis
Ostrea lurida
Protothaca laciniata
Protothaca staminea
Nuttallia nuttalli
Saxidomus nuttalli
Siliqua patula
Tagelus californianus
Tagelus subteres
Tellina carpenteri
Tellina modesta
Tresus nuttallii
Musculista senhousia
Spisula planulata
Solen rosaceus

Polychaete Worms
Armandia brevis
Axiothella rubrocincta
Anaitides sp.
Capitella capitata
Chaetopterus variopedatus
Diopatra splendidissima
Eunice sp.
Euzonus mucronata
Glycera dibranchiata
Glycera capitata
Goniada brunnea

Gastropods
Acteocina inculta
Aplysia californica
Assiminea californica
Bulla gouldiana
Cerithidea californica
Crepidula fornicata
Crepidula onyx
Cylichnella culcitella
Haminoea vesicula
Melampus olivaceus
Nassarius fossatus
Nassarius tegula
Navanax inermis
Olivella baetica
Olivella biplicata
Polinices lewisii
Serpulorbis squamigerus

Decapod Crustaceans
Callianassa californiensis
Callianassa gigas
Callianassa affinis
Cancer antennarius
Cancer productus
Emerita analoga
Hemigrapsus oregonensis
Loxorhynchus crispatus
Pachygrapsus crassipes
Pagurus samuelis
Pinnixa franciscana
Portunus xantusii
Scleroplax granulata
Malacoplax californiensis
Uca crenulata
Upogebia sp.
Pagurus hirsutiusculus

Amphipod Crustaceans
Ampithoe plumulosa
Eohaustorius washingtonianus
Jassa falcata

TABLE 7.2 (CONTINUED)
Invertebrate Species Identified in the Tijuana Estuary

Haploscoloplos elongatus	**Cirriped Crustaceans**
Hemipodus borealis	*Balanus amphitrite*
Laonice cirrata	*Balanus glandula*
Magelona pitelkai	
Nereis brandti	**Isopod Crustaceans**
Nephtys spp.	*Excirolana chiltoni*
Nephtys caecoides	
Nephtys californiensis	**Cnidaria**
Nephtys punctata	*Renilla kollikeri*
Notomastus tenuis	*Corymorpha palma*
Ophelia limacina	
Owenia fusiformis	**Platyhelminthes**
Pectinaria californiensis	*Stylochus* sp.
Polydora cornuta	*Stylochus franciscanus*
Polydora nuchalis	
Polydora spp.	**Turbellaria**
Prionospio cirrifera	Unidentified species
Prionospio spp.	
Scoloplos armiger	**Hemichordata**
Serpula vermicularis	*Saccoglossus* sp.
Spio filicornis	
Spiophanes missionensis	**Brachipoda**
Streblospio benedicti	*Glottidia albida*
Tharynx parvis	
Capitellidae	**Phoronida**
Goniadidae	*Phoronis architecta*
Lumbrineridae	
Magelonidae	**Nemertea**
Maldanidae	*Carinoma mutabilis*
Orbiniidae	Unidentified species
Phyllodocidae	

Source: Zedler, J.B., C.S. Nordby, and K.B. Kus. 1992. The Ecology of Tijuana Estuary, California: A National Estuarine Research Reserve. Technical Report, NOAA Office of Coastal Resource Management, Sanctuaries and Reserves Division, Washington, D.C.

The benthic invertebrate community in 1980 differed considerably in species composition from that in the early to mid-1970s. By 1980, for instance, the false mya had replaced the purple clam as the dominant bivalve in the estuary. Other species prominent in the 1970s were absent in 1980 benthic samples. These included the burrowing shrimp (*Callianassa gigas*), Washington clam (*Saxidomus nuttalli*), white sand clam (*Macoma secta*), bent-nose clam (*M. nasuta*), Carpenter's tellen (*Tellina carpenteri*), yellow clam (*Florimentis obesa*), and egg cockle (*Laevicardium substriatum*) (Zedler et al., 1992).

The distribution of bivalves in the estuary is dependent on sediment type. Hosmer (1977) correlated bivalve biomass to sediment grain size. Although the biomass of bivalves generally increased with increasing grain size, considerable variation was observed even among the dominant species. For example, the biomass and density of the purple clam peaked in coarse sand. The littleneck clam attained higher biomass and density in finer sediments but was found in a wide range of sediment types from silt and clay to coarse sand. The California jackknife clam inhabited a narrower range of sediment types than the littleneck clam, residing in fine to medium sand. Both biomass and density were greater in clams collected in fine sand.

Of the aforementioned dominant bivalve species, Hosmer (1977) registered a mean size of 71 mm for the purple clam, 27 mm for the California jackknife clam, and 22 mm for the littleneck clam. The mean sizes of the California jackknife clam and the littleneck clam changed significantly during later benthic sampling in the estuary (1986–1989). For example, Nordby and Zedler (1991) found the mean size of the littleneck clam to be about half of that reported by Hosmer (1977). The California jackknife clam, in turn, was larger than that reported by Hosmer (1977). In addition, no purple clams were collected in their samples. These changes reflect more systemwide shifts in the benthic invertebrate community structure occurring in the estuary during the 1978–1980 period, when floods caused mass mortality of many benthic invertebrate species.

Hydrographic disturbances after 1980, such as the floods of 1986, 1987, and 1988, which greatly reduced salinities, were also detrimental to benthic communities and were tolerated only by the most resilient species (i.e., short-lived forms with short generation times and protracted spawning). Nordby and Zedler (1991) demonstrated that over their sampling period from 1986 through 1988 the overwhelming trends of the benthic community in the estuary were toward lower species richness and abundance. As in the case of the benthic assemblages chronicled after the 1980 floods, younger and smaller animals predominated. Through periods of ongoing wastewater inflows, Nordby and Zedler (1991) ascertained that pollution-tolerant capitellids and spionids (mainly *Polydora nuchalis* and *P. cornuta*) were the dominant polychaetes. Collections of bivalves were more variable. *Tagelus californianus* was the dominant bivalve species in 1986, but it declined substantially in 1987, when *Cryptomya californica* and *Protothaca staminea* became abundant (Table 7.3). While *T. californicus* continued to decrease and *P. staminea* continued to increase in 1988, *C. californica* leveled off in numbers. Over the three-year period, bivalve densities dropped substantially from a high of more than 2500 individuals/m^2 in September 1986 to generally less than 1000 individuals/m^2 through December 1988. Most of the bivalves collected in the benthic samples were young individuals ranging from 0 to 1 year of age.

Fish

Table 7.4 shows the species of fish that have been identified in the Tijuana Estuary. Three resident species dominated the fish community in the estuary prior to the flooding events during the 1978–1980 period. These were the striped mullet (*Mugil cephalus*), California killifish (*Fundulus parvipinnis*), and longjaw mudsucker (*Gillichthys mirabilis*). Reduced salinity due to flooding in 1980 and hypersalinity due

TABLE 7.3
Annual Relative Abundance of the Dominant Channel Organisms Collected at the Tijuana Estuary (% of Total)

Organism	Year		
	1986–1987	1987–1988	1988–1989
Fishes			
Atherinops affinis	52%	14%	7%
Clevelandia ios	41%	58%	90%
Fundulus parvipinnis	4%	19%	1%
Gillichthys mirabilis	0%	<1%	<1%
Gambusia affinis	0%	0%	0%
Total fishes collected	20,888	4,976	54,301
Bivalves			
Tagelus californianus	73%	33%	27%
Protothaca staminea	19%	34%	42%
Macoma nasuta	2%	17%	19%
Cryptomya californica	0%	6%	4%
Total bivalves collected	658	490	651
Polychaetes			
Capitellidae	ND	33%	50%
Spionidae			
Boccardia spp.	ND	<1%	5%
Polydora spp.	ND	18%	20%
Nephtys spp.	ND	16%	1%
Pseudopolydora spp.	ND	0%	0%
Spionphanes missionensis	ND	0%	8%
Opheliidae			
Armandia brevis	ND	<1%	5%
Euzonus mucronata	ND	0%	0%
Unidentified taxa	ND	3%	0%
Total polychaetes collected	ND	276	1,422

Note: ND = no data.

Source: Nordby, C.S. and J.B. Zedler. 1991. *Estuaries* 14: 80–93.

to closure of the ocean inlet in 1984 caused important changes in the abundance, distribution, and diversity of fishes in the estuary. Topsmelt (*Atherinops affinis*) was the most abundant species collected during the 1980–1984 period. Striped mullet and California killifish were also relatively abundant after the 1980 flood and 1984 inlet closure, respectively. Species diversity decreased in response to these environmental changes. Thus, 29 species of fish were reported in the estuary prior to 1978 (Zedler et al., 1992), whereas only 21 species of fish were recorded during the 1986–1989 period (Nordby and Zedler, 1991).

TABLE 7.4
Species of Fish Recorded in the Tijuana Estuary

Atherinidae

Topsmelt — *Atherinops affinis*

Batrachoididae

Specklefin midshipman — *Porichthys myriaster*

Blennidae

Bay blenny — *Hypsoblennius gentilis*

Bothidae

Sanddabs — *Citharichthys* spp.
California halibut — *Paralichthys californicus*

Clupeidae

Pacific sardine — *Sardinops sagax caeruleus*

Cottidae

Scalyhead sculpin — *Artedius harringtoni*
Staghorn sculpin — *Leptocottus armatus*

Cynoglossidae

California tonguefish — *Symphurus atricauda*

Cyprinodontidae

California killifish — *Fundulus parvipinnis*

Dasyatididae

Round stingray — *Urolophus halleri*

Embiotocidae

Barred surfperch — *Amphistichus argenteus*
Shiner perch — *Cymatogaster aggregata*
Walleys surfperch — *Hyperprosopon argenteum*

Engraulidae

Deepbody anchovy — *Anchoa compressa*
Slough anchovy — *Anchoa delicatissima*
Northern anchovy — *Engraulis mordax*

Gobiidae

Yellowfin goby — *Acanthogobius flavimanus*
Arrow goby — *Clevelandia ios*
Longjaw mudsucker — *Gillichthys mirabilis*
Cheekspot goby — *Llypnus gilberti*
Bay goby — *Lepidogobius lepidus*
Shadow goby — *Quietula y-cauda*

Kyphosidae

Opaleye — *Girella nigricans*

TABLE 7.4 (CONTINUED)
Species of Fish Recorded in the Tijuana Estuary

Labridae
California sheepshead — *Semicossyphus pulcher*

Mugilidae
Striped mullet — *Mugil cephalus*

Myliobatidae
Bat ray — *Myliobatus californicus*

Ophidiidae
Basketweave cusk-eel — *Otophidium scrippsi*

Pleuronectidae
Diamond turbot — *Hypsopsetta guttulata*
C-O turbot — *Pleuronichthys coenosus*
Spotted turbot — *Pleuronichthys ritteri*
Hornyhead turbot — *Pleuronichthys verticalis*

Poeciliidae
Mosquitofish — *Gambusia affinis*

Rhinobatidae
Shovelnose guitarfish — *Rhinobatus productus*

Sciaenidae
White croaker — *Genyonemus lineatus*
California corbina — *Menticirrhus undulatus*
Queenfish — *Seriphus politus*

Scombridae
Pacific mackerel — *Scomber japonicus*

Serranidae
Kelp bass — *Paralabrax clathratus*
Spotted sandbass — *Paralabrax maculatofasciatus*
Barred sandbass — *Paralabrax nebulifer*

Sphyraenidae
California barracuda — *Sphyraena argentea*

Syngnathidae
Bay pipefish — *Syngnathus leptorhynchus*

Source: Zedler, J.B., C.S. Nordby, and K.B. Kus. 1992. The Ecology of Tijuana Estuary, California: A National Estuarine Research Reserve. Technical Report, NOAA Office of Coastal Resource Management, Sanctuaries and Reserves Division, Washington, D.C.

River flooding and wastewater inflows during the late 1980s resulted in further changes in the fish community. At this time, the arrow goby (*Clevelandia ios*) far outnumbered the other species in absolute abundance, comprising 75% of all the individuals collected in field samples. Nordby (1982) also asserted that goby larvae dominated other larval forms in the estuary. Topsmelt was the second most abundant species, accounting for 19% of all individuals. California killifish, in turn, constituted only 3% of the fish in the collections. While the arrow goby and topsmelt were co-dominant species in 1986, by 1989 the arrow goby had become the overwhelming numerical dominant. Table 7.3 shows the annual relative abundance (% of total) of the dominant fish species collected in the Tijuana Estuary during this time frame.

Over the 3-year period from 1986 to 1989, the number of fish species dropped from 14 to 6 in the estuary, reflecting the inability of some of the resident fishes to adapt to the acute salinity changes associated with river flooding and wastewater inflow. Zedler et al. (1992) disclosed that by 1991 the California killifish once again increased in abundance in the estuary to become a co-dominant form along with the arrow goby. Although there have been significant compositional variations in the fish community of the Tijuana Estuary, forage species have consistently dominated in terms of numerical abundance through the 1990s (Zedler, 2001).

Arrow gobies, as well as other forage species such as killifishes and anchovies, are important sources of food for recreational and commercial species. Some of the most important recreational and commercial species in the estuary include the California halibut (*Paralichthys californicus*), the white croaker (*Genyonemus lineatus*), sea bass (*Paralabrax* spp.), flounders (*Pleuronichthys* spp.), and walleys surfperch (*Hyperprosopon argenteum*). Resident species typically are more abundant than these fishes (Zedler et al., 1992).

Birds

Tidal creeks and channels are important resting and feeding areas for a variety of waterbird species that also frequent other habitats in the reserve (Table 7.5). Some of the commonly observed waterbirds in these aquatic habitats are herons and egrets (e.g., great egret, *Casmerodius albus*; great blue heron, *Ardea herodias*; snowy egret, *Egretta thula*; and reddish egret, *Egretta rufescens*), gulls and terns (e.g., California gull, *Larus californicus*; western gull, *L. occidentalis*; Forster's tern, *Sterna forsteri*; and black skimmer, *Rynchops niger*), waders (e.g., light-footed clapper rail, *Rallus longirostris levipes*), cormorants (e.g., double-crested cormorant, *Phalacrocorax auritus*), pelicans (brown pelican, *Pelecanus occidentalis*), and raptors (e.g., osprey, *Pandion haliaetus*). Forage fishes, including the topsmelt, California killifish, arrow goby, and northern anchovy (*Engraulis mordax*), are generally the preferred prey of these birds (Zedler et al., 1992; Zedler, 2001).

ANTHROPOGENIC IMPACTS

In contrast to some of the more pristine estuarine reserves such as the Jacques Cousteau NERR (see Chapter 3), the Tijuana River NERR has been affected by anthropogenic disturbances. Among the most significant disturbances are chronic

TABLE 7.5
Waterbird Species Observed at the Tijuana Estuary between October and April

Common Name	Scientific Name	Habitat[a]
Pelicans, Grebes, and Cormorants		
Pied-billed grebe	*Podilymbus podiceps*	I
Eared grebe	*Podiceps nigricollis*	I
Western grebe	*Aechmophorus occidentalis*	I
Double-crested cormorant	*Phalacrocorax auritus*	I, S
Brown pelican	*Pelecanus occidentalis*	I
Herons and Egrets		
Great blue heron	*Ardea herodias*	I, M
Great egret	*Casmerodius albus*	I, M, D
Little blue heron	*Egretta caerulea*	I, M, D
Reddish egret	*Egretta rufescens*	I
Snowy egret	*Egretta thula*	I, M, D
Waterfowl		
Canada goose	*Branta canadensis*	I
American wigeon	*Anas americana*	I, M, D
Gadwall	*Anas strepera*	I
Green-winged teal	*Anas crecca*	I, M
Mallard	*Anas platyrhynchos*	I
Northern pintail	*Anas acuta acuta*	I, M
Blue-winged teal	*Anas discors*	I
Cinnamon teal	*Anas cyanoptera*	I
Northern shoveler	*Anas clypeata*	I
Lesser scaup	*Aythya affinis*	I
Surf scoter	*Melanitta perspicillata*	I
Common goldeneye	*Bucephala clangula*	I
Bufflehead	*Bucephala albeola*	I, M
Red-breasted merganser	*Mergus serrator*	I
Ruddy duck	*Oxyura jamaicensis*	I
American coot	*Fulica americana*	I, S
Small Waders		
Spotted sandpiper	*Tringa macularia*	I, B
Black-bellied plover	*Pluvialis squatarola*	I, M, B, S, D
Killdeer	*Charadrius vociferus*	I, M, D
Semipalmated plover	*Charadrius semipalmatus*	I, B, D
Snowy plover	*Charadrius alexandrinus*	I, B, D
Ruddy turnstone	*Arenaria interpres*	I, M, S, D
Black turnstone	*Arenaria melanocephala*	I
Least sandpiper	*Calidris minutilla*	I, B, D
Red knot	*Calidris canutus*	I

(continued)

TABLE 7.5 (CONTINUED)
Waterbird Species Observed at the Tijuana Estuary between October and April

Common Name	Scientific Name	Habitat[a]
Western sandpiper	*Calidris mauri*	I, B, D
Dunlin	*Calidris alpina pacifica*	I, B, S, D
Sanderling	*Calidris alba*	I, B, S, D
Common snipe	*Gallinago gallinago*	M
Light-footed clapper rail	*Rallus longirostris levipes*	I, M
Large Waders		
Black-necked stilt	*Himantopus mexicanus*	I, M, S
American avocet	*Recurvirostra americana*	I, D
Willet	*Catoptrophorus semipalmatus*	I, M, B, S, D
Yellowlegs	*Tringa* spp.	I, M, D
Whimbrel	*Numenius phaeopus*	I, M, B, D
Long-billed curlew	*Numenius americanus*	I, M, S, D
Marbled godwit	*Limosa fedoa*	I, M, B, S, D
Dowitcher	*Limnodromus* sp.	I, M, S, D
Gulls and Terns		
Bonaparte's gull	*Larus philadelphia*	I
Heermann's gull	*Larus heermanni*	I, B, S
Ring-billed gull	*Larus delawarensis*	I, M, B, S, D
Mew gull	*Larus canus*	B
California gull	*Larus californicus*	I, S, D
Herring gull	*Larus argentatus*	I, S, D
Western gull	*Larus occidentalis*	I, M, B, S, D
Caspian tern	*Sterna caspia*	I, D
Royal tern	*Sterna maxima*	I, S
Elegant tern	*Sterna elegans*	I, S, D
Common tern	*Sterna hirundo*	S
Forster's tern	*Sterna forsteri*	I, B, S, D
Black skimmer	*Rynchops niger*	I

[a] Habitat: I = intertidal flats; M = salt marsh; B = beach; S = sandflat; D = dune.

Source: Zedler, J.B., C.S. Nordby, and K.B. Kus. 1992. The Ecology of Tijuana Estuary, California: A National Estuarine Research Reserve. Technical Report, NOAA Office of Coastal Resource Management, Sanctuaries and Reserves Division, Washington, D.C.

sewage wastewater inflows to the estuary from the City of Tijuana, which have averaged ~38 to 45 MLD. These inflows have been responsible for significant reductions in salinity and negative impacts on channel biota. During some periods (e.g., 1987–1988), sewage inflows amounted to ~83 MLD (Seamans, 1988). The chronic wastewater inflows have occasionally been augmented by sewage spills from broken pipelines. The net effect has been the periodic change of intermittent stream flow in the Tijuana River to essentially a permanently flowing system. However, this condition has improved in recent years (Jeffrey Crooks, Tijuana River National Estuarine Research Reserve, personal communication, 2003).

Watershed development and associated human activities are disturbing and destabilizing soils and accelerating erosion in land areas surrounding the river and estuary. The effects of development are evident throughout the watershed, and disturbed soils are easily mobilized by flooding events. The trampling of dune vegetation, destabilization of beach and dune sands, and storm activity facilitate erosion of the system (Zedler and Nordby, 1986). Runoff from farmlands may contribute to these problems. Land erosion and sediment influx to tidal creeks, channels, and lagoonal habitats raise turbidity levels and periodically bury benthic organisms in the estuary (Nordby, 1987). Sedimentation of channels is a major ecological impact in the estuary. Dams (i.e., Rodriquez, Barrett, and Morena dams) modulate downstream flows and protect the estuary from the influx of large volumes of sediment by regulating nearly 80% of the watershed area.

Inlet closure and subsequent sediment buildup have required the implementation of dredging operations to maintain navigable waterways. Some of the sediment enters the lower estuary from dune washovers that accompany major storms, but a larger volume of sediment derives from the surrounding watershed upriver and enters the system in largest quantities during floods. Dredging of the main estuary enhances tidal flushing and also increases turbidity as well as mortality of benthic organisms. Most of the biotic impacts caused by dredging are ephemeral, although recovery of the benthos can be protracted (Zedler et al., 1992).

The modification of stream flow is a major ecological stressor in the estuary. Aside from wastewater discharges, several other factors alter stream flow in the system. For example, the Barrett, Morena, and Rodriguez reservoirs trap stream flow and thus modify the volume of downstream discharges (Zedler et al., 1992).

Dikes and levees have been constructed in various areas of the reserve. Dike construction in lagoonal waters, for instance, has been used to create wastewater receiving ponds. Levees have been constructed along the Tijuana River flood plain for flood control. These structures, together with agricultural lands, have substantially altered the Tijuana flood plain. Agricultural and military activities were once concentrated throughout the southern half of the estuary. A military airport was constructed more than 50 years ago east of the estuary (Zedler et al., 1992).

Other human activities of significance that have directly damaged habitats include gravel extraction and the use of off-road vehicles. A primary goal of the Tijuana River NERR has been to promote restoration efforts, with the major focus on the restoration of tidal wetlands (see Zedler, 2001). Another concern is the maintenance of an open tidal inlet necessary to provide the habitat necessary to improve biodiversity in the estuary. The removal of sediment in the lower estuary and near the inlet will increase tidal flushing and restore tidal circulation favorable for the reestablishment of native species that have declined due to diminished tidal influence in response to repeated sedimentation events associated with storms and episodic flooding. Finally, restoration projects are stressing the importance of revegetating disturbed lands along perimeter areas of the Tijuana River to upland areas of the watershed. Resource management programs are vital to maintain the variety of habitats needed to support healthy communities of organisms in the system (Zedler et al., 1992).

SUMMARY AND CONCLUSIONS

The Tijuana Estuary represents one of the least fragmented salt marsh systems in Southern California. It is an estuary characterized by highly variable hydrologic conditions due to large fluxes in the volume of stream flow over an annual cycle and periodic flooding during the winter (wet) season. Human activities exacerbate natural perturbations in environmental conditions. For example, discharges of raw sewage and wastewaters from Tijuana, Mexico, as well as spills of sewage down border canyons, have delivered pathogens, nutrients, chemical contaminants, and additional fluid flow to the estuary. While the wastewater discharges in past years have provided for more consistent flow in the Tijuana River, reservoirs (i.e., Barrett, Morena, and Rodriguez reservoirs) have delayed flood flows, prolonged the period of wet-season flows, and reduced the total volume of stream flow. As a result, both monthly and annual flows of the Tijuana River have changed markedly, with coefficients of variation of 690 and 325%, respectively.

The neighboring watershed is heavily populated and characterized by considerable domestic and agricultural development. Construction and other disturbances have created unstable slopes and soils that are susceptible to erosion and runoff. Thus, large volumes of sediment are transported to tidal creeks, channels, and the lower estuary, especially during periods of heavy precipitation and flooding. Some areas of the estuary are clogged with sediment. Sedimentation has been catastrophic at the coast and is exacerbated by occasional dune washovers associated with storm surges. It has reduced tidal influence and has led to lower biodiversity. Periodic dredging near the coast has increased the tidal prism and improved circulation in the estuary.

Anthropogenic activities have had a significant impact on the estuarine biotic communities. For example, wastewater inflows during the 1980s markedly reduced salinities, causing mass mortality of benthic fauna and the alteration of fish assemblages. Species with younger individuals, shorter generation times, and protracted spawning seasons became the dominant taxa, replacing those with longer generation times and shorter spawning periods. Pollution-tolerant species (e.g., capitellids) attained peak abundance among the polychaete taxa. Bivalve densities and fish species richness were greatest at sites farthest from the source of the wastewaters.

Despite the array of anthropogenic impacts in the Tijuana River NERR, a wide range of habitats and biotic communities occurs in the system. Aside from extensive salt marshes, the Tijuana River NERR contains salt pannes, brackish marshes, riparian habitat, wetland–upland transition zones, intertidal flats, beaches and dunes, and tidal creeks and channels. The dominant vegetation in the salt marsh habitat includes the Pacific cordgrass (*Spartina foliosa*) in the lower marsh, perennial pickleweed (*Salicornia virginica*) in the mid-marsh plain, and glasswort (*S. subterminalis*), shoregrass (*Monanthocloe littoralis*), alkali heath (*Frankenia salina*), sea lavender (*Limonium californicum*), and Watson's saltbush (*Atriplex watsonii*) in the higher marsh. Algal mats are conspicuous under the canopy of salt marsh plants. Invertebrates commonly found on the marsh are the fiddler crab (*Uca crenulata*) and other species of crabs, as well as snails such as *Assiminea californica* and *Melampus olivaceus*. Insects are abundant and serve as food for birds. Wading birds are quite abundant near tidal creeks. Reptiles (lizards and snakes) and mammals (rabbits and squirrels) are frequently observed in the higher marsh. The smaller mammals are important prey for raptors (e.g., American kestrel, *Falco sparverius*; northern harrier, *Circus cyaneus*; and golden eagle, *Aquila chrysaetos*).

Salt pannes are evident in upper intertidal areas of salt marsh habitats. Much of the year they appear as denuded areas devoid of vegetation. During wet periods, the salt pannes often contain water derived from rainfall and tidal inundation. At this time, widgeon grass and insects usually proliferate. Waterfowl (e.g., northern pintails, *Anas acuta*) may be seen feeding here. At least two species of birds (California least tern, *Sterna antillarum browni*; and western snowy plover, *Charadrius alexandrinus nivosus*) are known to nest in this habitat. In summer, heat dessicates the salt pannes, leaving a salt-crusted surface. Beetles (e.g., *Bledius* spp. and *Cicindela* spp.) burrow into the surface sediments, and several species of reptiles and mammals may appear along the surface of the salt pannes.

Cattails (*Typha domingensis*) and bulrushes (*Scirpus californicus*) are the predominant emergent flora in the brackish marshes. Insects are abundant. Numerous species of birds (e.g., red-winged blackbirds, *Agelaius phoeniceus*; black-bellied plover, *Pluvialis squatarola*; and snowy egret, *Egretta thula*) are also reported here.

Riparian habitats consist of low shrub vegetation in low flood plain areas (e.g., salt bush, *Baccharis glutinosa*; and sandbar willow, *Salix hindsiana*) and forest vegetation (e.g., cottonwoods, *Populus fremontii*; and willows, *Salix* spp.) in higher woodlands. These habitats support some of the most diverse communities of birds, reptiles, and mammals in the reserve. Fish assemblages in these habitats are likewise highly diverse.

Vegetation in the wetland–transition habitat is comprised of a variety of wetland plants in a lowland community, coastal sage scrub in an upland community, and a rather abrupt boundary between them. Among abundant salt marsh plants are the salt grass (*Distichlis spicata*), perennial glasswort (*Salicornia subterminalis*), and sea lavender (*Limonium californicum*). Important upland plants include California salt bush (*Atriplex californica*), box-thorn (*Lycium californicum*), and several species of shrubs (e.g., lemonadeberry, *Rhus integrifolia*; and laurel sumac, *R. laurina*). Invasive species of note are the Australian salt bush (*Atriplex semibaccata*) and crystal ice plant (*Mesembryanthemum crystallinum*).

The wetland–transition habitat supports various herpetofauna (e.g., California king snakes, *Lampropeltis getulus californiae*; San Diego gopher snakes, *Pituophis melanoleucus annectens*; and side-blotched lizards, *Uta stansburiana*). Small mammals (mice, rats, squirrels, and rabbits) are relatively abundant. Larger mammalian species (e.g., striped skunks, *Mephitis mephitis*; and coyote, *Canis latrans*) also commonly appear in this habitat. A number of avian species (e.g., northern harrier, *Circus cyaneus*; and short-eared owl, *Asio flammeus*) actively prey on the smaller mammalian forms.

The dunes and beach habitat is one of the most physically stressed environments in the reserve; it is subjected to considerable wind and wave action, salt spray, and disturbances by humans. Some plant species of significance are the dune ragweed (*Ambrosia chamissonis*), sand verbena (*Abronia umbellata*), and sea rocket (*Cakile maritima*). Animal populations are sparse. Exceptions are tiger beetles (*Coelus* spp.) and ants. Lizards (e.g., silvery legless lizards, *Anniella pulchra pulchra*; and side-blotched lizards, *Uta stansburiana*) frequent the dunes, as do certain birds (e.g., Belding's Savannah sparrow, *Ammodramus sandwichensis beldingi*). Other avifauna that may be found in this habitat are the sanderling (*Calidris alba*), least sandpiper (*C. minutilla*), western sandpiper (*C. mauri*), willet (*Catoptrophorus semipalmatus*), whimbrel (*Numenius phaeopus*), marbled godwit (*Limosa fedoa*), dowitcher (*Limnodromus* sp.), and black-bellied plover (*Pluvialis squatarola*).

Mudflats and sandflats are valuable foraging habitats for crabs and numerous species of birds. These intertidal flats support amphipods, bivalves, polychaetes, gastropods, and other invertebrates that are important prey for wading birds that feed on the flats. Because of the rich food supply, more shorebird species use the intertidal flats than use any other habitat in the reserve.

Tidal creeks and channels form the principal aquatic habitats in the Tijuana Estuary NERR. Plant and animal taxa are well represented. Diatoms and dinoflagellates dominate the phytoplankton community, and benthic macroalgae (e.g., *Enteromorpha* sp. and *Ulva* sp.) constitute a drifting community. The benthic invertebrate community includes deposit-feeding and filter-feeding species whose distribution is strongly influenced by sediment type. In addition, the benthic fauna is highly susceptible to rapid and acute changes in salinity levels of the Tijuana Estuary. At times, perturbations in salinity have resulted in the extirpation of numerous benthic invertebrate species in the system. Mass mortality of benthic organisms has occurred during periods of hydrographic disturbances (e.g., floods). Such disturbances result in dramatic shifts in the structure of benthic communities that may persist for years.

Fish assemblages are dominated by forage species, such as the California killifish (*Fundulus parvipinnis*), arrow goby (*Clevelandia ios*), and topsmelt (*Atherinops affinis*). Several species are of recreational and commercial importance, such as sea bass (*Paralabrax* spp.), flounder (*Pleuronichthys* spp.), white croaker (*Genyonemus lineatus*), and California halibut (*Paralichthys californicus*). As in the case of the benthic invertebrate community, fish assemblages in the estuary have been impacted by variable salinity concentrations, with the species richness declining appreciably during periods of rapidly diminishing salinities.

Many waterbirds feed and rest along tidal creeks and channels of the estuary. Gulls and terns, herons and egrets, and cormorants and pelicans utilize these aquatic environments. They are often seen searching for forage fishes along the shoreline.

The Tijuana River NERR is a system heavily impacted by human activities and disturbances that have altered extensive habitat area. Environmental management programs are now focusing on protecting the health and viability of habitats in the system. Restoration projects have been initiated to revitalize impacted habitats. In addition, strategies have been devised to deal with physical and hydrologic disturbances in the reserve. The success of these habitat management programs depends on effective collaboration of state and federal government agencies, academic institutions, and the general public, as well as effective education initiatives that inform all individuals of the critical importance of this valuable system.

REFERENCES

Boland, J.M. 1981. Seasonal Abundances, Habitat Utilization, Feeding Strategies, and Interspecific Competition within a Wintering Shorebird Community and Their Possible Relationships with the Latitudinal Distribution of Shorebird Species. M.S. thesis, San Diego State University, San Diego, CA.

Cox, J.E. 1996. Management goals and functional boundaries of riparian forested wetlands. In: Mulamoottil, G., B.G. Warner, and E.A. McBean (Eds.). *Wetlands: Environmental Gradients, Boundaries, and Buffers.* Lewis Publishers, Boca Raton, FL, pp. 153–161.

Cox, G.W. and J.B. Zedler. 1986. The influence of mima mounds on vegetation patterns in the Tijuana Estuary salt marsh, San Diego County, California. *Bulletin of the Southern California Academy of Science* 85: 158–172.

Desmond, J.S., G.D. Williams, and G. Vivian-Smith. 2001. The diversity of habitats in southern California coastal wetlands. In: Zedler, J.B. (Ed.). *Handbook for Restoring Tidal Wetlands.* CRC Press, Boca Raton, FL, pp. 67–77.

Entrix Inc., PERL, and PWA, Ltd. 1991. Tijuana Estuary Tidal Restoration Program. Draft Environmental Impact Report/Environmental Impact Statement, California Coastal Conservancy and U.S. Fish and Wildlife Service, SCC, Oakland, CA, Volumes I–III.

Ferren, W.R., Jr. 1985. Carpinteria Salt Marsh: Environment, History, and Botanical Resources of a Southern California Estuary. Publication No. 4, The Herbarium, Department of Biological Sciences, University of California, Santa Barbara, CA.

Fong, P. 1986. Monitoring and Manipulation of Phytoplankton Dynamics in a Southern California Estuary. M.S. thesis, San Diego State University, San Diego, CA.

Ford, R.F., G. McGowen, and M.V. Needham. 1971. Biological inventory: investigations of fish, invertebrates, and marine grasses in the Tijuana River Estuary. In: *Environmental Impact Study for the Proposed Tijuana River Flood Control Channel.* Technical Report to Ocean Studies and Engineering, Long Beach, CA, pp. 39–64.

Hosmer, S.C. 1977. Pelecypod–Sediment Relationships at Tijuana Estuary. M.A. thesis, San Diego State University, San Diego, CA.

International Boundary and Water Commission (IBWC). 1976. Final Environmental Impact Statement. Technical Report, Tijuana River Flood Control Project, San Diego County, CA.

James, M.L. and J.B. Zedler. 2000. Dynamics of wetland and upland subshrubs at the salt marsh–coastal sage scrub ecotone. *American Midland Naturalist* 82: 81–99.

Levin, L., S. Talley, and J. Hewitt. 1998. Macrobenthos of *Spartina foliosa* (Pacific cordgrass) salt marshes in southern California: community structure and comparison to a Pacific mudflat and a *Spartina alterniflora* (Atlantic smooth cordgrass) marsh. *Estuaries* 21: 129–144.

NERRS. 2002. National Estuarine Research Reserve System: System-wide Monitoring Monitoring Program Deployment Plans for Phase 1 (Abiotic Factors). Technical Report, National Estuarine Research Reserve System, Silver Spring, MD.

Nordby, C.S. 1982. The Comparative Ecology of Ichthyoplankton within Tijuana Estuary and its Adjacent Nearshore Waters. M.S. thesis, San Diego State University, San Diego, CA.
Nordby, C.S. 1987. Response of channel organisms to estuarine closure and substrate disturbance. In: *Wetland and Riparian Systems of the American West*. Proceedings of Society of Wetland Scientists' Eighth Annual Meeting, Seattle, WA, pp. 318–321.
Nordby, C.S. and J.B. Zedler. 1991. Responses of fish and macrobenthic assemblages to hydrologic disturbances in Tijuana Estuary and Los Peñasquitos Lagoon, California. *Estuaries* 14: 80–93.
Peterson, C.H. 1975. Stability of species and of community for the benthos of two lagoons. *Ecology* 56: 958–965.
Rehse, M.A. 1981. Faunal Recovery in Tijuana Estuary. Unpublished Technical Report, San Diego State University, San Diego, CA.
Restore America's Estuaries. 2002. A National Strategy to Restore Coastal and Estuarine Habitat. Technical Report, Restore America's Estuaries, Arlington, VA.
Rudnicki, R. 1986. Dynamics of Macroalgae in Tijuana Estuary: Response to Simulated Wastewater Addition. M.S. thesis, San Diego State University, San Diego, CA.
Seamans, P. 1988. Wastewater creates a border problem. *Journal of the Water Pollution Control Federation* 60: 1799–1804.
Smith, S.H. 1974. The Growth and Mortality of the Littleneck Clam, *Protothaca staminea*, in Tia Juana Slough, California. M.S. thesis, San Diego State University, San Diego, CA.
Tiner, R.W. 1999. *Wetland Indicators: A Guide to Wetland Identification, Delineation, Classification, and Mapping*. Lewis Publishers, Boca Raton, FL.
Vivian-Smith, G. 2001. Developing a framework for restoration. In: Zedler, J.B. (Ed.). *Handbook for Restoring Tidal Wetlands*. CRC Press, Boca Raton, FL, pp. 39–88.
Winfield, T.P. 1980. Dynamics of Carbon and Nitrogen in a Southern California Salt Marsh. Ph.D. thesis, University of California, Riverside and San Diego State University, San Diego, CA.
Zedler, J.B. 1977. Salt marsh community structure in the Tijuana Estuary, California. *Estuarine and Coastal Marine Science* 5: 39–53.
Zedler, J.B. 1980. Algal mat productivity: comparisons in a salt marsh. *Estuaries* 3: 122–131.
Zedler, J.B. 1982. Salt marsh algal mat composition: spatial and temporal comparisons. *Bulletin of the Southern California Academy of Sciences* 81: 41–50.
Zedler, J.B. (Ed.). 2001. *Handbook for Restoring Tidal Wetlands*. CRC Press, Boca Raton, FL.
Zedler, J.B. and G.W. Cox. 1985. Characterizing wetland boundaries: a Pacific coast example. *Wetlands* 4: 43–55.
Zedler, J.B. and P.A. Beare. 1986. Temporal variability of salt marsh vegetation: the role of low salinity gaps and environmental stress. In: Wolfe, D. (Ed.). *Estuarine Variability*. Academic Press, New York, pp. 295–306.
Zedler, J.B. and C.S. Nordby. 1986. The Ecology of the Tijuana Estuary: An Estuarine Profile. Biological Report 85 (7.5), U.S. Fish and Wildlife Service, Washington, D.C.
Zedler, J.B., R. Koenigs, and W.P. Magdych. 1984. Review of Salinity and Predictions of Estuarine Responses to Lowered Salinity. Technical Report, State of California Water Resources Control Board, San Diego Association of Governments, San Diego, CA.
Zedler, J.B., C.S. Nordby, and K.B. Kus. 1992. The Ecology of Tijuana Estuary, California: A National Estuarine Research Reserve. Technical Report, NOAA Office of Coastal Resource Management, Sanctuaries and Reserves Division, Washington, D.C.

Index

Able, K.W., 69, 106, 109
Abronia umbellata, Tijuana River NERR, 243
Acantharchus pomotis, Jacques Cousteau NERR, 92–93
Acartia hudsonica, Delaware NERR, 136, 158
Acartia spp., Delaware NERR, 158
Acartia tonsa
    ACE Basin NERR, 194
    Delaware NERR, 135, 136, 158
    Jacques Cousteau NERR, 59, 98
    Weeks Bay NERR, 225
Accipiter cooperii, *see* Cooper's hawk
Accipiter striatus, *see* Sharp-shinned hawk
ACE Basin NERR
    anthropogenic impacts
        ditching/impounding, 204
        overview, 173, 203–205
        pollution, 173, 189, 203–204
    aquifer systems, 171, 173
    barrier islands, 173, 183
    barrier islands herpetofauna, 183
    Carolina bays, 175, 181
    coastal marine waters
        birds, 202
        fish, 200–201
        mammals, 202
        reptiles, 201–202
    conservation easements, 173
    description, 171–173
    designation, 171
    development, 203
    endangered/threatened species, 202–203
    estuary biotic communities
        benthic invertebrates, 195–197, 198
        decapod crustaceans list, 198
        fish, 197–200
        phytoplankton, 192–194
        zooplankton, 194–195
    estuary physical-chemical characteristics, 190–192
    land cover, 203
    land protection, 173
    land use cover, 174
    map, 172
    overview, 205
    pocosins, 181
    Snuggedy Swamp, 174
    St. Helena Sound, 190–191
    streamflow characteristics, 171, 172
    sub-basins, 175
    tidal influences, 191
    water quality characteristics, 191–192
    watershed communities (animal), 175–190
        amphibians/reptiles, 176–183
        amphibians/reptiles list, 177–180
        birds, 186–189
        insects, 189–190
        mammals, 183–186
        mammals list, 184–185
        overview, 175–176
    watershed communities (plant), 173–175
Acer rubrum
    Delaware NERR, 121, 152, 155
    Jacques Cousteau NERR, 71
Acipenser brevirostrum, ACE Basin NERR, 173, 200, 202
Acipenser oxyrhynchus, *see* Atlantic sturgeon
Acorus calamus, Jacques Cousteau NERR, 70
Acris crepitans crepitans, Jacques Cousteau NERR, 82
Acris gryllus, ACE Basin NERR, 181–182
Acteocina canaliculata, Jacques Cousteau NERR, 103
Actinastrum spp., Delaware NERR, 157
Actitis macularia, Delaware NERR, 161
Agalinis acuta, Waquoit Bay NERR, 38
Agardhiella subulata, Jacques Cousteau NERR, 95
Agelaius phoeniceus, *see* Red-winged blackbird
Agkistrodon piscivorus, ACE Basin NERR, 181, 182
Agriculture
    impact on ACE Basin NERR, 203–204
    impact on Tijuana River NERR, 255
    impact on Weeks Bay NERR, 227
    need for modified practices, xiii
Aix sponsa, Delaware NERR, 161, 163
Alder (*Alnus rugos*), Waquoit Bay NERR, 38
Alewife (*Alosa pseudoharengus*)
    Delaware NERR, 138
    Jacques Cousteau NERR, 94, 106
    Waquoit Bay NERR, 38
Alkali heath (*Frankenia salina*), Tijuana River NERR, 239, 242
Allen, D.E., 204

Alligator mississippiensis, ACE Basin NERR, 173, 182, 202
Alligator weed (*Alternathera philoxeroides*), Weeks Bay NERR, 220
Alnus rugos, Waquoit Bay NERR, 38
Alnus serrulata, Delaware NERR, 123
Alosa aestivalis, *see* Blueback herring
Alosa mediocris, *see* Hickory shad
Alosa pseudoharengus
    Delaware NERR, 138
    Jacques Cousteau NERR, 94, 106
    Waquoit Bay NERR, 38
Alosa sapidissima, *see* American shad
Alternathera philoxeroides, Weeks Bay NERR, 220
Ambrosia chamissonis, Tijuana River NERR, 243
Ambylopappus pusillus, Tijuana River NERR, 242
Ambystoma mabeei, 182
Ambystoma maculatum
    ACE Basin NERR, 181, 182
    Waquoit Bay NERR, 38
Ambystoma opacum, *see* Marbled salamander
Ambystoma talpoideum, ACE Basin NERR, 181
Ambystoma tigrinum tigrinum, Jacques Cousteau NERR, 82
Ameiurus catus, Delaware NERR, 140
Ameiurus natalis, Jacques Cousteau NERR, 92
Ameiurus nebulosus, Jacques Cousteau NERR, 92
Ameiurus spp., ACE Basin NERR, 204–205
Amelanchier candensis, Jacques Cousteau NERR, 82
American alligator (*Alligator mississippiensis*), ACE Basin NERR, 173, 182, 202
American avocet (*Recurvirostra americana*), Tijuana River NERR, 241, 244
American beach grass (*Ammophila breviligulata*), Jacques Cousteau NERR, 81
American beech (*Fagus grandifolia*), Delaware NERR, 121, 152
American black duck (*Anas rubripes*)
    Delaware NERR, 144, 163
    Jacques Cousteau NERR, 88
American brant (*Branta bernicla*), Jacques Cousteau NERR, 88
American brook lamprey (*Lampetra lamottie*), Jacques Cousteau NERR, 94
American crow (*Corvus brachyrhynchos*)
    ACE Basin NERR, 187
    Delaware NERR, 145
American eel (*Anguilla rostrata*)
    Delaware NERR, 138, 162
    Jacques Cousteau NERR, 93, 98, 106
    Waquoit Bay NERR, 42
American goldfinches (*Carduelis tristis*)
    ACE Basin NERR, 187
    Delaware NERR, 145, 161
American holly (*Ilex opaca*)
    Delaware NERR, 121, 152
    Jacques Cousteau NERR, 82
American kestrel (*Falco sparverius*)
    Delaware NERR, 145, 161
    Jacques Cousteau NERR, 89
    Tijuana River NERR, 240
American mannagrass (*Glyceria grandis*), Jacques Cousteau NERR, 67
American oystercatcher (*Haematopus palliatus*)
    ACE Basin NERR, 188
    Jacques Cousteau NERR, 87
American robin (*Turdus migratorius*), Delaware NERR, 145, 160
American shad (*Alosa sapidissima*)
    ACE Basin NERR, 200
    Delaware NERR, 138, 162
    Jacques Cousteau NERR, 94
American three-square (*Scirpus americanus*), Delaware NERR, 152
American toad (*Bufo americanus*), Waquoit Bay NERR, 38
American wigeon (*Anas americana*)
    Delaware NERR, 163
    Jacques Cousteau NERR, 88
Ammodramus sandwichensis beldingi, Tijuana River NERR, 239, 240, 243
Ammodytes sp., Jacques Cousteau NERR, 99–100, 109
Ammophila breviligulata
    Jacques Cousteau NERR, 81
    Waquoit Bay NERR, 41
Ammospiza caudacuta, *see* Sharp-tailed sparrow
Ammospiza maritima, *see* Seaside sparrow
Ampelisca abdita, Jacques Cousteau NERR, 100, 103
Ampelisca vadorum, ACE Basin NERR, 197
Ampelisca verrilli, Jacques Cousteau NERR, 103
Amphipods, ACE Basin NERR, 197
Amphiuma means, ACE Basin NERR, 182
Anabaena spp., Delaware NERR, 133
Anas acuta, 88
    Tijuana River NERR, 240
Anas americana, *see* American wigeon
Anas crecca, *see* Green-winged teal
Anas discors, Delaware NERR, 163
Anas platyrhynchos, *see* Mallard
Anas rubripes
    Delaware NERR, 144, 163
    Jacques Cousteau NERR, 88
Anas strepera, *see* Gadwall
Anchoa hepsetus, *see* Striped anchovy
Anchoa mitchilli, *see* Bay anchovy

Index

Andropogon virginicus var. *virginicus*, Jacques Cousteau NERR, 72
Anguilla rostrata, *see* American eel
Ankistrodesmus spp., Delaware NERR, 133, 157
Annelids
    Delaware NERR, 136, 137, 158, 159
    Waquoit Bay NERR, 44, 45
Anniella pulchra pulchra, Tijuana River NERR, 243
Annual pickleweed (*Salicornia bigelovii*), Tijuana River NERR, 239
Anthropogenic impacts, *see also specific impacts*
    ACE Basin NERR, 173, 189, 203–205
    categorization of, xi
    Delaware NERR, 128–133, 157
    leading to protection legislation, 1–2
    summary, xi–xiii
    summary table, xii
    Tijuana River NERR, 237, 240–242, 243, 252, 254–256, 255, 257
    Waquoit Bay NERR, 46, 48–51
    Weeks Bay NERR, 227–228
Apalone ferox, ACE Basin NERR, 182
Apeltes quadracus, *see* Fourspine stickleback
Aphredoderus sayanus, Jacques Cousteau NERR, 92
Aquila chrysaetos, Tijuana River NERR, 240
Arctic terns (*Sterna paradisaea*), Waquoit Bay NERR, 41
Arctostaphylos uva-ursi, *see* Bearberry
Ardea herodias, *see* Great blue heron
Arenaria interpres, *see* Ruddy turnstone
Argopecten irradians, Waquoit Bay NERR, 42, 45, 51
Arius felis, ACE Basin NERR, 199
Arrow arum (*Peltandra virginica*)
    Delaware NERR, 152
    Jacques Cousteau NERR, 70
    Weeks Bay NERR, 220
Arrow goby (*Clevelandia ios*), Tijuana River NERR, 249, 252
Arrow-grass (*Triglochin concinnum*), Tijuana River NERR, 239
Arrowheads (*Sagittaria engelmanniana*), Jacques Cousteau NERR, 67
Arrowheads (*Sagittaria latifolia*), Jacques Cousteau NERR, 67
Arrowheads (*Sagittaria spatulata*), Jacques Cousteau NERR, 67
Arrowheads (*Sagittaria* spp.), Jacques Cousteau NERR, 70
Arrow leaf (*Sagittaria lancifolia*), Weeks Bay NERR, 220
Arrow worms (*Sagitta* spp.), Jacques Cousteau NERR, 98

Artemisia californica, Tijuana River NERR, 242
Artemisia stelleriana, Waquoit Bay NERR, 41
Arthropods
    Delaware NERR, 136, 158, 159
    Waquoit Bay NERR, 42, 44, 45–46
Ashepoo-Combahee-Edisto Basin NERR, *see* ACE Basin NERR
Ashepoo River, *see also* ACE Basin NERR
    description, 171
Asian clam (*Potamocorbula amurensis*), as invasive, 25
Asio flammeus, *see* Short-eared owl
Asio otus, ACE Basin NERR, 188
Assiminea californica, Tijuana River NERR, 239
Atherinops affinis, Tijuana River NERR, 248, 252
Atlantic bumper (*Chloroscombrus chrysurus*), ACE Basin NERR, 199, 201
Atlantic croaker (*Micropogonias undulatus*)
    ACE Basin NERR, 199
    Delaware NERR, 140
    Jacques Cousteau NERR, 94, 106
    Weeks Bay NERR, 226
Atlantic menhaden (*Brevoortia tyrannus*)
    ACE Basin NERR, 199
    Delaware NERR, 138, 140, 160, 162
    Jacques Cousteau NERR, 94, 98, 106, 109
Atlantic rock crab (*Cancer irroratus*), Jacques Cousteau NERR, 104
Atlantic silverside (*Menida menidia*)
    ACE Basin NERR, 199, 201
    Delaware NERR, 138, 140
    Jacques Cousteau NERR, 98, 106, 109
Atlantic sturgeon (*Acipenser oxyrhynchus*)
    ACE Basin NERR, 200
    Delaware NERR, 138, 140
Atlantic tomcod (*Microgadus tomcod*), Waquoit Bay NERR, 46
Atlantic white cedar (*Chamaecyparis thyoides*), Jacques Cousteau NERR, 71, 82
Atriplex leucophylla, Tijuana River NERR, 243
Atriplex patula, Jacques Cousteau NERR, 66
Atriplex pusillus, Tijuana River NERR, 242
Atriplex semibaccata, Tijuana River NERR, 242
Atriplex watsonii, Tijuana River NERR, 239, 242
Aureococcus anophagefferens, Jacques Cousteau NERR, 96–97
Australian salt bush (*Atriplex semibaccata*), Tijuana River NERR, 242
Awned meadow beauty (*Rhexia aristosa*), Jacques Cousteau NERR, 72
Aythra valisneria, *see* Canvasback
Aythya affinis, *see* Lesser scaup
Aythya americana, Delaware NERR, 163
Aythya marila, *see* Greater scaup

B

Baccharis glutinosa, Tijuana River NERR, 241
Baccharis halimifolia, Delaware NERR, 124–125
Baccharus halmifolia, Weeks Bay NERR, 220
Bain, M.B., 225, 226
Bairdiella chrysoura, see Silver perch
Balaena glacialis, Delaware NERR, 145
Balaenoptera physalus, Delaware NERR, 145
Balanus improvisus, ACE Basin NERR, 196
Balanus spp., Waquoit Bay NERR, 42, 45
Bald cypress (*Taxodium distichum*), ACE Basin NERR, 173, 174
Bald eagle (*Haliaeetus leucocephalus*)
   ACE Basin NERR, 202
   Delaware NERR, 144, 145
   Jacques Cousteau NERR, 89
Banded killifish (*Fundulus diaphanus*), Jacques Cousteau NERR, 94
Banded sunfish (*Enneacanthus obesus*), Jacques Cousteau NERR, 92, 93
Banded water snake (*Nerodia fasciata*), ACE Basin NERR, 182
Barnacle larvae (cirripedes), ACE Basin NERR, 194
Barnacles (*Balanus* spp.), Waquoit Bay NERR, 42, 45
Barn owls (*Tyto alba*), ACE Basin NERR, 188
Barn swallow (*Hirundo rustica*), Delaware NERR, 161
Barred owl (*Strix varia*), Delaware NERR, 144
Barrier islands, ACE Basin NERR, 173
Batea catharinensis, ACE Basin NERR, 197
Batis maritima, Tijuana River NERR, 239
Bats, Weeks Bay NERR, 223
Bault, E.I., 225–226
Bay anchovy (*Anchoa mitchilli*)
   ACE Basin NERR, 199, 201
   Delaware NERR, 138, 140
   Jacques Cousteau NERR, 94, 99, 106, 109
   Weeks Bay NERR, 226
Bayberry (*Myrica pensylvanica*)
   Jacques Cousteau NERR, 71, 72, 82
   Waquoit Bay NERR, 41
Bay scallop (*Argopecten irradians*), Waquoit Bay NERR, 42, 45, 51
Beach grass (*Ammophila breviligulata*), Waquoit Bay NERR, 41
Beach heather (*Hudsonia tomentosa*)
   Jacques Cousteau NERR, 82
   Waquoit Bay NERR, 41
Beach pea (*Lathyrus japonicus* var. *glaber*), Waquoit Bay NERR, 41
Beach pea (*Lathyrus maritimus*), Jacques Cousteau NERR, 81

Beach plum (*Prunus maritima*)
   Jacques Cousteau NERR, 82
   Waquoit Bay NERR, 41
Bearberry (*Arctostaphylos uva-ursi*)
   Jacques Cousteau NERR, 81
   Waquoit Bay NERR, 37
Beaver (*Castor canadensis*)
   ACE Basin NERR, 185
   Delaware NERR, 145, 162
   Jacques Cousteau NERR, 86
Belding's Savannah sparrow (*Ammodramus sandwichensis beldingi*), Tijuana River NERR, 239, 240, 243
Bell, S.S., 196
Belted-kingfisher (*Ceryle alcyon*), Weeks Bay NERR, 223
Benthic habitat mapping, 20–21
Benthic invertebrates categorization, 195
Bent-nose clam (*Macoma nasuta*), Tijuana River NERR, 247
Best management practices (BMPs), 227, 228
Betula populifolia, Jacques Cousteau NERR, 71
Biddulphia spp., Delaware NERR, 133
Bidens cernus, Delaware NERR, 156
Bidens laevis, Jacques Cousteau NERR, 70
Big brown bat (*Eptesicus fuscus*)
   ACE Basin NERR, 185
   Jacques Cousteau NERR, 86
Big cordgrass (*Spartina cynosuroides*)
   Delaware NERR, 121, 152
   Jacques Cousteau NERR, 66
   Weeks Bay NERR, 220
Biotic impacts/stressors, types, xi
Bird's foot violet (*Viola pedata*), Waquoit Bay NERR, 38
Black-and-white warbler (*Mniotilta varia*)
   ACE Basin NERR, 188
   Jacques Cousteau NERR, 92
Black-backed gulls (*Larus marinus*), Waquoit Bay NERR, 41
Black-banded sunfish (*Enneacanthus chaetodon*), Jacques Cousteau NERR, 93
Black-bellied plover (*Pluvialis squatarola*)
   Jacques Cousteau NERR, 87
   Tijuana River NERR, 241, 244
Black bullhead (*Ictalurus melas*), Jacques Cousteau NERR, 94
Blackcheek tonguefish (*Symphurus plagiusa*), Jacques Cousteau NERR, 109
Black cherry (*Prunus serotina*)
   Delaware NERR, 121, 152
   Jacques Cousteau NERR, 82
Black crappie (*Pomoxis nigromaculatus*), Jacques Cousteau NERR, 94

Black-crowned night heron (*Nycticorax nycticorax*), Jacques Cousteau NERR, 88
Black drum (*Pogonias cromis*), Delaware NERR, 140, 162
Black grass (*Juncus gerardii*), Jacques Cousteau NERR, 66
Black gum (*Nyssa sylvatica*)
  Delaware NERR, 121, 152, 155
  Jacques Cousteau NERR, 71
Black haw (*Viburnum prunifolium*), Delaware NERR, 121
Blackjack oak (*Quercus marilandica*), Jacques Cousteau NERR, 81
Black-necked stilt (*Himantopus mexicanus*), Tijuana River NERR, 241
Black needle rush (*Juncus roemerianus*), Weeks Bay NERR, 220
Blacknose dace (*Rhinichthys atratulus*), Jacques Cousteau NERR, 94
Black oak (*Quercus velutina*), Jacques Cousteau NERR, 80, 81
Black racer (*Coluber constrictor constrictor*), ACE Basin NERR, 181
Black rail (*Laterallus jamaicensis*), Delaware NERR, 144
Black rat snake (*Elaphe obsoleta*), Delaware NERR, 140, 160
Black rush (*Juncus gerardi*), Waquoit Bay NERR, 39
Black scoter (*Melanitta nigra*), Delaware NERR, 144
Black skimmer (*Rynchops niger*)
  ACE Basin NERR, 188, 189
  Jacques Cousteau NERR, 87
Black swamp snake (*Seminatrix pygaea*), ACE Basin NERR, 181
Black vulture (*Coragyps atratus*), ACE Basin NERR, 187
Bladderworts (*Utricularia* spp.), Jacques Cousteau NERR, 72
Blarina brevicauda, see Short-tailed shrew
Bledius spp., Tijuana River NERR, 240
Blueback herring (*Alosa aestivalis*)
  ACE Basin NERR, 200
  Delaware NERR, 138
  Jacques Cousteau NERR, 94
  Waquoit Bay NERR, 38, 42
Blueberry (*Vaccinium* spp.), Jacques Cousteau NERR, 80, 82
Blue crab (*Callinectes sapidus*)
  ACE Basin NERR, 196, 197
  Delaware NERR, 138, 162
  Waquoit Bay NERR, 41, 45, 51
  Weeks Bay NERR, 227
Bluefish (*Pomatomus salatrix*)
  Delaware NERR, 140, 162
  Jacques Cousteau NERR, 109
  Waquoit Bay NERR, 46
Bluegill (*Lepomis macrochirus*), Jacques Cousteau NERR, 92, 94
Blue-green algae
  Delaware NERR, 157
  Tijuana River NERR, 245
Blue grosbeak (*Guiraca caerulea*), Delaware NERR, 145
Blue jay (*Cyanocitta cristata*), Delaware NERR, 145, 160
Bluespotted sunfish (*Enneacanthus gloriosus*), Jacques Cousteau NERR, 93
Blue-winged teal (*Anas discors*), Delaware NERR, 163
Bluntnose minnow (*Pimephales notatus*), Jacques Cousteau NERR, 93
Bluntscale bulrush (*Scirpus smithii* var. *smithii*), Jacques Cousteau NERR, 70
Boat-tailed grackle (*Quiscalus major*)
  ACE Basin NERR, 187
  Delaware NERR, 144, 160
Bobcat (*Lynx rufus*), ACE Basin NERR, 186
Bobolink (*Dolichonyx oxyzivorus*), Delaware NERR, 145, 161
Bog Pond, Waquoit Bay NERR, 35
Bog turtle (*Clemmys muhlenbergii*), Jacques Cousteau NERR, 85
Bologna, P.A.X., 95
Boothe, B.B., 196
Bottlenose porpoise (*Tursiops truncatus*)
  ACE Basin NERR, 202
  Delaware NERR, 145
Bourne Pond, Waquoit Bay NERR, 35
Box-thorn (*Lycium californicum*), Tijuana River NERR, 242
Boyking's lobelia (*Lobelia boykinii*), Jacques Cousteau NERR, 72
Bracken fern (*Pteridium aquilinum*)
  ACE Basin NERR, 174
  Jacques Cousteau NERR, 71, 72, 80
Brania clavata, Jacques Cousteau NERR, 103
Branta bernicla, Jacques Cousteau NERR, 88
Branta canadensis, Jacques Cousteau NERR, 88
Brevoortia patronus, ACE Basin NERR, 226, 227
Brevoortia tyrannus, see Atlantic menhaden
Bricker, S.B., xi, 16
Bridled shiner (*Notropis bifrenatus*), Jacques Cousteau NERR, 94
Broadhead skink (*Eumeces laticeps*), ACE Basin NERR, 176
Broad-winged hawk (*Buteo platypterus*), Jacques Cousteau NERR, 89, 91

Brook trout (*Salmo fontinalis*), Jacques Cousteau NERR, 94
Broom crowberry (*Cormea conradii*), Jacques Cousteau NERR, 81
Brown bullhead (*Ameiurus nebulosus*), Jacques Cousteau NERR, 92
Brown creeper (*Certhia americana*), Delaware NERR, 161
Brown macroalgae (*Fucus* spp.), Waquoit Bay NERR, 43
Brown macroalgae (*Laminaria agardhii*), Waquoit Bay NERR, 43
Brown macroalgae (*Petroderma maculiform*?), Waquoit Bay NERR, 4?
Brown macroalgae (*Pseudolithoderma* spp.), Waquoit Bay NERR, 43
Brown macroalgae (*Ralfsia* spp.), Waquoit Bay NERR, 43
Brown macroalgae, Waquoit Bay NERR, 43
Brown pelican (*Pelecanus occidentalis*), ACE Basin NERR, 189
Brown thrasher (*Toxostoma rufum*), Delaware NERR, 145
Brown-tide alga (*Aureococcus anophageferens*), Jacques Cousteau NERR, 96–97
Brown trout (*Salmo gairdneri*), Jacques Cousteau NERR, 94
Brown water snake (*Nerodia taxispilota*), ACE Basin NERR, 182
Brush mouse (*Peromyscus boyli*), Tijuana River NERR, 243
Bubo virginianus, see Great horned owl
Bucephala albeola, see Bufflehead
Bucephala clangula, Jacques Cousteau NERR, 88
Bufflehead (*Bucephala albeola*)
  Delaware NERR, 144, 163
  Jacques Cousteau NERR, 88
Bufo americanus, Waquoit Bay NERR, 38
Bufo quercicus, ACE Basin NERR, 181
Bufo woodhousii fowleri, Jacques Cousteau NERR, 82
Bullfrog (*Rana catesbeiana*)
  ACE Basin NERR, 182
  Delaware NERR, 140, 160
  Jacques Cousteau NERR, 82
Bullhead catfish (*Ameiurus* spp.), ACE Basin NERR, 204–205
Bullhead lilies (*Nuphar variegatum*), Jacques Cousteau NERR, 72
Bulrush (*Scirpus* spp.), Jacques Cousteau NERR, 67
Buntings, ACE Basin NERR, 188, 189
Burdick, D.M., 49, 51
Burger, J., 88

Burrowing shrimp (*Callianassa gigas*), Tijuana River NERR, 247
Burying amphipods (*Corophium* sp.), Waquoit Bay NERR, 40
Busycon canaliculatum, Waquoit Bay NERR, 45
Busycon carica, Waquoit Bay NERR, 45
Buteo jamaicensis, see Red-tailed hawk
Buteo lagopus, Delaware NERR, 144–145
Buteo lineatus, Delaware NERR, 144–145
Buteo platypterus, Jacques Cousteau NERR, 89, 91
Butorides virescens, Delaware NERR, 161
Button bush (*Cephalanthus occidentalis*)
  Delaware NERR, 123
  Jacques Cousteau NERR, 70
Button sedge (*Carex bullata*), Jacques Cousteau NERR, 72

C

Cactus mouse (*Peromyscus eremicus*), Tijuana River NERR, 243
Cakile edentula, Jacques Cousteau NERR, 81
Cakile maritima, Tijuana River NERR, 243
Calder, D.R., 196
Caleb Pond, Waquoit Bay NERR, 35
Calidris alba, see Sanderling
Calidris alpina, see Dunlin
Calidris canutus, see Red knot
Calidris fuscicollis, Jacques Cousteau NERR, 87
Calidris mauri, see Western sandpiper
Calidris minutilla, see Least sandpiper
Calidris pusilla, Jacques Cousteau NERR, 87
California brown pelican (*Pelecanus occidentalis*), Tijuana River NERR, 237
California Department of Parks and Recreation, 237
California halibut (*Paralichthys californicus*), Tijuana River NERR, 252
California horn snail (*Cerithidea californica*), Tijuana River NERR, 239, 244
California jackknife clam (*Tagelus californianus*), Tijuana River NERR, 245, 248
California killifish (*Fundulus parvipinnis*), Tijuana River NERR, 239, 248–249, 252
California king snake (*Lampropeltis getulus californiae*), Tijuana River NERR, 242
California least tern (*Sterna antillarum browni*), Tijuana River NERR, 237, 240, 243–244

Index

California salt bush (*Atriplex pusillus*), Tijuana River NERR, 242
Callianassa gigas, Tijuana River NERR, 247
Callinectes sapidus, Jacques Cousteau NERR, 69
Callinectes sapidus, see Blue crab
Calycomonas ovalis, Jacques Cousteau NERR, 97
Camissonia cheiranthifolia spp. *suffruticosa*, Tijuana River NERR, 243
Campbell Scientific Weather Station, 11, 12
Canada goose (*Branta canadensis*), Jacques Cousteau NERR, 88
Canada lily (*Lilium canadense*), Delaware NERR, 156
Canby's dropwort (*Oxypolyis canbyi*), ACE Basin NERR, 202
Cancer irroratus, Jacques Cousteau NERR, 104
Canis latrans, see Coyote
Canvasback (*Aythra valisneria*)
  Delaware NERR, 163
  Jacques Cousteau NERR, 88
Capitella spp., Waquoit Bay NERR, 46
Capitellids (*Heteromastus filiformis*), Waquoit Bay NERR, 41
Carabid beetle (*Tachys corax*), Tijuana River NERR, 240
Carassius auratus, Jacques Cousteau NERR, 94
Carcinus maenas, as invasive, 25
Carcinus maenas, Waquoit Bay NERR, 46
Cardinalis cardinalis, see Northern cardinal
Carduelis tristis, see American goldfinches
Caretta caretta, see Loggerhead turtle
Carex bullata, Jacques Cousteau NERR, 72
Carex exilis, Jacques Cousteau NERR, 72
Carex kobomugi, Jacques Cousteau NERR, 81
Carolina bays, 175, 181
Carolina chickadee (*Parus carolinensis*)
  Delaware NERR, 145, 160
  Jacques Cousteau NERR, 92
Carolina wren (*Thryothorus ludovicianus*)
  ACE Basin NERR, 187, 188
  Jacques Cousteau NERR, 92
Carp (*Cyprinus carpio*)
  Delaware NERR, 140
  Jacques Cousteau NERR, 94
Carpenter frog (*Rana virgatipes*)
  ACE Basin NERR, 181
  Jacques Cousteau NERR, 82
Carpenter's tellen (*Tellina carpenteri*), Tijuana River NERR, 247
Carphophis amoenus amoenus, Jacques Cousteau NERR, 85
Carpobrotus edulis, Tijuana River NERR, 243
Casmerodius albus, Delaware NERR, 161
Casmerodius albus, see Great egret
Castelli, P.M., 88

Castor canadensis, see Beaver
Cathartes aura, see Turkey vulture
Catoptrophorus semipalmatus, see Willet
Catostomus commersoni, see White sucker
Cattails, (*Typha* spp.), Delaware NERR, 124–125, 152
Cattails (*Typha angustifolia*)
  Delaware NERR, 125
  Jacques Cousteau NERR, 70
  Weeks Bay NERR, 220
Cattails (*Typha glauca*), Jacques Cousteau NERR, 70
Cattails (*Typha latifolia*), Delaware NERR, 125
Cattle egret (*Casmerodius albus*), Delaware NERR, 161
CCX Fiberglass Products Plant, 173
Celtis occidentalis, Jacques Cousteau NERR, 82
Cemophora coccinea, Jacques Cousteau NERR, 85
Centralized Data Management Office (CDMO)
  function, 4–5, 11
  location, 4
Centropages hamatus, Jacques Cousteau NERR, 98
Centropages typicus, Jacques Cousteau NERR, 98
Centropristis striata, Jacques Cousteau NERR, 106
Cephalanthus occidentalis, see Button bush
Ceramium fastigiatum, Jacques Cousteau NERR, 95
Ceratium spp., Delaware NERR, 133
Cerithidea californica, Tijuana River NERR, 239, 244
Certhia americana, Delaware NERR, 161
Ceryle alcyon, Weeks Bay NERR, 223
Chaetomorpha spp., Jacques Cousteau NERR, 95
Chaetopsis spp., ACE Basin NERR, 190
Chain pickerel (*Esox niger*), Jacques Cousteau NERR, 92, 93
Chamaecyparis thyoides, Jacques Cousteau NERR, 71, 82
Chamaedaphne calyculata, Jacques Cousteau NERR, 71
Channel catfish (*Ictalurus punctatus*)
  Delaware NERR, 140
  Jacques Cousteau NERR, 94
Chant, R.J., 62
Charadrius alexandrinus nivosus, Tijuana River NERR, 240, 243, 244
Charadrius melodus, see Piping plovers
Charadrius semipalmatus, see Semipalmated plover
Charadrius vociferus, see Killdeer
Charadrius wilsonia, ACE Basin NERR, 188
Chelonia mydas, see Green sea turtle

Chelydra serpentina, see Snapping turtle
Chemical contaminants, see also specific types
   ACE Basin NERR, 203–204
   major classes, 22
   monitoring, 22
   overview, xi, xii–xiii
   Waquoit Bay NERR, 49
Chen caerulescens, Jacques Cousteau NERR, 88
Chestnut oak (*Quercus prinus*), Jacques Cousteau NERR, 80, 81
Chicken turtle (*Deirochelys reticularia*), ACE Basin NERR, 182
Chinese mitten crap (*Eriocheir sinensis*), as invasive, 25
Chipping sparrow (*Spizella passerina*), Delaware NERR, 145
Chironomidae, Delaware NERR, 136
Chlamydomonas spp., Delaware NERR, 133
Chlorella spp.
   Delaware NERR, 133
   Jacques Cousteau NERR, 97
Chloroscombrus chrysurus, ACE Basin NERR, 199, 201
Chroomonas amphioxiea, Jacques Cousteau NERR, 97
Chroomonas minuta, Jacques Cousteau NERR, 97
Chroomonas vectensis, Jacques Cousteau NERR, 97
Chrysemys picta, Jacques Cousteau NERR, 85
Chrysemys rubriventris, see Red-bellied turtle
Cicindela spp., Tijuana River NERR, 240
Cinnamon fern (*Osmunda cinnamonea*), Jacques Cousteau NERR, 72
Circus cyaneus, see Northern harrier
Cistothorus palustris, see Marsh wren
Cladium jamaicense, Weeks Bay NERR, 220
Cladium marascoides, Waquoit Bay NERR, 39
Cladocerans, Weeks Bay NERR, 225
Cladonia spp., Waquoit Bay NERR, 37
Cladophora vagabunda, Waquoit Bay NERR, 43, 51
Clam worms (*Nereis virens*), Waquoit Bay NERR, 41
Clangula hyemalis, see Oldsquaw
Clapper rails (*Rallus longirostris*), Jacques Cousteau NERR, 91
Clemmys guttata, Jacques Cousteau NERR, 85
Clemmys insculpta, Jacques Cousteau NERR, 85
Clemmys muhlenbergii, Jacques Cousteau NERR, 85
Clethra alnifolia, see Sweet pepperbush
Clethrionomys gapperi, Jacques Cousteau NERR, 86
Clevelandia ios, Tijuana River NERR, 249, 252
Clupea harengus, Jacques Cousteau NERR, 106

Cnemidophorus sexlineatus, ACE Basin NERR, 176, 180
Coastal Change Analysis Program (C-CAP), land cover/habitat change database, 19
Coastal Services Center, 6
Coastal Training Program (CTP), 5
   function, 28
Coastal Zone Management,CZM
Coastal Zone Management Act (CZMA) of 1972, importance, 2
Coast horned lizard (*Phrynosoma coronatum blainvillei*), Tijuana River NERR, 243
Coast sedge (*Carex exilis*), Jacques Cousteau NERR, 72
Code of Federal Regulations 15 CFR, 4, 7–8
Codium fragile
   Jacques Cousteau NERR, 95
   Waquoit Bay NERR, 42, 43
Coelus globosus, Tijuana River NERR, 243
Coelus hirticollis gravida, Tijuana River NERR, 243
Coelus latesignata latesignata, Tijuana River NERR, 243
Colinus virginianus, Delaware NERR, 145
Collins, J.T., 176
Colonial nesting birds, Jacques Cousteau NERR, 87–88, 89
Coluber constrictor constrictor
   ACE Basin NERR, 181
   Jacques Cousteau NERR, 85
Combahee River, see also ACE Basin NERR
   description, 171
Comely shiner (*Notropis amoenus*), Jacques Cousteau NERR, 94
Commercially/recreationally important species
   Delaware NERR, 140, 162–163
   Tijuana River NERR, 252
   Weeks Bay NERR, 227
Common cattail (*Typha latifolia*), Waquoit Bay NERR, 38
Common goldeneye (*Bucephala clangula*), Jacques Cousteau NERR, 88
Common grackle (*Quiscalus quiscula*), Delaware NERR, 145, 161
Common merganser (*Mergus merganser*)
   Delaware NERR, 144
   Jacques Cousteau NERR, 88
Common reed (*Phragmites australis*)
   Delaware NERR, 124, 152
   effects overview, 24
   Jacques Cousteau NERR, 66, 67, 69, 70
   Waquoit Bay NERR, 38
   Weeks Bay NERR, 220

Index

Common shiner (*Notropis cornustus*), Jacques Cousteau NERR, 94
Common tern (*Sterna hirundo*)
  Waquoit Bay NERR, 41
  Weeks Bay NERR, 223
Comptonia peregrina, *see* Sweetfern
Conant, R., 176
Condylura cristata, Jacques Cousteau NERR, 86
Conservation easements, 173
Consumption advisories, Delaware NERR, 131–132
Contopus virens, Jacques Cousteau NERR, 91
Conway, T.M., 63
Cooney, T.W., 171
Cooperative Institute for Coastal and Estuarine Environmental Technology (CICEET), 23
Cooper's hawk (*Accipiter cooperii*)
  ACE Basin NERR, 187
  Delaware NERR, 144
Copepod nauplii
  ACE Basin NERR, 194, 195
  Delaware NERR, 157–158
Copepods, Weeks Bay NERR, 225
Copperhead (*Agkistrodon controtrix*), ACE Basin NERR, 181, 182
Coragyps atratus, ACE Basin NERR, 187
Cordylanthus maritimus ssp. *maritimus*, Tijuana River NERR, 237, 240
Coreopsis rosea, Jacques Cousteau NERR, 72
Cormea conradii, Jacques Cousteau NERR, 81
Cormorants, ACE Basin NERR, 202
Cormorants (*Phalacrocorax* spp.), Jacques Cousteau NERR, 87
Corn snake (*Elaphe guttata guttata*)
  ACE Basin NERR, 180
  Jacques Cousteau NERR, 85
Cornus amomum, Jacques Cousteau NERR, 70
Cornus spp., Delaware NERR, 123
Corophium cylindricum, Jacques Cousteau NERR, 100
Corophium sp.
  Delaware NERR, 136, 158
  Waquoit Bay NERR, 40
Corvus brachyrhynchos
  ACE Basin NERR, 187
  Delaware NERR, 145
Corvus ossifragus, ACE Basin NERR, 187
Cossura sp., Jacques Cousteau NERR, 103
Costa, J.E., 51
Cotton mouse (*Peromyscus gossypinus*), ACE Basin NERR, 185
Cottonmouth (*Agkistrodon piscivorus*), ACE Basin NERR, 181, 182
Cottonwoods (*Populus fremontii*), Tijuana River NERR, 241
Coturnicops noveboracensis, Delaware NERR, 144
Coull, B.C., 196
Cowwheat (*Melampyrum lineare*), Jacques Cousteau NERR, 81
Cox, G.W., 242
Coyote (*Canis latrans*)
  ACE Basin NERR, 183
  Tijuana River NERR, 243
Cranberry bogs
  Jacques Cousteau NERR, 72
  Waquoit Bay NERR, 39
Crangon septemspinosa
  Jacques Cousteau NERR, 98, 104
  Waquoit Bay NERR, 46
Crassostrea virginica, Delaware NERR, 136
Crassostrea virginica, *see* Eastern oyster
Crawford, R.E., 51
Creek chub (*Semotilus atromaculatus*), Jacques Cousteau NERR, 93
Creek chubsucker (*Erimyzon oblongus*), Jacques Cousteau NERR, 93
Crepidula fornicata, Waquoit Bay NERR, 45
Crepidula plana, Waquoit Bay NERR, 45
Cressa truxillensis, Tijuana River NERR, 242
Crotalus adamanteus, ACE Basin NERR, 180
Crotalus horridus horridus, *see* Timber rattlesnake
Cryptomya californica, 245, 247, 248
Cryptotis parva, *see* Least shrew
Crystal ice plant (*Mesembryanthemum crystallinum*), Tijuana River NERR, 242
Cunner (*Tautogolabrus adspersus*), Jacques Cousteau NERR, 100
Curly-grass fern (*Schizaea pusilla*), Jacques Cousteau NERR, 71
Cyanocitta cristata, Delaware NERR, 145, 160
Cyathura polita
  Delaware NERR, 158
  Jacques Cousteau NERR, 100
Cyclops spp., Delaware NERR, 135, 158
Cyclotella sp., Jacques Cousteau NERR, 97
Cygnus columbianus, Jacques Cousteau NERR, 88
Cylindrotheca closterium
  ACE Basin NERR, 194
  Jacques Cousteau NERR, 97
Cynoscion nebulosus, ACE Basin NERR, 199
Cynoscion regalis, *see* Weakfish
Cyprinodon variegatus, Delaware NERR, 138
Cyprinus carpio, *see* Carp

CZM
    as federal government assessment program, xiii
    NERRS relationship, 3–4

D

Dangleberry (*Gaylussacia frondosa*), Jacques Cousteau NERR, 71, 72
Dardeau, M.R., 223, 225
Data loggers
    Jacques Cousteau NERR, 63, 64
    SWMP, 10–11, 12
Data management
    data needs, 6
    SWMP improvements, 11
Davis, L.V., 190
Davis, R.B., 193–194
Deer mouse (*Peromyscus maniculatus*), Tijuana River NERR, 243
Deirochelys reticularia, ACE Basin NERR, 182
Delaware NERR
    description, 119
    designation, 119
    map, 120
    overview, 163–165
Delaware NERR/Lower St. Jones River Reserve
    anthropogenic impacts, 128–133
        habitat alteration, 132–133
        nutrient loading, 128
        pollution, 128–132
        toxic substances list, 129
    aquatic habitat, 125
    biotic communities
        amphibians, 140
        benthic fauna, 136–138
        benthic fauna density, 136, 137–138
        birds, 140–145, 146–150
        finfish, 138–140
        mammals, 145, 151
        phytoplankton, 133–134
        plankton density, 135
        reptiles, 140
        zooplankton, 134–136
    description, 119, 120–121
    development, 120, 128, 131
    land use, 120–121, 151, 157
    plant list, 122–123
    upland vegetation, 121
    water quality, 125–128
        dissolved oxygen, 125–126, 127
        salinity, 125, 126
        specific conductivity, 126
        turbidity, 127–128
        water depth, 127, 128
        water pH, 126, 127
        water temperature, 126
    watershed, 120–121
    wetland vegetation, 121, 123–125
    wetland vegetation cover, 124
    wetland vegetation/marsh plants list, 122–123
Delaware NERR/Upper Blackbird Creek Reserve
    anthropogenic impact
        habitat alteration, 157
        pollution, 157
    aquatic habitat, 156
    biotic communities
        amphibians, 160
        benthic fauna/density, 158–159
        birds, 160–161
        commercially/recreationally important species, 140, 162–163
        finfish, 160
        mammals, 161–162
        phytoplankton, 157
        plant list, 153–155
        reptiles, 160
        zooplankton, 157–158
    description, 119, 121
    upland vegetation, 151–152
    water quality/variables, 126–127, 156
    watershed, 151
    wetland vegetation, 152–156
Delaware's Department of Natural Resources and Environmental Control, 131
Dendroica coronata, Delaware NERR, 145, 161
Dendroica discolor, Delaware NERR, 145, 161
Dendroica petechia, *see* Yellow warbler
Dentroica pinus, Jacques Cousteau NERR, 92
Deposit feeders, Tijuana River NERR, 245
Dermo (*Perkinsus marinus*), Delaware NERR, 162
Dermochelys coriacea, ACE Basin NERR, 201
Desmognathus fuscus, Jacques Cousteau NERR, 82
Desmond, J.S., 242
Diaphanosoma spp., Delaware NERR, 158
Diatoms
    ACE Basin NERR, 193–194
    Delaware NERR, 133, 157
    Jacques Cousteau NERR, 97
    Tijuana River NERR, 245
Didelphis marsupialis, *see* Opossum
Dindo, J.J., 223
Dinoflagellates
    ACE Basin NERR, 193
    Jacques Cousteau NERR, 97
    Tijuana River NERR, 245

Diopatra cuprea, Jacques Cousteau NERR, 103–104
Dipodomys agilis, Tijuana River NERR, 243
Dissolved oxygen, *see also specific sites*
  summary analysis, 13
Distichlis spicata, *see* Salt grass; Spike grass
Dolichonyx oxyzivorus, Delaware NERR, 145, 161
Dolphins, ACE Basin NERR, 202
Dorosoma cepedianum, Jacques Cousteau NERR, 94
Dotted smartweed (*Polygonum punctatum*), Jacques Cousteau NERR, 70
Double-crested cormorant (*Phalacrocorax auritus*), Delaware NERR, 144
Dover Air Force Base Superfund site, 132
Dover Gas Light Company Superfund site, 132, 133
Dow, C.L., 63
Dowitcher (*Limmodromus* sp.), Tijuana River NERR, 241, 244
Downy woodpecker (*Picoides pubescens*), Delaware NERR, 145
Dragonflies, ACE Basin NERR, 190
Dreissena polymorpha, as invasive, 25
Drosera capillaris, Jacques Cousteau NERR, 71, 72
Drosera spp., Weeks Bay NERR, 220
Drowned beaked-rush (*Rhynchospora inundata*), Jacques Cousteau NERR, 72
Dudley, B.W., 196
Dumetella carolinensis, *see* Gray catbird
Dune primrose (*Camissonia cheiranthifolia* spp. *suffruticosa*), Tijuana River NERR, 243
Dune ragweed (*Ambrosia chamissonis*), Tijuana River NERR, 243
Dunlin (*Calidris alpina*)
  Jacques Cousteau NERR, 87
  Waquoit Bay NERR, 41
Durand, J.B., 98, 100, 106
Dusky-footed woodrat (*Neotoma fuscipes*), Tijuana River NERR, 243
Dusty miller (*Artemisia stelleriana*), Waquoit Bay NERR, 41
Dwarf salamander (*Eurycea quadridigitata*), ACE Basin NERR, 182
Dwarf white bladderwort (*Utricularia olivacea*), Jacques Cousteau NERR, 72

E

Eagles, ACE Basin NERR, 187

Eastern box turtle (*Terrapene carolina*), Jacques Cousteau NERR, 85
Eastern brook trout (*Salvelinus fontinalis*), Waquoit Bay NERR, 38
Eastern chipmunk (*Tamius striatus*), Jacques Cousteau NERR, 86
Eastern coachwhip (*Masticophis flagellum*), ACE Basin NERR, 183
Eastern cottontail (*Sylvilagus floridanus*)
  ACE Basin NERR, 183
  Delaware NERR, 145, 161
  Jacques Cousteau NERR, 86
Eastern diamondback rattlesnake (*Crotalus adamanteus*), ACE Basin NERR, 180
Eastern fence lizard (*Sceloporus undulatus*), ACE Basin NERR, 176
Eastern garter snake (*Thanmnophis sirtalis*)
  ACE Basin NERR, 180
  Jacques Cousteau NERR, 85
Eastern glass lizard (*Ophisaurus ventralis*), ACE Basin NERR, 181
Eastern harvest mouse (*Reithrodontomys humilus*), ACE Basin NERR, 183
Eastern hognose snake (*Heterodon platyrhinos*), Jacques Cousteau NERR, 85
Eastern king snake (*Lampropeltis getula getula*), Jacques Cousteau NERR, 84
Eastern meadowlark (*Sturnella magna*), Delaware NERR, 161
Eastern mole (*Scalopus aquaticus*)
  ACE Basin NERR, 186
  Jacques Cousteau NERR, 86
Eastern mudminnow (*Umbra pygmaea*), Jacques Cousteau NERR, 92, 93
Eastern mud turtle (*Kinosternon subrubrum subrubrum*)
  ACE Basin NERR, 182
  Delaware NERR, 140, 160
  Jacques Cousteau NERR, 85
Eastern oyster (*Crassostrea virginica*)
  ACE Basin NERR, 197
  Delaware NERR, 162
  Weeks Bay NERR, 227
Eastern painted turtle (*Chrysemys picta*), Jacques Cousteau NERR, 85
Eastern pipistrelle (*Pipistrellus subflavus*), Jacques Cousteau NERR, 86
Eastern ribbon snake (*Thamnophis sauritus*), Jacques Cousteau NERR, 84
Eastern screech owl (*Otus asio*)
  Delaware NERR, 145
  Jacques Cousteau NERR, 89, 91
Eastern spadefoot frog (*Scaphiopus holbrooki*), Jacques Cousteau NERR, 82

Eastern tiger salamander (*Ambystoma tigrinum tigrinum*), Jacques Cousteau NERR, 82
Eastern wood peewee (*Contopus virens*), Jacques Cousteau NERR, 91
Eastern woodrat (*Neotoma floridana*), ACE Basin NERR, 185
Eastern worm snake (*Carphophis amoenus amoenus*), Jacques Cousteau NERR, 85
Echinoderms
  Jacques Cousteau NERR, 103–104
  Waquoit Bay NERR, 42, 44
Ectinosoma spp., Delaware NERR, 158
Edisto River, *see also* ACE Basin NERR
  description, 171
Edotea triloba, Delaware NERR, 136
Education, Outreach, and Interpretation Program, *see also* Environmental education
  description, 26
  objectives, 26–27
Eelgrass (*Zostera marina*)
  importance, 43, 45
  Jacques Cousteau NERR, 94, 95–96, 109
  Waquoit Bay NERR, 43, 45
Egg cockle (*Laevicardium substriatum*), Tijuana River NERR, 247
Egrets
  ACE Basin NERR, 188
  Jacques Cousteau NERR, 87–88
Egretta caerulea, Jacques Cousteau NERR, 88
Egretta thula, *see* Snowy egret
Egretta tricolor, Jacques Cousteau NERR, 88
Eidson, J.P., 191
Elaphe guttata guttata, *see* Corn snake
Elaphe obsoleta, Delaware NERR, 140, 160
Eleocharis equisetoides, Jacques Cousteau NERR, 72
Elops saurus, ACE Basin NERR, 200
Endangered/threatened species
  ACE Basin NERR, 202
  definition, 203
  Jacques Cousteau NERR, 90–91
  Tijuana River NERR, 237, 239–240
Enis directus, Delaware NERR, 136
Enneacanthus chaetodon, Jacques Cousteau NERR, 93
Enneacanthus gloriosus, Jacques Cousteau NERR, 93
Enneacanthus obesus, Jacques Cousteau NERR, 92, 93
Ensis directus, Waquoit Bay NERR, 45
Enteromorpha spp., Waquoit Bay NERR, 43
Environmental education
  NERRS, 5–6, 7, 25–28

Strategic Education Plan, 26
Environmental Monitoring and Assessment Program, xiii
Epifauna, 195
Epigaea repens, Jacques Cousteau NERR, 81
Eptesicus fuscus, *see* Big brown bat
Eremophila alpestris, Tijuana River NERR, 240
Eretmochelys imbricata, ACE Basin NERR, 201
Erimyzon oblongus, Jacques Cousteau NERR, 93
Eriocaulon parkeri, Jacques Cousteau NERR, 70
Eriocheir sinensis, as invasive, 25
Eriogonum fasciculatum, Tijuana River NERR, 242
Erosion
  ACE Basin NERR, 204
  Tijuana River NERR, 255
  Weeks Bay NERR, 227
Esox americanus, Jacques Cousteau NERR, 93
Esox niger, Jacques Cousteau NERR, 92, 93
Estuaries
  commercial interests summary, 1
  recreational interests summary, 1
Estuaries and Clean Waters Act (2000), 24
Estuarine-dependent species statistics, 1
Estuarine Reserves Division (ERD), role, 3
Etheostoma fusiforme, Jacques Cousteau NERR, 92
Etheostoma olmetedi, Jacques Cousteau NERR, 92, 93
Etropus microstomus, Jacques Cousteau NERR, 109
Euglena/Eutrepta spp., Jacques Cousteau NERR, 97
Euglenoids, Jacques Cousteau NERR, 72
Eumeces fasciatus, Jacques Cousteau NERR, 82
Eumeces laticeps, ACE Basin NERR, 176
Eumida sanguinea, Jacques Cousteau NERR, 103
European green crab (*Carcinus maenas*), as invasive, 25
European sicklegrass (*Parapholis incurva*), Tijuana River NERR, 242
Eurycea bislineata, *see* Northern two-lined salamander
Eurycea longicauda, ACE Basin NERR, 181
Eurycea quadridigitata, ACE Basin NERR, 182
Eurytemora affinis
  Delaware NERR, 135, 136, 158
  Jacques Cousteau NERR, 59, 98
Euterpina acutifrons, ACE Basin NERR, 194
Eutrophication, *see also* Nutrient loading
  effects, 16
  severity, 16
  Waquoit Bay NERR, 49
Evening bat (*Nycticeius humeralis*), ACE Basin NERR, 183

Index

Exogene dispar, Jacques Cousteau NERR, 103
Exotic species, *see* Invasive/exotic species

F

Fagus grandifolia, Delaware NERR, 121, 152
Falco columbarius, see Merlin
Falco peregrinus, see Peregrine falcon
Falco sparverius, see American kestrel
Fallfish (*Semotilus corporalis*), Jacques Cousteau NERR, 93
False mya (*Cryptomya californica*), 245, 247, 248
Federal Geographical Data Committee, 5
Ferns, Weeks Bay NERR, 220
Ferren, W.R., Jr., 242
Fetterbush (*Leucothoe racemosa*), Jacques Cousteau NERR, 71, 72
Fiddler crabs (*Uca crenulata*), Tijuana River NERR, 239, 244
Fiddler crabs (*Uca* spp.)
  ACE Basin NERR, 193
  Delaware NERR, 136, 138
Field sparrow (*Spizella pusilla*), Delaware NERR, 145, 160
Filinia spp., Delaware NERR, 157–158
Filter feeders, Tijuana River NERR, 245
Finback whale (*Balaenoptera physalus*), Delaware NERR, 145
Fish crows (*Corvus ossifragus*), ACE Basin NERR, 187
Fish River, Weeks Bay NERR, 218
Five-lined skink (*Eumeces fasciatus*), Jacques Cousteau NERR, 82
Flagellates, ACE Basin NERR, 194
Flathead catfish (*Ictalurus furcatus*), ACE Basin NERR, 204–205
Flathead minnow (*Pimephales promelas*), Jacques Cousteau NERR, 94
Flat Pond, Waquoit Bay NERR, 35
Fleshy jaumea (*Jaumea carnosa*), Tijuana River NERR, 239
Floating heart (*Nymphoides cordata*), Jacques Cousteau NERR, 72
Florida cooter (*Pseudemys floridana*), ACE Basin NERR, 182
Florida softshell (*Apalone ferox*), ACE Basin NERR, 182
Florimentis obesa, Tijuana River NERR, 247
Flounders (*Pleuronichthys* spp.), Tijuana River NERR, 252
Flycatchers, ACE Basin NERR, 187
Forage fishes, Waquoit Bay NERR, 43
Ford, R.F., 245
Forster's tern (*Sterna forsteri*), Delaware NERR, 161
Forsythe National Wildlife Refuge, 66, 81
Fourspine stickleback (*Apeltes quadracus*)
  Jacques Cousteau NERR, 94, 99, 106, 109
  Waquoit Bay NERR, 46
Four-toed salamander (*Hemidactylium scutatum*), Jacques Cousteau NERR, 82
Fowler's toad (*Bufo woodhousii fowleri*), Jacques Cousteau NERR, 82
Foxes, Weeks Bay NERR, 223
Frankenia grandifolia, Tijuana River NERR, 242
Frankenia palmeri, Tijuana River NERR, 242
Frankenia salina, Tijuana River NERR, 239, 242
Fraxinus pennsylvanica, Delaware NERR, 121, 152, 155
Free-tailed bat (*Tadarida brasiliensis*), ACE Basin NERR, 185
Fucus spp., Waquoit Bay NERR, 43
Fundulus diaphanus, Jacques Cousteau NERR, 94
Fundulus heteroclitus, see Mummichog
Fundulus majalis, Jacques Cousteau NERR, 106
Fundulus parvipinnis, Tijuana River NERR, 239, 248–249, 252

G

Gadwall (*Anas strepera*)
  Delaware NERR, 163
  Jacques Cousteau NERR, 88
Gammarus sp., Delaware NERR, 136, 158
Gasterosteus aculeatus, see Three-spined stickleback
Gastropods, Jacques Cousteau NERR, 103–104
Gaultheria procumbens, Jacques Cousteau NERR, 71, 72, 81
Gavia spp., Jacques Cousteau NERR, 87
Gavia stelata, Delaware NERR, 144
Gaylussacia baccata
  Jacques Cousteau NERR, 71, 72
  Waquoit Bay NERR, 37
Gaylussacia dumosa, Jacques Cousteau NERR, 72
Gaylussacia frondosa, Jacques Cousteau NERR, 71, 72
Gaylussacia spp., Jacques Cousteau NERR, 80
Geist, M.A., 42
Gem clam (*Gemma gemma*)
  Delaware NERR, 136
  Jacques Cousteau NERR, 103
  Waquoit Bay NERR, 40
Genyonemus lineatus, Tijuana River NERR, 252
Geographic Information System, *see* GIS

Geothlypis trichas, see Yellowthroat
Geukensia demissa, Waquoit Bay NERR, 42, 45
Geukensia demissa, see Ribbed mussel
Geum virginiana, Delaware NERR, 156
Gillichthys mirabilis, Tijuana River NERR, 248
GIS
  habitat information database, 11
  use, 6
  watershed land use mapping, 18–19
Gizzard shad (*Dorosoma cepedianum*), Jacques Cousteau NERR, 94
Glassworts (*Salicornia* spp.), Waquoit Bay NERR, 39
Glassworts (*Salicornia subterminalis*), Tijuana River NERR, 239, 242
Glaucomys volans, Jacques Cousteau NERR, 86
Global sea level rise
  Delaware NERR, 132
  effects, 21–22
  monitoring, 21–22
  Waquoit Bay NERR, 41–42
Global sea surface temperature, 22
Globose dune beetle (*Coelus globosus*), Tijuana River NERR, 243
Glossy crayfish snake (*Regina regida*), ACE Basin NERR, 182
Glossy ibis (*Plegadis falcinellus*)
  Delaware NERR, 161
  Jacques Cousteau NERR, 88
Glyceria dibranchiata, Delaware NERR, 136
Glyceria grandis, Jacques Cousteau NERR, 67
Glycinde solitaria, Jacques Cousteau NERR, 103
Goatsrue (*Tephrosia virginiana*), Jacques Cousteau NERR, 81
Gobies (*Gobiosoma* spp.), Jacques Cousteau NERR, 98, 109
Golden crest (*Lophiola aurea*), Jacques Cousteau NERR, 72
Golden eagle (*Aquila chrysaetos*), Tijuana River NERR, 240
Golden mouse (*Ochrotomys nuttalli*), ACE Basin NERR, 185
Golden shiner (*Notemigonus crysoleucas*), Jacques Cousteau NERR, 92, 94
Goldfish (*Carassius auratus*), Jacques Cousteau NERR, 94
Gracilaria tikvahiae, see Red macroalga
Grackles, ACE Basin NERR, 189
Grammarus spp., Delaware NERR, 138
Granivores (birds), ACE Basin NERR, 186, 187, 188, 189
Grasshoppers, ACE Basin NERR, 190
Grass shrimp (*Palaemonetes* spp.)
  ACE Basin NERR, 197
  Delaware NERR, 138, 158

Grass shrimp (*Palaemonetes vulgaris*), Waquoit Bay NERR, 46
Gray, I.E., 190
Gray birch (*Betula populifolia*), Jacques Cousteau NERR, 71
Gray catbird (*Dumetella carolinensis*)
  Delaware NERR, 145, 160–161
  Jacques Cousteau NERR, 92
Gray fox (*Urocyon cinereoargenteus*)
  ACE Basin NERR, 185
  Delaware NERR, 151, 161
  Jacques Cousteau NERR, 86
Gray seals (*Halichoerus grypus*), Delaware NERR, 145
Gray squirrel (*Sciurus carolinensis*)
  ACE Basin NERR, 183
  Delaware NERR, 145, 151, 161
  Jacques Cousteau NERR, 86
Gray treefrogs (*Hyla chrysoscelis*), Jacques Cousteau NERR, 82
Gray treefrogs (*Hyla versicolor*), Jacques Cousteau NERR, 82
Great Bay Boulevard Wildlife Management Area, 66
Great blue heron (*Ardea herodias*)
  Delaware NERR, 161
  Tijuana River NERR, 239
  Weeks Bay NERR, 223
Great egret (*Casmerodius albus*)
  Jacques Cousteau NERR, 87, 88
  Tijuana River NERR, 239
Greater scaup (*Aythya marila*)
  Delaware NERR, 163
  Jacques Cousteau NERR, 88
Greater siren (*Siren lacertina*), ACE Basin NERR, 182
Greater yellowlegs (*Tringa melanoleuca*)
  Delaware NERR, 161
  Jacques Cousteau NERR, 87
  Waquoit Bay NERR, 41
Great horned owl (*Bubo virginianus*)
  Delaware NERR, 144, 145
  Jacques Cousteau NERR, 89, 91
Grebes, ACE Basin NERR, 202
Green algae, Delaware NERR, 157
Green algae (*Enteromorpha* spp.), Waquoit Bay NERR, 43
Green algae
  Jacques Cousteau NERR, 72
  Waquoit Bay NERR, 43
Green ash (*Fraxinus pennsylvanica*), Delaware NERR, 121, 152, 155
Green-backed heron (*Butorides virescens*), Delaware NERR, 161

Index

Green crabs (*Carcinus maenus*), Waquoit Bay NERR, 46
Green frog (*Rana clamatans*), Waquoit Bay NERR, 38
Green frog (*Rana clamitans melanota*)
  Delaware NERR, 140, 160
  Jacques Cousteau NERR, 82
Green macroalga (*Cladophora vagabunda*), Waquoit Bay NERR, 43, 51
Green sea turtle (*Chelonia mydas*)
  ACE Basin NERR, 201
  Delaware NERR, 140
Green-winged teal (*Anas crecca*)
  Delaware NERR, 161, 163
  Jacques Cousteau NERR, 88
Groundsel bush (*Baccharis halimifolia*), Delaware NERR, 124–125
Ground skink (*Scinella lateralis*)
  ACE Basin NERR, 176
  Jacques Cousteau NERR, 82, 84
Grouseberry (*Gaylussacia dumosa*), Jacques Cousteau NERR, 72
Guinardia spp., Delaware NERR, 133
Guiraca caerulea, Delaware NERR, 145
Gulf menhaden (*Brevoortia patronus*), ACE Basin NERR, 226, 227
Gulf whiting (*Menticirrhus littoralis*), ACE Basin NERR, 201
Gulls
  ACE Basin NERR, 202
  Delaware NERR, 144
  Jacques Cousteau NERR, 87

H

Habitat alteration
  ACE Basin NERR, 204
  anthropogenic impact types, xi
  Delaware NERR, 132–133
  effects, 18
  land cover/habitat change database, 19
Habitat restoration
  example reserves, 24
  funding, 24
  monitoring, 23
  restoration science, 23
  Tijuana River NERR, 24, 237, 255
Hackberry (*Celtis occidentalis*), Jacques Cousteau NERR, 82
Haematopus palliatus, see American oystercatcher
Hagen, S.M., 69
Hake (*Urophycis* spp.), Jacques Cousteau NERR, 109

Halectinosoma winonae, ACE Basin NERR, 196
Hales, L.S., 103–104
Haliaeetus leucocephalus, see Bald eagle
Halichoerus grypus, Delaware NERR, 145
Halicyclops fosteri, Delaware NERR, 135, 136, 158
Hamblin Pond, Waquoit Bay NERR, 35
Haplopappus venetus, Tijuana River NERR, 242
Haplosporidium nelsoni, Delaware NERR, 162
Harbor seals (*Phoca vitulina*), Delaware NERR, 145
Hardhead catfish (*Arius felis*), ACE Basin NERR, 199
Hard-shelled clam (*Mercenaria mercenaria*), Waquoit Bay NERR, 40, 42, 45
Harp seals (*Pagophilus groenlandicus*), Delaware NERR, 145
Hastings, R.W., 92
Hawks, ACE Basin NERR, 187, 188
Hawksbill turtle (*Eretmochelys imbricata*), ACE Basin NERR, 201
Heavy metals
  ACE Basin NERR, 204
  Delaware NERR, 128, 129, 130
  Weeks Bay NERR, 227
Helonias bullata, Jacques Cousteau NERR, 71
Hemidactylium scutatum, Jacques Cousteau NERR, 82
Hemigrapsus oregonensis, Tijuana River NERR, 239, 244
Herons, Jacques Cousteau NERR, 88
Herring gulls (*Larus argentatus*)
  ACE Basin NERR, 188
  Waquoit Bay NERR, 41
Heteromastus filiformis
  ACE Basin NERR, 197
  Delaware NERR, 136
  Waquoit Bay NERR, 41
Heterosigma carterae, Jacques Cousteau NERR, 97
Hibiscus moscheutos, Jacques Cousteau NERR, 70
Hibiscus palustris, Delaware NERR, 121
Hickory shad (*Alosa mediocris*)
  ACE Basin NERR, 200
  Jacques Cousteau NERR, 94
Highbush blueberry (*Vaccinium corymbosum*), Jacques Cousteau NERR, 72, 82
High-priority SWMP initiatives
  anthropogenic impacts understanding, 16
  benthic community surveys, 21
  benthic habitat mapping, 19–21
  biomonitoring, 17–18
  chlorophyll fluorescence, 17
  description, 15

nutrient monitoring, 16–17
specific water quality problems, 15–16
water quality monitoring, 15–16
watershed land use mapping, 18–19
Himantopus mexicanus, Tijuana River NERR, 241
Hippolyte pleuracanthus, Waquoit Bay NERR, 45
Hirundo rustica, Delaware NERR, 161
Hispid cotton rat (*Sigmodon hispidus*), ACE Basin NERR, 183
Hogchoker (*Trinectes maculatus*)
 ACE Basin NERR, 199
 Jacques Cousteau NERR, 94, 100
Holoplankton, 194
Homo sapiens, Jacques Cousteau NERR, 86
Hooded merganser (*Lophodytes cucullatus*), Jacques Cousteau NERR, 88
Horned grebe (*Podiceps auritus*), Delaware NERR, 144
Horned lark (*Eremophila alpestris*), Tijuana River NERR, 240
Horned pondweed (*Zanniuchellia palustris*), Jacques Cousteau NERR, 67
Horseshoe crab (*Limulus polyphemus*)
 Delaware NERR, 162–163
 Waquoit Bay NERR, 40–41, 46
Hosmer, S.C., 245, 248
Hottentot-fig (*Carpobrotus edulis*), Tijuana River NERR, 243
House mouse (*Mus musculus*)
 ACE Basin NERR, 186
 Jacques Cousteau NERR, 86
House sparrow (*Passer domesticus*), Delaware NERR, 145
House wren (*Troglodytes aedon*), Delaware NERR, 160, 161
Howes, B.G., 39
Huckleberry (*Gaylussacia* spp.), Jacques Cousteau NERR, 80
Huckleberry/black huckleberry (*Gaylussacia baccata*)
 Jacques Cousteau NERR, 71, 72
 Waquoit Bay NERR, 37
Hudsonian godwit (*Lemosa haemastica*), Jacques Cousteau NERR, 87
Hudsonia tomentosa, *see* Beach heather
Human coastal population
 as estuary stressor, xi, xiii
 statistics, 1
Humans (*Homo sapiens*), Jacques Cousteau NERR, 86
Hump-back whale (*Megaptera novaeangliae*), Delaware NERR, 145
Hunchak-Kariouk, K., 63

Hybognathus nuchalis, Jacques Cousteau NERR, 93
Hydrodictyon spp., Delaware NERR, 133, 157
Hyla andersonii, Jacques Cousteau NERR, 82
Hyla chrysoscelis, Jacques Cousteau NERR, 82
Hyland, J.L., 196
Hyla versicolor, Jacques Cousteau NERR, 82
Hylocichla mustilina, *see* Wood thrush
Hypericum gentianoides, Jacques Cousteau NERR, 82
Hyperprosopon argenteum, Tijuana River NERR, 252
Hypoxia analysis of 22 sites, 14

I

Ichthyoplankton, Jacques Cousteau NERR, 99–100
Ictalurus catus, ACE Basin NERR, 199
Ictalurus furcatus, ACE Basin NERR, 204–205
Ictalurus melas, Jacques Cousteau NERR, 94
Ictalurus punctatus, *see* Channel catfish
Icterus galbula, Delaware NERR, 145, 160
Ilex glabra, Jacques Cousteau NERR, 80
Ilex opaca, *see* American holly
Ilex spp., Weeks Bay NERR, 220
Ilex verticillata, *see* Winterberry
Ilyanassa obsoleta
 ACE Basin NERR, 197
 Delaware NERR, 138
Ilyanassa sp., Delaware NERR, 136
Impatiens capensis, Jacques Cousteau NERR, 70
Indigo bunting (*Passerina cyanea*)
 ACE Basin NERR, 187
 Delaware NERR, 145
Infauna, 195
Inkberry (*Ilex glabra*), Jacques Cousteau NERR, 80
Inland silverside (*Menidia beryllina*), ACE Basin NERR, 200
Insectivores (birds), ACE Basin NERR, 186, 187, 188, 189
Insects, Tijuana River NERR, 239
Institute of Marine and Coastal Sciences, Rutgers University, 62
International Boundary and Water Commission, 245
Introduced species, *see* Invasive/exotic species
Invasive/exotic species
 effects, 24, 25
 examples, 24, 25
 statistics, 24–25
 Tijuana River NERR, 235, 242, 243
Iridoprocne bicolor, Jacques Cousteau NERR, 91

Index

Ironcolor shiner (*Notropis chalybaeus*), Jacques Cousteau NERR, 92
Isoetes riparia, Jacques Cousteau NERR, 70
Itea virginica, Weeks Bay NERR, 220
Iva frutescens, see Marsh elder

J

Jacques Cousteau NERR
  Atlantic Flyway, 86
  climatic conditions, 61
  description, 59–60
  designation, 59
  development, 59, 81
  education, 27–28
  eelgrass decline, 95–96
  endangered/threatened species list/table, 90–91
  environmental setting, 61–62
  estuarine biotic communities (animal)
    benthic fauna, 100–104
    benthic invertebrates list/table, 101–103
    benthic invertebrates salinity distribution, 105
    finfish, 104, 106–110
    fish/decapods list/table, 107–109
    zooplankton, 98–100
    zooplankton monthly mean abundance, 99
  estuarine biotic communities (animal) benthic macroinvertebrates density map, 104
  estuarine biotic communities (plant)
    benthic flora, 94–96
    phytoplankton, 96–97
  grasses, 72
  Holocene barrier island complex, 61
  macroalgae, 95–96
  map, 60
  Mullica River Basin/tributaries, 61
  Mullica River-Great Bay Estuary
    description, 61, 62
    water quality, 62–65, 66, 67, 68–69, 70
  overview, 110–111
  partners, 60
  picoplankton blooms, 96–97
  Pine Barrens species, 82, 84, 85–86
  SAV loss, 95–96
  seagrass distribution map, 96
  sediments, 62
  species overview, 59–60
  tidal water bodies, 61–62
  wasting disease (from *Labyrinthula zosterae*), 95–96
  water circulation, 62
  water quality
    dissolved oxygen, 64, 67, 68
    seasonal pH, 64, 68
    seasonal salinity, 64, 66
    seasonal water temperature, 65
    turbidity levels, 64, 69
    water depth, 64–65, 70
  water quality monitoring sites, 64
  watershed biotic communities (animal)
    amphibians/anurans list/table, 85
    amphibians/reptiles, 82–85
    amphibians/reptiles list/table, 83–84
    birds, 86–89, 91–92
    fish, 92–94
    fish list/table, 93
    mammals, 85–86
  watershed biotic communities, endangered/threatened species list/table, 90–91
  watershed biotic communities (plant)
    Atlantic white cedar swamp forests, 71
    barrier island plant communities, 81–82
    brackish tidal marshes, 66–69
    broadleaf swamp forests, 71
    freshwater marshes, 69–70
    herbaceous wetland communities, 71, 72
    lowland plant communities, 71–80
    lowland plant list/table, 73–80
    pine transition forest, 71–72
    pitch pine lowland forests, 71
    salt marshes, 65–66
    shrubby wetland communities, 71, 72
    upland plant communities, 80–81
Japanese sedge (*Carex kobomugi*), Jacques Cousteau NERR, 81
Jassa falcata, Jacques Cousteau NERR, 98
Jaumea carnosa, Tijuana River NERR, 239
Jehu Pond, Waquoit Bay NERR, 35
Jivoff, P., 106, 109
Juncus acutus, Tijuana River NERR, 241
Juncus gerardi, Waquoit Bay NERR, 39
Juncus gerardii, Jacques Cousteau NERR, 66
Juncus roemerianus
  ACE Basin NERR, 173
  Weeks Bay NERR, 220
Juniperus virginiana, see Red cedar

K

Kalmia augustifolia, see Sheep laurel
Kalmia latifolia, Jacques Cousteau NERR, 80, 81
Kangaroo rat (*Dipodomys agilis*), Tijuana River NERR, 243
Katodinium rotundatum, Jacques Cousteau NERR, 97

Kemp's Ridley turtle (*Lepidochelys kempii*)
    ACE Basin NERR, 201
    Delaware NERR, 140
Kennish, M.J., xi, xii–xiii
Kettle hole ponds, Waquoit Bay NERR, 39
Killdeer (*Charadrius vociferus*)
    Delaware NERR, 161
    Tijuana River NERR, 241
Killifishes, ACE Basin NERR, 200
Kingfish (*Menticirrhus saxatilis*), Jacques Cousteau NERR, 94, 106
Kinosternon subrubrum subrubrum, see Eastern mud turtle
Knob-styled dogwood (*Cornus amomum*), Jacques Cousteau NERR, 70
Knott, D.M., 194
Knotted spikerush (*Eleocharis equisetoides*), Jacques Cousteau NERR, 72
Kraeuter, J.N., 190

L

Labidocera aestiva, Jacques Cousteau NERR, 98
Lady crabs (*Ovalipes ocellatus*)
    Jacques Cousteau NERR, 104
    Waquoit Bay NERR, 42
Ladyfish (*Elops saurus*), ACE Basin NERR, 200
Laevicardium substriatum, Tijuana River NERR, 247
Lagodon rhomboides, ACE Basin NERR, 201
Laminaria agardhii, Waquoit Bay NERR, 43
Lampetra lamottie, Jacques Cousteau NERR, 94
Lampropeltis getula getula, Jacques Cousteau NERR, 84
Lampropeltis getulus californiae, Tijuana River NERR, 242
Landscaping (alternative landscaping), xiii
Lanius ludovicianus, Tijuana River NERR, 240
Largemouth bass (*Micropterus salmoides*), Jacques Cousteau NERR, 92, 94
Larus argentatus, see Herring gulls
Larus atricilla, see Laughing gull
Larus delawarensis, Waquoit Bay NERR, 41
Larus marinus, Waquoit Bay NERR, 41
Larus spp., Delaware NERR, 144, 161
Lasiurus borealis, Delaware NERR, 151, 161–162
Laterallus jamaicensis, Delaware NERR, 144
Lathrop, R.G., 63
Lathyrus japonicus var. *glaber*, Waquoit Bay NERR, 41
Lathyrus maritimus, Jacques Cousteau NERR, 81
Laughing gull (*Larus atricilla*)
    ACE Basin NERR, 188, 189

    Delaware NERR, 161
    Weeks Bay NERR, 223
Laurel oak (*Quercus laurifolia*), Weeks Bay NERR, 220
Least Bell's vireo (*Vireo belli pusilius*), Tijuana River NERR, 237
Least sandpiper (*Calidris minutilla*)
    Jacques Cousteau NERR, 87
    Tijuana River NERR, 241, 244
    Waquoit Bay NERR, 41
Least shrew (*Cryptotis parva*)
    Delaware NERR, 151, 161
    Jacques Cousteau NERR, 86
Least tern (*Sterna antillarum*)
    ACE Basin NERR, 188, 202
    Delaware NERR, 144
    Jacques Cousteau NERR, 87
    Waquoit Bay NERR, 41
Leatherback turtle (*Dermochelys coriacea*), ACE Basin NERR, 201
Leatherleaf (*Chamaedaphne calyculata*), Jacques Cousteau NERR, 71
Leiophyllum buxifolium, Jacques Cousteau NERR, 81
Leiostomus xanthurus, see Spot
Leitoscoloplos robustus, Jacques Cousteau NERR, 103
Lemonadeberry (*Rhus integrifolia*), Tijuana River NERR, 243
Lemosa fedoa, see Marbled godwit
Lemosa haemastica, Jacques Cousteau NERR, 87
Lepidochelys kempii, see Kemp's Ridley turtle
Lepomis auritus, Jacques Cousteau NERR, 94
Lepomis gibbosus, Jacques Cousteau NERR, 92
Lepomis macrochirus, Jacques Cousteau NERR, 92, 94
Leptastacus spp., Delaware NERR, 135
Lepus californicus, Tijuana River NERR, 240
Lesser golden plover (*Pluvialis dominica*), Jacques Cousteau NERR, 87
Lesser scaup (*Aythya affinis*)
    Delaware NERR, 144, 163
    Jacques Cousteau NERR, 88
Lesser yellowlegs (*Tringa flavipes*), Waquoit Bay NERR, 41
Leucothoe racemosa, Jacques Cousteau NERR, 71, 72
Lewitus, A.J., 194
Liatris scariosa var. *novae-angliae*, Waquoit Bay NERR, 38
Libinia dubia, Waquoit Bay NERR, 46
Libinia emarginata, Jacques Cousteau NERR, 104
Lichens (*Cladonia* spp.), Waquoit Bay NERR, 37

Index

Light-footed clapper rail (*Rallus longirostris levipes*), Tijuana River NERR, 237, 244
Lilium canadense, Delaware NERR, 156
Limmodromus sp., Tijuana River NERR, 241, 244
Limnodromus griseus, *see* Short-billed dowitcher
Limonium californicum, Tijuana River NERR, 239, 242
Limonium nashii, Waquoit Bay NERR, 39
Limulus polyphemus, *see* Horseshoe crab
Lined shore crab (*Pachygrapsus crassipes*), Tijuana River NERR, 239, 244
Liquidambar styraciflua, *see* Sweet gum
Liriodendron tulipifera, Delaware NERR, 121
Little blue heron (*Egretta caerulea*), Jacques Cousteau NERR, 88
Little blue-stem (*Schizachyrium scoparium*), Waquoit Bay NERR, 37–38
Little brown myotis (*Myotis lucifugus*), Jacques Cousteau NERR, 86
Little grass frog (*Pseudacris ocularis*), ACE Basin NERR, 181
Little ice plant (*Mesembryanthemum nodilforum*), Tijuana River NERR, 242
Littleneck clam (*Protothaca staminea*), Tijuana River NERR, 245, 248
Littorina irrorata, ACE Basin NERR, 197
Live oak (*Quercus virginiana*), Weeks Bay NERR, 220
Lizardfish (*Synodus foetens*), Jacques Cousteau NERR, 109
Lobelia boykinii, Jacques Cousteau NERR, 72
Loblolly pines (*Pinus taeda*)
  ACE Basin NERR, 174
  Weeks Bay NERR, 220
Loggerhead shrike (*Lanius ludovicianus*), Tijuana River NERR, 240
Loggerhead turtle (*Caretta caretta*)
  ACE Basin NERR, 173, 201–202
  Delaware NERR, 140
Long-billed curlew (*Numenius americanus*), Tijuana River NERR, 239, 244
Long-eared owls (*Asio otus*), ACE Basin NERR, 188
Longjaw mudsucker (*Gillichthys mirabilis*), Tijuana River NERR, 248
Longleaf pine (*Pinus palustris*)
  ACE Basin NERR, 174
  Weeks Bay NERR, 220
Long's bulrush (*Scirpus longii*), Jacques Cousteau NERR, 72
Long-tailed weasel (*Mustela frenata*)
  ACE Basin NERR, 185
  Delaware NERR, 151, 161
  Jacques Cousteau NERR, 86

Tijuana River NERR, 243
Long-Term Ecosystem Observatory, 59, 100
Loons (*Gavia* spp.), Jacques Cousteau NERR, 87
Lophiola aurea, Jacques Cousteau NERR, 72
Lophodytes cucullatus, Jacques Cousteau NERR, 88
Lowbush blueberry (*Vaccinium angustifolium*), Waquoit Bay NERR, 37
Lowland broomsedge (*Andropogon virginicus* var. *virginicus*), Jacques Cousteau NERR, 72
Lunatia heros, Waquoit Bay NERR, 45
Lutra canadensis, *see* River otter
Lycium californicum, Tijuana River NERR, 242
Lynx rufus, ACE Basin NERR, 186
Lyonia mariana, Jacques Cousteau NERR, 72
Lythrum salicaria, Jacques Cousteau NERR, 70

M

Mabee's salamander (*Ambystoma mabeei*), 182
Macoma nasuta, Tijuana River NERR, 247
Macoma secta, Tijuana River NERR, 247
Macrofauna, 195
Macrozooplankton, 194
Magnolia River, Weeks Bay NERR, 218
Magnolia virginiana, Jacques Cousteau NERR, 71
Mahoney, J.B., 96–97
Malaclemys terrapin terrrapin, *see* Northern diamondback terrapin
Mallard (*Anas platyrhynchos*)
  Delaware NERR, 161
  Jacques Cousteau NERR, 88
Malpass, W., 42
Many-lined salamander (*Stereochilus marginatus*), ACE Basin NERR, 181
Marbled godwit (*Lemosa fedoa*)
  Jacques Cousteau NERR, 87
  Tijuana River NERR, 239, 244
Marbled salamander (*Ambystoma opacum*)
  ACE Basin NERR, 181
  Jacques Cousteau NERR, 82
Marcus, J.M., 204
Margined madtom (*Noturus insignis*), Jacques Cousteau NERR, 94
Marine Activities, Resources, and Education (MARE) Summer Institute, 27–28
Marine Resources Research Institute, SCDNR, 199
Marion, K.R., 223
Marmota monax, Jacques Cousteau NERR, 86
Marsh crabs (*Sesarma reticulatum*), Delaware NERR, 138

Marsh elder (*Iva frutescens*)
  Delaware NERR, 123
  Jacques Cousteau NERR, 66
  Waquoit Bay NERR, 39
  Weeks Bay NERR, 220
Marsh fleabane (*Pluchea purpurascens*), Jacques Cousteau NERR, 66
Marshpepper smartweed (*Polygonum hydropiper*), Delaware NERR, 121, 152
Marsh rabbit (*Sylvilagus palustris*)
  ACE Basin NERR, 185, 186
  Weeks Bay NERR, 223
Marsh wren (*Cistothorus palustris*)
  Delaware NERR, 144, 160
  Jacques Cousteau NERR, 91
Masked shrew (*Sorex cinereus*)
  Delaware NERR, 151, 161
  Jacques Cousteau NERR, 86
Masticophis flagellum, ACE Basin NERR, 183
Mathews, T., 203, 204
Meadow jumping mouse (*Zapus hudsonius*)
  Delaware NERR, 145, 161
  Jacques Cousteau NERR, 86
Meadow vole (*Microtus pennsylvanicus*), Jacques Cousteau NERR, 86
Mediomastus ambiseta, Waquoit Bay NERR, 46
Mediomastus sp., Jacques Cousteau NERR, 103
Medium-priority SWMP initiatives
  chemical contaminant monitoring, 22
  description, 15
  global sea level rise, 21–22
Megafauna, 195
Megaptera novaeangliae, Delaware NERR, 145
Meiofauna, 195
Melampus bidentatus, Delaware NERR, 138
Melampus olivaceus, Tijuana River NERR, 239
Melampyrum lineare, Jacques Cousteau NERR, 81
Melanerpes carolinus, Delaware NERR, 145
Melanitta nigra, Delaware NERR, 144
Melanitta perspicillata, Delaware NERR, 144
Melanitta spp., Jacques Cousteau NERR, 88
Melinna cristata, Jacques Cousteau NERR, 103
Melita nitida, ACE Basin NERR, 197
Melosira spp., Delaware NERR, 133, 157
Melospiza melodia, *see* Song sparrow
Menida menidia, *see* Atlantic silverside
Menidia beryllina, ACE Basin NERR, 200
Menticirrhus saxatilis, Jacques Cousteau NERR, 94, 106
Mephitis mephitis, *see* Striped skunk
Mercenaria mercenaria, Waquoit Bay NERR, 40, 42, 45
Mergus merganser, *see* Common merganser

Mergus serrator, *see* Red-breasted merganser
Merlin (*Falco columbarius*)
  ACE Basin NERR, 187
  Jacques Cousteau NERR, 89
Meroplankton, 194
Mesembryanthemum crystallinum, Tijuana River NERR, 242
Mesembryanthemum nodilforum, Tijuana River NERR, 242
Mesozooplankton, 194
Metabolic rates, analysis of 22 sites, 14
Meteorological data, overview, 12–13
Microcystis spp., Delaware NERR, 133
Microfauna, 195
Microgadus tomcod, Waquoit Bay NERR, 46
Microphthalmus sezelkowii, Jacques Cousteau NERR, 103
Micropogonias undulatus, *see* Atlantic croaker
Micropterus salmoides, Jacques Cousteau NERR, 92, 94
Microtus pennsylvanicus, Jacques Cousteau NERR, 86
Microtus pinetorum, *see* Pine vole
Microzooplankton, 194
Miller-Way, T.M., 217
Mimic glass lizard (*Ophisaurus mimicus*), ACE Basin NERR, 176
Mimus polyglottos
  ACE Basin NERR, 187
  Delaware NERR, 145, 161
  Jacques Cousteau NERR, 92
Mink (*Mustela vison*)
  ACE Basin NERR, 185
  Delaware NERR, 163
  Jacques Cousteau NERR, 86
Mitchella repens, Jacques Cousteau NERR, 71
Mnemiopsis leidyi, Jacques Cousteau NERR, 98
Mniotilta varia, *see* Black-and-white warbler
Mockingbird (*Mimus polyglottos*), Jacques Cousteau NERR, 92
Moles, ACE Basin NERR, 186
Mole salamander (*Ambystoma talpoideum*), ACE Basin NERR, 181
Molgula manhattensis
  ACE Basin NERR, 196
  Waquoit Bay NERR, 42
Mollusca, Delaware NERR, 136, 137
Mollusks, Waquoit Bay NERR, 44, 45
Monanthocloe littoralis, Tijuana River NERR, 239, 242
Monitoring, *see also* SWMP
  baseline monitoring program, 4
  components, 9
  importance, xiii

Index   281

Moon snails (*Lunatia heros*), Waquoit Bay NERR, 45
Moon snails (*Polinices duplicatus*), Waquoit Bay NERR, 45
Morone americana, see White perch
Morone saxatilis, see Striped bass
Morus bassanus, Delaware NERR, 144
Moser, F.C., 100, 103
Mountain laurel (*Kalmia latifolia*), Jacques Cousteau NERR, 80, 81
Mountford, K., 97
Mourning dove (*Zenaida macroura*)
  ACE Basin NERR, 187, 188
  Delaware NERR, 145, 161
MSX (*Haplosporidium nelsoni*), Delaware NERR, 162
Mud crabs (*Neopanope texana*), Waquoit Bay NERR, 46
Mud snails (*Ilyanassa obsoleta*), Delaware NERR, 138
Mud sunfish (*Acantharchus pomotis*), Jacques Cousteau NERR, 92–93
Mugil cephalus, see Striped mullet
Mugil curema, ACE Basin NERR, 201
Muhlenbergia torreyana, Jacques Cousteau NERR, 72
Mulina lateralis, Delaware NERR, 136, 197
Multiflora rose (*Rosa multiflora*), Jacques Cousteau NERR, 82
Mummichog (*Fundulus heteroclitus*)
  Delaware NERR, 138, 160
  Jacques Cousteau NERR, 69, 94, 98, 106
Muskrat (*Ondatra zibethicus*)
  Delaware NERR, 145, 162, 163
  Jacques Cousteau NERR, 86
Mus musculus, see House mouse
Mustela frenata, see Long-tailed weasel
Mustela vison, see Mink
Mya arenaria
  Delaware NERR, 136
  Waquoit Bay NERR, 40, 42, 43
Mycteria americana, ACE Basin NERR, 173, 202
Myotis lucifugus, Jacques Cousteau NERR, 86
Myrica cerifera, ACE Basin NERR, 173, 174
Myrica gale, Waquoit Bay NERR, 39
Myrica pensylvanica, see Bayberry
Myriophyllum tenellum, Jacques Cousteau NERR, 72
Mysid shrimp (*Neomysis americana*), Delaware NERR, 135
Mytilus edulis, Waquoit Bay NERR, 45

N

Nadeau, R.J., 98, 100, 106
Nannochloris atomus, Jacques Cousteau NERR, 96–97
Narragansett Bay NERR, restoration, 24
Narrow-leaved cattail (*Typha angustifolia*), Jacques Cousteau NERR, 67
Nassarius obsoletes, Jacques Cousteau NERR, 103
National Coastal Assessment Program, xiii
National Estuaries Day, 27
National Estuarine Research Reserve System, *see* NERRS
National Estuarine Sanctuary Program
  change to NERRS, 2
  establishment, 2
National Estuary Program (NEP), xiii
National Marine Fisheries Service National Habitat Program, xiii
National Oceanic and Atmospheric Administration, *see* NOAA
National Pollution Discharge Elimination System (NPDES), 131, 203
National Status and Trends (NS&T) Program, xiii
National Strategy for Coastal Habitat Restoration, 24
National Wetlands Inventory, xiii
Nematodes, ACE Basin NERR, 196
Nemertinea
  Delaware NERR, 136, 137, 158, 159
  Waquoit Bay NERR, 44
Neomysis americana
  Delaware NERR, 135, 136, 158
  Jacques Cousteau NERR, 98
Neopanope texana
  Jacques Cousteau NERR, 98
  Waquoit Bay NERR, 46
Neotoma floridana, ACE Basin NERR, 185
Neotoma fuscipes, Tijuana River NERR, 243
Neotropical migrants, Jacques Cousteau NERR, 92
Nereis spp., Waquoit Bay NERR, 46
Nereis succinea, ACE Basin NERR, 197
Nereis virens, Waquoit Bay NERR, 41
Nerodia fasciata, ACE Basin NERR, 182
Nerodia sipedon, see Northern water snake
Nerodia taxispilota, ACE Basin NERR, 182
NERRS
  categories of federal awards, 8
  description, xiii–xiv
  education/outreach, 25–28
  functional elements, 7
  functions overview, 4–5
  funding, 4

linking programs, 5–6
list of major components, 6
list of sites, 3
management, 3, 7
map of site locations, 2
mission, 6–9
nominations, 8–9
overview, 28–29
overview of sites, 2–3
partnerships, 4, 7
program components overview, 9–29
  habitat restoration, 23–24
  invasive/exotic species, 24–25
  monitoring/research, 9–22
  special high-priority initiatives, 23–28
reasons for creating, xiv
reserve description, 3
role of reserve, 4
SWMP, 4
NERRS Action Plan, 6
NERRS Strategic Plan, 6
New England blazing star (*Liatris scariosa* var. *novae-angliae*), Waquoit Bay NERR, 38
New Jersey chorus frog (*Pseudacris triseriata kalmi*), Jacques Cousteau NERR, 82
New Jersey Department of Environmental Protection, 59
Nichols, T.C., 88
Nitocra lacustrus, ACE Basin NERR, 196
Nitzschia spp.
  ACE Basin NERR, 194
  Delaware NERR, 157
  Jacques Cousteau NERR, 97
NOAA
  Estuarine Reserves Division (ERD), 3, 6
  role with NERRS, 2, 3, 4
Nodding bur-marigold (*Bidens cernus*), Delaware NERR, 156
Nonpoint source pollution, Delaware NERR, 131
Nordby, C.S., 248, 252
North Brigantine State Natural Area, 81
Northern black racer (*Coluber constrictor constrictor*), Jacques Cousteau NERR, 85
Northern bobwhite (*Colinus virginianus*), Delaware NERR, 145
Northern cardinal (*Cardinalis cardinalis*)
  ACE Basin NERR, 187
  Delaware NERR, 145
Northern cricket frog (*Acris crepitans crepitans*), Jacques Cousteau NERR, 82
Northern diamondback terrapin (*Malaclemys terrapin terrapin*)
  Delaware NERR, 140, 160, 163

Jacques Cousteau NERR, 85
Northern dusky salamander (*Desmognathus fuscus*), Jacques Cousteau NERR, 82
Northern fence lizard (*Sceloporus undulatus hyacinthinus*), Jacques Cousteau NERR, 82, 84
Northern gannet (*Morus bassanus*), Delaware NERR, 144
Northern gannet (*Sula bassanus*), Jacques Cousteau NERR, 87
Northern harrier (*Circus cyaneus*)
  ACE Basin NERR, 187
  Delaware NERR, 144, 161
  Jacques Cousteau NERR, 89
  Tijuana River NERR, 240
Northern mockingbird (*Mimus polyglottos*)
  ACE Basin NERR, 187
  Delaware NERR, 145, 161
Northern oriole (*Icterus galbula*), Delaware NERR, 145, 160
Northern parula warbler (*Parula americana*)
  Delaware NERR, 160
  Jacques Cousteau NERR, 91
Northern pine snake (*Pituophis melanoleucus melanoleucus*), Jacques Cousteau NERR, 85
Northern pintail (*Anas acuta*), 88
  Tijuana River NERR, 240
Northern pipefish (*Syngnathus fuscus*), Jacques Cousteau NERR, 98, 106, 109
Northern puffer (*Sphoeroides maculatus*), Jacques Cousteau NERR, 106
Northern red salamander (*Pseudotriton ruber*), Jacques Cousteau NERR, 82
Northern right whale (*Balaena glacialis*), Delaware NERR, 145
Northern scarlet snake (*Cemophora coccinea*), Jacques Cousteau NERR, 85
Northern spring peeper (*Pseudacris crucifer crucifer*)
  Delaware NERR, 140
  Jacques Cousteau NERR, 82
Northern two-lined salamander (*Eurycea bislineata*)
  Delaware NERR, 140, 160
  Jacques Cousteau NERR, 82
Northern water snake (*Nerodia sipedon*)
  Delaware NERR, 140, 160
  Jacques Cousteau NERR, 84
North Inlet/Winyah Bay NERR, CDMO, 4–5
Norway rat (*Rattus norvegicus*)
  ACE Basin NERR, 183
  Jacques Cousteau NERR, 86
Notemigonus crysoleucas, Jacques Cousteau NERR, 92, 94

Index

Notholca spp., Delaware NERR, 157–158
Notomastus lateris, Jacques Cousteau NERR, 100
Notophthalmus viridescens viridescens, Waquoit Bay NERR, 38
Notropis amoenus, Jacques Cousteau NERR, 94
Notropis analostanus, Jacques Cousteau NERR, 94
Notropis bifrenatus, Jacques Cousteau NERR, 94
Notropis chalybaeus, Jacques Cousteau NERR, 92
Notropis cornustus, Jacques Cousteau NERR, 94
Notropis hudsonius, Jacques Cousteau NERR, 94
Noturus gyrinus, Jacques Cousteau NERR, 93
Noturus insignis, Jacques Cousteau NERR, 94
Nucula sp., Jacques Cousteau NERR, 103
Numenius americanus, Tijuana River NERR, 239, 244
Numenius phaeopus, Tijuana River NERR, 244
Nuphar lutea, Delaware NERR, 152
Nuphar variegatum, Jacques Cousteau NERR, 72
Nurphur advena, Jacques Cousteau NERR, 70
Nutrient loading, *see also* Eutrophication
    Delaware NERR, 128
    overview, xi
    Waquoit Bay NERR, 48–49
    Weeks Bay NERR, 227
Nutrient monitoring, commencement, 11–12
Nuttall's pondweed (*Potamogeton epihydrus*), Jacques Cousteau NERR, 67
Nycticeius humeralis, ACE Basin NERR, 183
Nycticorax nycticorax, Jacques Cousteau NERR, 88
Nycticorax violaceus, Jacques Cousteau NERR, 88
Nymphaea odorata, Jacques Cousteau NERR, 72
Nymphoides cordata, Jacques Cousteau NERR, 72
Nyssa sylvatica, *see* Black gum

O

Oak toad (*Bufo quercicus*), ACE Basin NERR, 181
Oceanites oceanicus, Jacques Cousteau NERR, 87
Ochrotomys nuttalli, ACE Basin NERR, 185
Odocoileus virginianus, *see* White-tailed deer
O'Herron, J.C.,, III, 138, 140
Oithona colcarva, Delaware NERR, 136
Oithona similis, Jacques Cousteau NERR, 59
Oithona spp., Delaware NERR, 158
Old-field mouse (*Peromyscus polionotus*), ACE Basin NERR, 183

Oldsquaw (*Clangula hyemalis*)
    Delaware NERR, 144
    Jacques Cousteau NERR, 88
Oligochaeta, Delaware NERR, 136
Olney three-square bulrush (*Scirpus americanus*), Jacques Cousteau NERR, 66–67
Olsen, P.S., 96–97
Omnivores (birds), ACE Basin NERR, 186, 187, 188, 189
Onaclea sensibilis, Jacques Cousteau NERR, 70
Ondatra zibethicus, *see* Muskrat
Opheodrys aestivus, *see* Rough green snake
Ophisaurus attenuatus, ACE Basin NERR, 176, 181
Ophisaurus mimicus, ACE Basin NERR, 176
Ophisaurus ventralis, ACE Basin NERR, 181
Opossum (*Didelphis marsupialis*)
    ACE Basin NERR, 183
    Jacques Cousteau NERR, 86
    Tijuana River NERR, 243
Opossum shrimp (*Neomysis americana*), Delaware NERR, 136, 158
Opsanus tau, Jacques Cousteau NERR, 106
Orach (*Atriplex patula*), Jacques Cousteau NERR, 66
Orange jewelweed (*Impatiens capensis*), Jacques Cousteau NERR, 70
Orchelimum fidicinium, ACE Basin NERR, 190
Orchestra grillus, Delaware NERR, 138
Oryzomys palustris, *see* Rice rat
Oscillatoria spp., Delaware NERR, 133
Osmunda cinnamonea, Jacques Cousteau NERR, 72
Osprey (*Pandion haliaetus*)
    Delaware NERR, 144, 145
    Jacques Cousteau NERR, 89
Otus asio, *see* Eastern screech owl
Ovalipes ocellatus, *see* Lady crabs
Ovenbird (*Seiurus aurocapillus*)
    Delaware NERR, 145, 161
    Jacques Cousteau NERR, 92
Overfishing, xi, xii
Owls, ACE Basin NERR, 186, 187, 188
Oxypolyis canbyi, ACE Basin NERR, 202
Oyster drill (*Urosalpinx cinera*), Delaware NERR, 162
Oyster reefs/beds, 197
Oyster toadfish (*Opsanus tau*), Jacques Cousteau NERR, 106

P

Pachygrapsus crassipes, Tijuana River NERR, 239, 244

Pacific cordgrass (*Spartina foliosa*), Tijuana River NERR, 239
Pagophilus groenlandicus, Delaware NERR, 145
PAHs
 ACE Basin NERR, 204
 Delaware NERR, 129
 Weeks Bay NERR, 227
Painted bunting (*Passerina cirus*), ACE Basin NERR, 188
Palaemonetes spp.
 ACE Basin NERR, 197
 Delaware NERR, 138, 158
 Jacques Cousteau NERR, 69
Palmettos, ACE Basin NERR, 173
Pandion haliaetus, see Osprey
Panicum wrightianum, Jacques Cousteau NERR, 72
Panopeus herbstii, Jacques Cousteau NERR, 98
Paracalanus crassirostris, Jacques Cousteau NERR, 98
Paracalanus parva, Jacques Cousteau NERR, 98
Paracaprella tenuis, ACE Basin NERR, 197
Paralabrax spp., Tijuana River NERR, 252
Paralichthys californicus, Tijuana River NERR, 252
Paralichthys dentatus
 Delaware NERR, 140, 162
 Jacques Cousteau NERR, 106
Paralichthys spp., Weeks Bay NERR, 227
Parapholis incurva, Tijuana River NERR, 242
Paraphosux sp., Jacques Cousteau NERR, 1034
Paraprionospio pinnata, ACE Basin NERR, 197
Parasites on oysters, 162
Pardosa ramulosa, Tijuana River NERR, 240
Parker's pipewort (*Eriocaulon parkeri*), Jacques Cousteau NERR, 70
Partridge berry (*Mitchella repens*), Jacques Cousteau NERR, 71
Parula americana, see Northern parula warbler
Parus carolinensis, see Carolina chickadee
Parvocalanus crassirostris, ACE Basin NERR, 194
Passer domesticus, Delaware NERR, 145
Passerina cirus, ACE Basin NERR, 188
Passerina cyanea see Indigo bunting
PCBs
 ACE Basin NERR, 204
 Delaware NERR, 129, 131
 Weeks Bay NERR, 227
Pelecanus occidentalis
 ACE Basin NERR, 189
 Tijuana River NERR, 237
Pelicans, ACE Basin NERR, 202
Peltandra virginica, see Arrow arum
Penaeus setiferus, ACE Basin NERR, 196

Penaeus spp.
 ACE Basin NERR, 197
 Weeks Bay NERR, 227
Pennock, J.R., 223, 225
Perca flavescens, Jacques Cousteau NERR, 92, 94
Peregrine falcon (*Falco peregrinus*)
 ACE Basin NERR, 202
 Delaware NERR, 144
 Jacques Cousteau NERR, 89
 Tijuana River NERR, 237
Perennial glasswort (*Salicornia subterminalis*), Tijuana River NERR, 242
Perennial glasswort (*Salicornia virginica*), Jacques Cousteau NERR, 66
Perennial pickleweed (*Salicornia virginica*), Tijuana River NERR, 239, 242
Perkinsus marinus, Delaware NERR, 162
Peromyscus boyli, Tijuana River NERR, 243
Peromyscus eremicus, Tijuana River NERR, 243
Peromyscus fallax, Tijuana River NERR, 243
Peromyscus gossypinus, ACE Basin NERR, 185
Peromyscus leucopus, see White-footed mouse
Peromyscus maniculatus, Tijuana River NERR, 243
Peromyscus polionotus, ACE Basin NERR, 183
Pesticides, Delaware NERR, 129, 131
Peterson, C.H., 245
Petrels, ACE Basin NERR, 202
Petroderma maculiforme, Waquoit Bay NERR, 43
Phalacrocorax auritus, Delaware NERR, 144
Phalacrocorax spp., Jacques Cousteau NERR, 87
Phoca vitulina, Delaware NERR, 145
Phragmites australis, see Common reed
Phrynosoma coronatum blainvillei, Tijuana River NERR, 243
Phytoplankton, categorization, 193
Pickerel frog (*Rana palustris*), Jacques Cousteau NERR, 82
Pickerel weed (*Ponderia cordata*)
 Delaware NERR, 121, 152
 Jacques Cousteau NERR, 70
 Weeks Bay NERR, 220
Picoides borealis, ACE Basin NERR, 202
Picoides pubescens, Delaware NERR, 145
Pig frog (*Rana grylio*), ACE Basin NERR, 182
Pimephales notatus, Jacques Cousteau NERR, 93
Pimephales promelas, Jacques Cousteau NERR, 94
Pineapple weed (*Ambylopappus pusillus*), Tijuana River NERR, 242
Pine Barrens treefrog (*Hyla andersonii*), Jacques Cousteau NERR, 82
Pinelands Commission, Jacques Cousteau NERR, 59

Index

Pine snake (*Pituophis melanoleucus melanoleucus*), ACE Basin NERR, 180
Pine vole (*Microtus pinetorum*)
  Delaware NERR, 145, 161
  Jacques Cousteau NERR, 86
Pine warbler (*Dentroica pinus*), Jacques Cousteau NERR, 92
Pineweed (*Hypericum gentianoides*), Jacques Cousteau NERR, 82
Pinfish (*Lagodon rhomboides*), ACE Basin NERR, 201
Pinus echinata, *see* Shortleaf pine
Pinus elliottii, Weeks Bay NERR, 220
Pinus palustris
  ACE Basin NERR, 174
  Weeks Bay NERR, 220
Pinus rigida, *see* Pitch pines
Pinus spp., ACE Basin NERR, 173
Pinus taeda, *see* Loblolly pines
Pipilo erythrophthalmus, Delaware NERR, 145, 161
Piping plovers (*Charadrius melodus*)
  ACE Basin NERR, 188
  Waquoit Bay NERR, 41
Pipistrellus subflavus, Jacques Cousteau NERR, 86
Pirate perch (*Aphredoderus sayanus*), Jacques Cousteau NERR, 92
Pitcher plants (*Sarracenia purpurea*), Jacques Cousteau NERR, 71, 72
Pitch pines (*Pinus rigida*)
  Jacques Cousteau NERR, 71, 80, 81, 82
  Waquoit Bay NERR, 36–37, 38
Pituophis melanoleucus annectens, Tijuana River NERR, 242
Pituophis melanoleucus melanoleucus
  ACE Basin NERR, 180
  Jacques Cousteau NERR, 85
Plant hoppers, ACE Basin NERR, 190
Platyhelminthes
  Delaware NERR, 136, 137
  Waquoit Bay NERR, 44
Plegadis falcinellus, *see* Glossy ibis
Plethodon cinereus, *see* Red-backed salamander
Pleuronichthys spp., Tijuana River NERR, 252
Pluchea purpurascens, Jacques Cousteau NERR, 66
Pluvialis dominica, Jacques Cousteau NERR, 87
Pluvialis squatarola, *see* Black-bellied plover
Pocosins, 181
Podiceps auritus, Delaware NERR, 144
Pogonias cromis, Delaware NERR, 140, 162
Poison ivy (*Rhus radicans*), Waquoit Bay NERR, 41
Poison ivy (*Toxicodendrom radicans*), Weeks Bay NERR, 220
Polinices duplicatus, Waquoit Bay NERR, 45
Pollachius virens, Waquoit Bay NERR, 46
Pollack (*Pollachius virens*), Waquoit Bay NERR, 46
Polychaetes, Delaware NERR, 136
Polychaetes (*Diopatra cuprea*), Jacques Cousteau NERR, 103–104
Polychaetes, Waquoit Bay NERR, 42, 46
Polycirrus hematodes, Jacques Cousteau NERR, 103
Polydora spp., Waquoit Bay NERR, 46
Polydori ligni, Jacques Cousteau NERR, 100, 103
Polygonum hydropiper, Delaware NERR, 121, 152
Polygonum punctatum, Jacques Cousteau NERR, 70
Polypogon monspeliensis, Tijuana River NERR, 242
Pomatomus salatrix, *see* Bluefish
Pomatomus saltatrix, Jacques Cousteau NERR, 106
Pomoxis nigromaculatus, Jacques Cousteau NERR, 94
Pompano (*Trachinotus carolinus*), ACE Basin NERR, 201
Ponderia cordata, *see* Pickerel weed
Populus fremontii, Tijuana River NERR, 241
Porzana carolina, *see* Sora
Post oak (*Quercus stellata*), Jacques Cousteau NERR, 81
Potamocorbula amurensis, as invasive, 25
Potamogeton epihydrus, Jacques Cousteau NERR, 67
Potamogeton perfoliatus, Jacques Cousteau NERR, 67
Potamogeton pusillus, Jacques Cousteau NERR, 67
Prairie warbler (*Dendroica discolor*), Delaware NERR, 145, 161
Prionospio spp., Waquoit Bay NERR, 46
Procyon lotor, *see* Raccoon
Progne subis, Jacques Cousteau NERR, 91
Prokelisia marginata, ACE Basin NERR, 190
Prorocentrum minimum, Jacques Cousteau NERR, 97
Protected Area Geographic Information System (PAGIS), 11
Prothonotary warbler (*Protonotaria citrea*), Delaware NERR, 160
Protonotaria citrea, Delaware NERR, 160
Protothaca staminea, Tijuana River NERR, 245, 248
Prunus maritima, *see* Beach plum

Prunus serotina, see Black cherry
Pseudacris crucifer crucifer, see Northern spring peeper
Pseudacris ocularis, ACE Basin NERR, 181
Pseudacris triseriata, ACE Basin NERR, 182
Pseudacris triseriata kalmi, Jacques Cousteau NERR, 82
Pseudemys concinna, ACE Basin NERR, 182
Pseudemys floridana, ACE Basin NERR, 182
Pseudobradya pulchella, ACE Basin NERR, 196
Pseudocalanus minutus, Jacques Cousteau NERR, 98
Pseudodiaptomus coronatus, Jacques Cousteau NERR, 98
Pseudodiaptomus pelagicus, Delaware NERR, 135, 136
Pseudolithoderma spp., Waquoit Bay NERR, 43
Pseudopleuronectes americanus, see Winter flounder
Pseudotriton ruber, Jacques Cousteau NERR, 82
Pteridium aquilinum, see Bracken fern
Ptilocherirus pinquis, Jacques Cousteau NERR, 103
Puffinus griseus, Jacques Cousteau NERR, 87
Pumpkinseed (*Lepomis gibbosus*), Jacques Cousteau NERR, 92
Purple clam (*Sanquinolaria nuttalli*), Tijuana River NERR, 245, 247, 248
Purple loosestrife (*Lythrum salicaria*), Jacques Cousteau NERR, 70
Purple martin (*Progne subis*), Jacques Cousteau NERR, 91
Pyramimonas spp., Jacques Cousteau NERR, 97

Q

Quercus alba, see White oak
Quercus coccinea, Jacques Cousteau NERR, 80, 81
Quercus falcata, see Southern red oak
Quercus ilicifolia, see Scrub oak
Quercus laurifolia, Weeks Bay NERR, 220
Quercus marilandica, see Blackjack oak
Quercus phellos, Jacques Cousteau NERR, 82
Quercus prinus, Jacques Cousteau NERR, 80, 81
Quercus rubra, Delaware NERR, 152
Quercus spp., ACE Basin NERR, 173, 174
Quercus stellata, Jacques Cousteau NERR, 81
Quercus velutina, Jacques Cousteau NERR, 80, 81
Quercus virginiana, Weeks Bay NERR, 220
Quiscalus major, see Boat-tailed grackle
Quiscalus quiscula, Delaware NERR, 145, 161

R

Rabbitfoot beardgrass (*Polypogon monspeliensis*), Tijuana River NERR, 242
Rabbits, Tijuana River NERR, 243
Raccoon (*Procyon lotor*)
  ACE Basin NERR, 183
  Delaware NERR, 151, 161
  Jacques Cousteau NERR, 86
  Weeks Bay NERR, 223
Rails
  ACE Basin NERR, 188
  Delaware NERR, 144, 161
Rainbow trout (*Salmo trutta*), Jacques Cousteau NERR, 94
Raja eglanteria, Jacques Cousteau NERR, 109
Ralfsia spp., Waquoit Bay NERR, 43
Rallus limicola, Jacques Cousteau NERR, 91
Rallus longirostris, Jacques Cousteau NERR, 91
Rallus longirostris levipes, Tijuana River NERR, 237, 244
Rallus spp., Delaware NERR, 144, 161
Rana catesbeiana, see Bullfrog
Rana clamatans, Waquoit Bay NERR, 38
Rana clamitans melanota
  Delaware NERR, 140, 160
  Jacques Cousteau NERR, 82
Rana grylio, ACE Basin NERR, 182
Rana heckscheri, ACE Basin NERR, 182
Rana palustris, Jacques Cousteau NERR, 82
Rana sylvatica, see Wood frog
Rana utricularia, see Southern leopard frog
Rana virgatipes, see Carpenter frog
Raptors
  ACE Basin NERR, 186, 187, 188, 189
  Delaware NERR, 144–145, 161
  Jacques Cousteau NERR, 89, 91
  Tijuana River NERR, 243, 252
Rathkea octopunctata, Jacques Cousteau NERR, 98
Rattus norvegicus, see Norway rat
Razor clams (*Ensis directus*), Waquoit Bay NERR, 45
Recurvirostra americana, Tijuana River NERR, 241, 244
Red-backed salamander (*Plethodon cinereus*)
  Delaware NERR, 140, 160
  Jacques Cousteau NERR, 82
Red-backed vole (*Clethrionomys gapperi*), Jacques Cousteau NERR, 86
Red bat (*Lasiurus borealis*), Delaware NERR, 151, 161–162
Red-bellied turtle (*Chrysemys rubriventris*)
  Delaware NERR, 140, 160

Jacques Cousteau NERR, 85
Red-bellied woodpecker (*Melanerpes carolinus*), Delaware NERR, 145
Redbelly water snake (*Nerodia erythrogastor*), ACE Basin NERR, 182
Red-breasted merganser (*Mergus serrator*)
 Delaware NERR, 144, 163
 Jacques Cousteau NERR, 88
Redbreasted sunfish (*Lepomis auritus*), Jacques Cousteau NERR, 94
Red cedar (*Juniperus virginiana*)
 ACE Basin NERR, 173
 Delaware NERR, 121, 152
 Jacques Cousteau NERR, 82
 Weeks Bay NERR, 220
Red-cockaded woodpecker (*Picoides borealis*), ACE Basin NERR, 202
Red drum (*Sciaenops ocellatus*), ACE Basin NERR, 199
Red-eyed vireo (*Vireo olivaceus*)
 ACE Basin NERR, 188
 Delaware NERR, 145, 161
 Jacques Cousteau NERR, 92
Redfin pickerel (*Esox americanus*), Jacques Cousteau NERR, 93
Red fox (*Vulpes vulpes*)
 ACE Basin NERR, 183
 Delaware NERR, 151, 161
 Jacques Cousteau NERR, 86
Redhead (*Aythya americana*), Delaware NERR, 163
Redhead grass (*Potamogeton perfoliatus*), Jacques Cousteau NERR, 67
Red knot (*Calidris canutus*)
 Jacques Cousteau NERR, 87
 Tijuana River NERR, 244
Red macroalga (*Gracilaria tikvahiae*)
 Jacques Cousteau NERR, 95
 Waquoit Bay NERR, 43, 51
Red maple (*Acer rubrum*)
 Delaware NERR, 121, 152, 155
 Jacques Cousteau NERR, 71
Red oak (*Quercus rubra*), Delaware NERR, 152
Red-shouldered hawk (*Buteo lineatus*), Delaware NERR, 144–145
Red-spotted newt (*Notophthalmus viridescens viridescens*), Waquoit Bay NERR, 38
Red squirrel (*Tamiasciurus hudsonicus*), Jacques Cousteau NERR, 86
Redstart (*Setophaga ruticilla*), Jacques Cousteau NERR, 91
Red-tailed hawk (*Buteo jamaicensis*)
 ACE Basin NERR, 187
 Delaware NERR, 144, 145, 161

Red-throated loon (*Gavia stelata*), Delaware NERR, 144
Red-winged blackbird (*Agelaius phoeniceus*)
 ACE Basin NERR, 187
 Delaware NERR, 144, 160
 Tijuana River NERR, 241
Regina regida, ACE Basin NERR, 182
Rehse, M.A., 245
Reithrodontomys humilus, ACE Basin NERR, 183
Reithrodontomys megalotis, Tijuana River NERR, 243
Remotely operated vehicles, 20–21
Remote sensing, 19
REMUS (Remote Environmental Monitoring UnitS), 20–21
Research
 importance, xiii
 NERRS, 4–5, 7
Restoration, *see* Habitat restoration
Rhexia aristosa, Jacques Cousteau NERR, 72
Rhinichthys atratulus, Jacques Cousteau NERR, 94
Rhododendron viscosum, Jacques Cousteau NERR, 71, 72
Rhus integrifolia, Tijuana River NERR, 243
Rhus laurina, Tijuana River NERR, 242
Rhus radicans, Waquoit Bay NERR, 41
Rhynchospora inundata, Jacques Cousteau NERR, 72
Rhynchospora nitens, Jacques Cousteau NERR, 72
Ribbed mussel (*Geukensia demissa*)
 ACE Basin NERR, 197
 Delaware NERR, 138
Rice rat (*Oryzomys palustris*)
 ACE Basin NERR, 186
 Jacques Cousteau NERR, 86
Richard Stockton College of New Jersey, Jacques Cousteau NERR, 59
Ring-billed gulls (*Larus delawarensis*), Waquoit Bay NERR, 41
Rithropanopeus harrisii, Jacques Cousteau NERR, 69
Riverbank guillwort (*Isoetes riparia*), Jacques Cousteau NERR, 70
River cooter (*Pseudemys concinna*), ACE Basin NERR, 182
River frog (*Rana heckscheri*), ACE Basin NERR, 182
River otter (*Lutra canadensis*)
 ACE Basin NERR, 186
 Delaware NERR, 145, 162, 163
 Jacques Cousteau NERR, 86
Robinson, C.L., 225, 226
Rosa multiflora, Jacques Cousteau NERR, 82

Rosa palustris, Jacques Cousteau NERR, 70
Rosa rugosa, see Salt spray rose
Roseate terns (*Sterna dougalli*), Waquoit Bay NERR, 41
Rose tickseed (*Coreopsis rosea*), Jacques Cousteau NERR, 72
Rotifers, ACE Basin NERR, 194, 195
Rough avens (*Geum virginiana*), Delaware NERR, 156
Rough green snake (*Opheodrys aestivus*)
　ACE Basin NERR, 182
　Jacques Cousteau NERR, 85
Rough-legged hawk (*Buteo lagopus*), Delaware NERR, 144–145
ROVER (Remote Observation Vehicle Earth Resources), benthic habitat mapping, 21
Royal tern (*Sterna maxima*)
　ACE Basin NERR, 189
　Weeks Bay NERR, 223
Ruddy turnstone (*Arenaria interpres*)
　Jacques Cousteau NERR, 87
　Waquoit Bay NERR, 41
Rufous-sided towhee (*Pipilo erythrophthalmus*), Delaware NERR, 145, 161
Ruppia maritima, see Widgeon grass
Rushes (*Juncus roemerianus*), ACE Basin NERR, 173
Rutgers University, Jacques Cousteau NERR, 59, 62, 63, 100
Rynchops niger, see Black skimmer

S

Sabatia stellaris, Jacques Cousteau NERR, 66
Sabellaria vulgaris, ACE Basin NERR, 197
Sage Lot Pond, Waquoit Bay NERR, 35
Sagittaria engelmanniana, Jacques Cousteau NERR, 67
Sagittaria lancifolia, Weeks Bay NERR, 220
Sagittaria latifolia, Jacques Cousteau NERR, 67
Sagittaria spatulata, Jacques Cousteau NERR, 67
Sagittaria spp., Jacques Cousteau NERR, 70
Sagitta spp., Jacques Cousteau NERR, 98
Salicornia bigelovii
　Jacques Cousteau NERR, 66
　Tijuana River NERR, 239
Salicornia europea, Jacques Cousteau NERR, 66
Salicornia spp.
　ACE Basin NERR, 173
　Waquoit Bay NERR, 39
Salicornia subterminalis, Tijuana River NERR, 239, 242

Salicornia virginica
　Jacques Cousteau NERR, 66
　Tijuana River NERR, 239, 242
Salinity
　ACE Basin NERR, 191–192
　analysis of 22 sites, 13
　Weeks Bay NERR, 218–219
Salix hindsiana, Tijuana River NERR, 241
Salix spp.
　Delaware NERR, 121, 152
　Tijuana River NERR, 241
　Waquoit Bay NERR, 38
Salmo fontinalis, Jacques Cousteau NERR, 94
Salmo gairdneri, Jacques Cousteau NERR, 94
Salmo trutta, Jacques Cousteau NERR, 94
Salt bush (*Atriplex leucophylla*), Tijuana River NERR, 243
Salt bush (*Baccharis glutinosa*), Tijuana River NERR, 241
Salt grass (*Distichlis spicata*)
　Delaware NERR, 121, 152
　Tijuana River NERR, 239, 242
　Weeks Bay NERR, 220
Salt marsh bird's beak (*Cordylanthus maritimus* ssp. *maritimus*), Tijuana River NERR, 237, 240
Salt marsh hay (*Spartina patens*), Waquoit Bay NERR, 39
Salt marsh pink (*Sabatia stellaris*), Jacques Cousteau NERR, 66
Salt marsh snails (*Melampus bidentatus*), Delaware NERR, 138
Salt-meadow cordgrass (*Spartina patens*)
　Delaware NERR, 121, 152
　Jacques Cousteau NERR, 66
　Weeks Bay NERR, 220
Salt spray rose (*Rosa rugosa*)
　Jacques Cousteau NERR, 82
　Waquoit Bay NERR, 41
Saltwort (*Batis maritima*), Tijuana River NERR, 239
Saltwort grass (*Salicornia bigelovii*), Jacques Cousteau NERR, 66
Saltworts (*Salicornia* spp.), ACE Basin NERR, 173
Salvelinus fontinalis, Waquoit Bay NERR, 38
Samphir (*Salicornia europea*), Jacques Cousteau NERR, 66
Sandbar willow (*Salix hindsiana*), Tijuana River NERR, 241
Sand bay shrimp (*Crangon septemspinosa*), Jacques Cousteau NERR, 104
Sand dollars (*Echinarachnius parma*), Delaware NERR, 136

Sand dune tiger beetle (*Coelus latesignata latesignata*), Tijuana River NERR, 243
Sanderling (*Calidris alba*)
  ACE Basin NERR, 188
  Jacques Cousteau NERR, 87
  Tijuana River NERR, 244
  Waquoit Bay NERR, 41
San Diego, 237
San Diego gopher snake (*Pituophis melanoleucus annectens*), Tijuana River NERR, 242
San Diego pocket mouse (*Peromyscus fallax*), Tijuana River NERR, 243
Sandifer, P.A., 176
Sand lance (*Ammodytes* sp.), Jacques Cousteau NERR, 99–100, 109
Sand myrtle (*Leiophyllum buxifolium*), Jacques Cousteau NERR, 81
Sandplain gerardia (*Agalinis acuta*), Waquoit Bay NERR, 38
Sand shrimp (*Crangon septemspinosa*), Waquoit Bay NERR, 46
Sand verbena (*Abronia umbellata*), Tijuana River NERR, 243
Sandy beach tiger beetle (*Coelus hirticollis gravida*), Tijuana River NERR, 243
Sanger, D., 14, 203
Sanquinolaria nuttalli, Tijuana River NERR, 245, 247, 248
Sansbury, C.E., 190
Sarracenia purpurea, Jacques Cousteau NERR, 71, 72
Sarsia spp., 98
Sassafras (*Sassafras albidum*)
  Delaware NERR, 121, 152
  Jacques Cousteau NERR, 71, 81
Satinfin shiner (*Notropis analostanus*), Jacques Cousteau NERR, 94
Saw grass (*Cladium jamaicense*), Weeks Bay NERR, 220
Scalopus aquaticus, *see* Eastern mole
Scaphiopus holbrooki, Jacques Cousteau NERR, 82
Scarlet oak (*Quercus coccinea*), Jacques Cousteau NERR, 80, 81
SCE&G Canadys Power Station, 173
Sceloporus undulatus, ACE Basin NERR, 176
Sceloporus undulatus hyacinthinus, Jacques Cousteau NERR, 82, 84
Scenedesmus spp., 133, 157
Schizachyrium scoparium, Waquoit Bay NERR, 37–38
Schizaea pusilla, Jacques Cousteau NERR, 71
Schizopera knabeni, ACE Basin NERR, 196

Schreiber, R.A., 223
Sciaenops ocellatus, ACE Basin NERR, 199
Scinella lateralis, *see* Ground skink
Scirpus americanus
  Delaware NERR, 152
  Jacques Cousteau NERR, 66–67
Scirpus longii, Jacques Cousteau NERR, 72
Scirpus pungens, Jacques Cousteau NERR, 70
Scirpus smithii var. *smithii*, Jacques Cousteau NERR, 70
Scirpus spp., Jacques Cousteau NERR, 67
Sciurus carolinensis, *see* Gray squirrel
Scoloplos robustus, Jacques Cousteau NERR, 100
Scophthalmus aquosus, *see* Windowpane
Scoters, ACE Basin NERR, 202
Scoters (*Melanitta* spp.), Jacques Cousteau NERR, 88
Scott, G.I., 204
Scottolana spp., Delaware NERR, 158
Scrub oak (*Quercus ilicifolia*)
  Jacques Cousteau NERR, 80, 81
  Waquoit Bay NERR, 36–37, 38
Scud (*Corophium* sp.), Delaware NERR, 158
Sea bass (*Paralabrax* spp.), Tijuana River NERR, 252
Sea blite (*Suaeda esteroa*), Tijuana River NERR, 239
Sea herring (*Clupea harengus*), Jacques Cousteau NERR, 106
Sea lavender (*Limonium californica*), Tijuana River NERR, 239, 242
Sea lavender (*Limonium nashii*), Waquoit Bay NERR, 39
Sea lettuce (*Ulva lactuca*)
  Jacques Cousteau NERR, 95
  Waquoit Bay NERR, 43, 45
Sea level rise, *see* Global sea level rise
Sea myrtle (*Baccharus halmifolia*), Weeks Bay NERR, 220
Sea rocket (*Cakile edentula*), Jacques Cousteau NERR, 81
Sea rocket (*Cakile maritima*), Tijuana River NERR, 243
Seaside goldenrod (*Solidago sempervirens*)
  Jacques Cousteau NERR, 66, 81
  Waquoit Bay NERR, 41
Seaside sparrow (*Ammospiza maritima*)
  Delaware NERR, 144, 160
  Jacques Cousteau NERR, 91
Sea squirts (*Molgula manhattensis*), Waquoit Bay NERR, 42
Sedberry, G.R., 201
Sedges, Jacques Cousteau NERR, 72
Sedimentation-erosion table (SET), assessment, 22

Sediment profile imagery (SPI), benthic habitat
    mapping, 20
Seiurus aurocapillus, see Ovenbird
Seminatrix pygaea, ACE Basin NERR, 181
Semipalmated plover (*Charadrius semipalmatus*)
    Jacques Cousteau NERR, 87
    Tijuana River NERR, 241
    Waquoit Bay NERR, 41
Semipalmated sandpiper (*Calidris pusilla*),
    Jacques Cousteau NERR, 87
Semotilus atromaculatus, Jacques Cousteau
    NERR, 93
Semotilus corporalis, Jacques Cousteau NERR,
    93
Sensitive fern (*Onaclea sensibilis*), Jacques
    Cousteau NERR, 70
Septic system problems
    Delaware NERR, 132
    Weeks Bay NERR, 227–228
Serviceberry (*Amelanchier candensis*), Jacques
    Cousteau NERR, 82
Sesarma reticulatum, Delaware NERR, 138
Setophaga ruticilla, Jacques Cousteau NERR, 91
Sharp, J.H., 132
Sharp-shinned hawk (*Accipiter striatus*)
    ACE Basin NERR, 187
    Delaware NERR, 144–145, 161
    Jacques Cousteau NERR, 89
Sharp-tailed sparrow (*Ammospiza caudacuta*)
    Delaware NERR, 144
    Jacques Cousteau NERR, 91
Sheep laurel (*Kalmia augustifolia*)
    Jacques Cousteau NERR, 71, 72, 81, 82
    Waquoit Bay NERR, 38–39
Sheepshead minnow (*Cyprinodon variegatus*),
    Delaware NERR, 138
Shoregrass (*Monanthocloe littoralis*), Tijuana
    River NERR, 239
Short, F.T., 49, 51
Short-beaked bald-rush (*Rhynchospora nitens*),
    Jacques Cousteau NERR, 72
Short-billed dowitcher (*Limnodromus griseus*)
    Jacques Cousteau NERR, 87
    Waquoit Bay NERR, 41
Short-eared owl (*Asio flammeus*)
    Delaware NERR, 145
    Jacques Cousteau NERR, 89
Shortleaf pine (*Pinus echinata*), 80
    ACE Basin NERR, 174
Shortnose sturgeon (*Acipenser brevirostrum*),
    ACE Basin NERR, 173, 200, 202
Short-tailed shrew (*Blarina brevicauda*)
    ACE Basin NERR, 185
    Delaware NERR, 151, 161
    Jacques Cousteau NERR, 86

Shrews, ACE Basin NERR, 186
Shrimp
    ACE Basin NERR, 196, 197
    Weeks Bay NERR, 227
Side-blotched lizard (*Uta stansburiana*), Tijuana
    River NERR, 242, 243
Sigmodon hispidus, ACE Basin NERR, 183
Silver perch (*Bairdiella chrysoura*)
    ACE Basin NERR, 200
    Jacques Cousteau NERR, 106, 109
Silvery legless lizard (*Anniella pulchra pulchra*),
    Tijuana River NERR, 243
Silvery minnow (*Hybognathus nuchalis*), Jacques
    Cousteau NERR, 93
Siren lacertina, ACE Basin NERR, 182
Site profile, description, 9
Six-lined racerunner (*Cnemidophorus
    sexlineatus*), ACE Basin NERR, 176,
    180
Skate (*Raja eglanteria*), Jacques Cousteau NERR,
    109
Skeletonema costatum
    ACE Basin NERR, 193–194
    Jacques Cousteau NERR, 97
Skeletonema spp., Delaware NERR, 157
Skimmers, ACE Basin NERR, 202
Slash pines (*Pinus elliottii*), Weeks Bay NERR,
    220
Slender glass lizard (*Ophisaurus attenuatus*),
    ACE Basin NERR, 176, 181
Slender pondweed (*Potamogeton pusillus*),
    Jacques Cousteau NERR, 67
Slender water-milfoil (*Myriophyllum tenellum*),
    Jacques Cousteau NERR, 72
Slipper shells (*Crepidula fornicata*), Waquoit Bay
    NERR, 45
Slipper shells (*Crepidula plana*), Waquoit Bay
    NERR, 45
Small-mouth flounder (*Etropus microstomus*),
    Jacques Cousteau NERR, 109
Smith, S.H., 245
Smooth alder (*Alnus serrulata*), Delaware NERR,
    123
Smooth bur-marigold (*Bidens laevis*), Jacques
    Cousteau NERR, 70
Smooth cordgrass (*Spartina alterniflora*)
    ACE Basin NERR, 173, 202
    Delaware NERR, 121, 138, 152
    invasive/exotic species effects on, 24, 25
    Jacques Cousteau NERR, 66
    Waquoit Bay NERR, 39
    Weeks Bay NERR, 220
Smooth earth snake (*Virginia valeriae*), ACE
    Basin NERR, 181
Snapping turtle (*Chelydra serpentina*)

Index **291**

ACE Basin NERR, 182
Delaware NERR, 140, 160, 163
Jacques Cousteau NERR, 85
Snow goose (*Chen caerulescens*), Jacques Cousteau NERR, 88
Snowy egret (*Egretta thula*)
  Delaware NERR, 161
  Jacques Cousteau NERR, 87, 88
  Tijuana River NERR, 241
Soft-shelled clam (*Mya arenaria*), Waquoit Bay NERR, 40, 42, 43
Solidago sempervirens, *see* Seaside goldenrod
Songbirds, Jacques Cousteau NERR, 91
Song sparrow (*Melospiza melodia*)
  Delaware NERR, 145, 160
  Jacques Cousteau NERR, 89
  Tijuana River NERR, 240
Sooty shearwaters (*Puffinus griseus*), Jacques Cousteau NERR, 87
Sora (*Porzana carolina*)
  Delaware NERR, 144
  Jacques Cousteau NERR, 91
Sorex cinereus, *see* Masked shrew
South Carolina Department of Natural Resources, 176
South Carolina Marine Resources Research Institute, 11
South Carolina slimy salamander (*Plethodon variolatus*), ACE Basin NERR, 182
Southeast Area Monitoring and Assessment Program, 201
Southeastern crowned snake (*Tantilla coronata*), ACE Basin NERR, 181
Southern bog lemming (*Synaptomys cooperi*), Jacques Cousteau NERR, 86
Southern cricket frog (*Acris gryllus*), ACE Basin NERR, 181–182
Southern flying squirrel (*Glaucomys volans*), Jacques Cousteau NERR, 86
Southern leopard frog (*Rana utricularia*)
  ACE Basin NERR, 182
  Delaware NERR, 140, 160
  Jacques Cousteau NERR, 82
Southern red oak (*Quercus falcata*)
  ACE Basin NERR, 174
  Delaware NERR, 121
  Jacques Cousteau NERR, 81, 82
South Slough NERR
  as first reserve, 2
  restoration, 24
Sparrows, ACE Basin NERR, 187, 188, 189
Spartina alterniflora, *see* Smooth cordgrass
Spartina cynosuroides, *see* Big cordgrass
Spartina foliosa
  invasive/exotic species effects, 25

  Tijuana River NERR, 239
Spartina patens
  Delaware NERR, 121, 152
  Jacques Cousteau NERR, 66
  Waquoit Bay NERR, 39
  Weeks Bay NERR, 220
Spartina salt marshes, Jacques Cousteau NERR, 65–66
Spatterdock (*Nurphur advena*), Jacques Cousteau NERR, 70
Spermophilus beechyii, Tijuana River NERR, 240
Sphaerosyllis spp., Jacques Cousteau NERR, 103
Sphagnum moss, Jacques Cousteau NERR, 71, 72, 81
Sphagnum sp., Waquoit Bay NERR, 39
Sphoeroides maculatus, Jacques Cousteau NERR, 106
Spider crab (*Libinia dubia*), Waquoit Bay NERR, 46
Spider crab (*Libinia emarginata*), Jacques Cousteau NERR, 104
Spike grass (*Distichlis spicata*)
  Jacques Cousteau NERR, 66
  Waquoit Bay NERR, 39
Spionids, Tijuana River NERR, 248
Spisula solidissima
  Delaware NERR, 136
  Jacques Cousteau NERR, 103–104
Spizella passerina, Delaware NERR, 145
Spizella pusilla, Delaware NERR, 145, 160
Sponges, Waquoit Bay NERR, 45
Spot (*Leiostomus xanthurus*)
  ACE Basin NERR, 199
  Delaware NERR, 138, 140, 160, 162
  Jacques Cousteau NERR, 106, 109
Spotted hake (*Urophycis regia*), ACE Basin NERR, 199
Spotted salamander (*Ambystoma maculatum*), ACE Basin NERR, 181, 182
Spotted sandpiper (*Actitis macularia*), Delaware NERR, 161
Spotted seatrout (*Cynoscion nebulosus*), ACE Basin NERR, 199
Spotted shiner (*Notropis hudsonius*), Jacques Cousteau NERR, 94
Spotted turtle (*Clemmys guttata*), Jacques Cousteau NERR, 85
Spring rush (*Juncus acutus*), Tijuana River NERR, 241
Squirrels
  Tijuana River NERR, 243
  Weeks Bay NERR, 223
Staggerbush (*Lyonia mariana*), Jacques Cousteau NERR, 72

Star drum (*Stellifer lanceolatus*), ACE Basin NERR, 199, 201
Star-nosed mole (*Condylura cristata*), Jacques Cousteau NERR, 86
Stearns, D.E., 225
Stellifer lanceolatus, ACE Basin NERR, 199, 201
Stenotomus chrysops, Waquoit Bay NERR, 46
Stenotomus spp., ACE Basin NERR, 201
Stenotomus versicolor, Delaware NERR, 140, 162
Stereochilus marginatus, ACE Basin NERR, 181
Sterna antillarum, see Least tern
Sterna antillarum browni, Tijuana River NERR, 237, 240, 243–244
Sterna dougalli, Waquoit Bay NERR, 41
Sterna forsteri, Delaware NERR, 161
Sterna hirundo, see Common tern
Sterna maxima, see Royal tern
Sterna paradisaea, Waquoit Bay NERR, 41
Sternotherus odoratus, see Stinkpot
Stewardship, NERRS, 5–6, 7
Stinkpot (*Sternotherus odoratus*)
    ACE Basin NERR, 182
    Jacques Cousteau NERR, 85
Strategic Education Plan, 26
    guiding principles, 26
Streblospio benedicti
    ACE Basin NERR, 197
    Delaware NERR, 136
Stressors, see Anthropogenic impacts
Striped anchovy (*Anchoa hepsetus*)
    ACE Basin NERR, 201
    Jacques Cousteau NERR, 106
Striped bass (*Morone saxatilis*)
    ACE Basin NERR, 200
    Delaware NERR, 140, 162
    Jacques Cousteau NERR, 94
    Waquoit Bay NERR, 46
Striped killifish (*Fundulus majalis*), Jacques Cousteau NERR, 106
Striped mullet (*Mugil cephalus*)
    ACE Basin NERR, 199
    Tijuana River NERR, 248–249
    Weeks Bay NERR, 226
Striped skunk (*Mephitis mephitis*)
    ACE Basin NERR, 183
    Delaware NERR, 151, 161
    Jacques Cousteau NERR, 86
    Tijuana River NERR, 243
Strix varia, Delaware NERR, 144
Structural controls, xiii
Sturnella magna, Delaware NERR, 161
Sturnella neglecta, Tijuana River NERR, 240
Suaeda esteroa, Tijuana River NERR, 239
Sula bassanus, Jacques Cousteau NERR, 87

Summer flounder (*Paralichthys dentatus*), Delaware NERR, 140, 162
Sundew (*Drosera capillaris*), Jacques Cousteau NERR, 71, 72
Superfund sites, Delaware NERR, 132–133
Supersaturation events, summary analysis, 14
Surf clams (*Spisula solidissima*), Delaware NERR, 136
Surf scoter (*Melanitta perspicillata*), Delaware NERR, 144
Sutton, C.C., 136
Swallows, ACE Basin NERR, 187, 189
Swamp azalea (*Rhododendron viscosum*), Jacques Cousteau NERR, 71, 72
Swamp darter (*Etheostoma fusiforme*), Jacques Cousteau NERR, 92
Swamp pink (*Helonias bullata*), Jacques Cousteau NERR, 71
Swamp rose (*Rosa palustris*), Jacques Cousteau NERR, 70
Swamp rose mallow (*Hibiscus moscheutos*), Jacques Cousteau NERR, 70
Swamp rose mallow (*Hibiscus palustris*), Delaware NERR, 121
Sweetbay magnolia (*Magnolia virginiana*), Jacques Cousteau NERR, 71
Sweetfern (*Comptonia peregrina*)
    Jacques Cousteau NERR, 80, 81
    Waquoit Bay NERR, 37
Sweet flag (*Acorus calamus*), Jacques Cousteau NERR, 70
Sweet gale (*Myrica gale*), Waquoit Bay NERR, 39
Sweet gum (*Liquidambar styraciflua*)
    ACE Basin NERR, 174
    Delaware NERR, 121, 152, 155
Sweet pepperbush (*Clethra alnifolia*)
    Delaware NERR, 123
    Jacques Cousteau NERR, 71, 82
SWMP
    abiotic factors, 10
    ACE Basin NERR, 191–192
    biological monitoring, 10
    components, 10, 12–15
    data loggers, 10–11, 12
    description, 4
    development, 10–12
    high-priority initiatives, 15–21
    Jacques Cousteau NERR, 63, 64
    medium-priority initiatives, 15, 21–22
    mission, 9–10
    overview, 28–29
    standardization, 10, 11
    trends monitoring, 13–15
    value, 10
    watershed/land use classifications, 10, 19

Index

Sylvilagus audubonii sactidiegi, Tijuana River NERR, 240
Sylvilagus floridanus, see Eastern cottontail
Sylvilagus palustris, see Marsh rabbit
Symphurus plaguisa, Jacques Cousteau NERR, 109
Synaptomys cooperi, Jacques Cousteau NERR, 86
Syngnathus fuscus, Jacques Cousteau NERR, 98, 106, 109
Synodus foetens, Jacques Cousteau NERR, 109
Synthesis of Water Quality Data: National Estuarine Research Reserve System-wide Monitoring Program, A (2002), 11
Synthesis of Water Quality Data from the National Estuarine Research Reserve's System-wide Monitoring Program, A (2001), 11
System-wide Monitoring Program, *see* SWMP
Szedlmayer, S.T., 106, 109

T

Tachycineta bicolor, Delaware NERR, 161
Tachys corax, Tijuana River NERR, 240
Tadarida brasiliensis, ACE Basin NERR, 185
Tadpole madtom (*Noturus gyrinus*), Jacques Cousteau NERR, 93
Tagelus californianus, Tijuana River NERR, 245, 248
Tamiasciurus hudsonicus, Jacques Cousteau NERR, 86
Tamius striatus, Jacques Cousteau NERR, 86
Tantilla coronata, ACE Basin NERR, 181
Tautoga onitis
  Delaware NERR, 162
  Jacques Cousteau NERR, 106
Tautogolabrus adspersus, Jacques Cousteau NERR, 100
Taxodium distichum, ACE Basin NERR, 173, 174
Teal, J.M., 39, 190
Tellina carpenteri, Tijuana River NERR, 247
Temora longicornis, Jacques Cousteau NERR, 98
Tephrosia virginiana, Jacques Cousteau NERR, 81
Terns
  ACE Basin NERR, 202
  Jacques Cousteau NERR, 87
Terrapene carolina, Jacques Cousteau NERR, 85
Tesselated darter (*Etheostoma olmetedi*), Jacques Cousteau NERR, 92, 93
Tetraedron spp., Delaware NERR, 157
Tetragnatha laboriosa, Tijuana River NERR, 240
Thalassiosira spp., ACE Basin NERR, 194

Thamnophis sauritus, Jacques Cousteau NERR, 84
Thanmnophis sirtalis, see Eastern garter snake
Threatened species, *see also* Endangered/threatened species
  ACE Basin NERR, 202
  definition, 203
Three-lined salamander (*Eurycea longicauda*), ACE Basin NERR, 181
Three-spined stickleback (*Gasterosteus aculeatus*)
  Jacques Cousteau NERR, 94
  Waquoit Bay NERR, 46
Three-square bulrush (*Scirpus pungens*), Jacques Cousteau NERR, 70
Thryothorus ludovicianus, see Carolina wren
Tier II parameters, 16
Tier I parameters, 16
Tijuana River NERR
  algal mats, 238
  anthropogenic impacts
    brackish marshes, 240–241
    development, 237, 255
    dunes/beach habitat, 243
    gravel extraction, 255
    overview, 241–242, 252, 254–256, 257
    sedimentation, 255, 256
    sewage-contaminated inflows, 235, 237, 252, 254, 256
    stream flow modification, 255
  commercially/recreationally important species, 252
  description, 235–237
  designation, 235
  endangered species/protection, 237, 239–240
  estuary/aquatic habitat
    benthic invertebrates, 245–248
    benthic invertebrates size change, 248
    birds, 252, 253–254, 258
    birds/waterbirds list, 253–254
    fish, 248–252, 258
    fish list, 250–251
    invertebrate list, 246–247
    plants, 245
    species annual relative abundance, 249
    tidal creeks/channels, 244–252, 258
  flood effects, 248–249, 252
  habitat list/area, 236
  invasive/exotic species, 235, 242, 243
  invertebrate list, 246–247
  map, 236
  overview, 256–258
  restoration, 24, 237, 255
  salinity, 238
  water quality monitoring sites, 237

watershed habitat
   brackish marsh, 240–241, 257
   dunes/beach, 243–244, 257–258
   intertidal flats, 244, 258
   map, 238
   overview, 237
   riparian habitat, 241–242, 257
   salt marsh, 238–240
   salt pannes, 240, 257
   wetland-upland transition, 242–243, 257
Timber rattlesnake (*Crotalus horridus horridus*)
   ACE Basin NERR, 181
   Jacques Cousteau NERR, 84
Toads, ACE Basin NERR, 181
Topsmelt (*Atherinops affinis*), Tijuana River NERR, 248, 252
Torrey's droopseed (*Muhlenbergia torreyana*), Jacques Cousteau NERR, 72
Tortanus discaudatus, Jacques Cousteau NERR, 98
Toxicodendrom radicans, Weeks Bay NERR, 220
Toxostoma rufum, Delaware NERR, 145
Trachemys scripta, ACE Basin NERR, 182
Trachinotus carolinus, ACE Basin NERR, 201
Trailing arbutus (*Epigagea repens*), Jacques Cousteau NERR, 81
Treefrogs, ACE Basin NERR, 181
Tree swallow (*Iridoprocne bicolor*), Jacques Cousteau NERR, 91
Tree swallow (*Tachycineta bicolor*), Delaware NERR, 161
Trichechus manatus, ACE Basin NERR, 202
Tri-colored heron (*Egretta tricolor*), Jacques Cousteau NERR, 88
Trident red maple (*Acer rubrum*), Jacques Cousteau NERR, 71
Triglochin concinnum, Tijuana River NERR, 239
Trinectes maculatus, see Hogchoker
Tringa flavipes, Waquoit Bay NERR, 41
Tringa melanoleuca, see Greater yellowlegs
Tringa spp., Tijuana River NERR, 244
Troglodytes aedon, Delaware NERR, 160, 161
Troglodytes troglodytes, ACE Basin NERR, 188
Tube-building amphipod (*Ampelisca abdita*), Jacques Cousteau NERR, 100, 103
Tuckerton Seaport, Jacques Cousteau NERR, 60
Tufted titmouse (*Parus bicolor*), Delaware NERR, 145
Tulip tree (*Liriodendron tulipifera*), Delaware NERR, 121
Tundra swan (*Cygnus columbianus*), Jacques Cousteau NERR, 88
Turbellarians, Delaware NERR, 136

Turbonilla sp., Jacques Cousteau NERR, 100
Turdus migratorius, Delaware NERR, 145, 160
Turkey-beard (*Xerophyllum asphodeloides*), Jacques Cousteau NERR, 71, 72
Turkey vulture (*Cathartes aura*)
   ACE Basin NERR, 187
   Delaware NERR, 145, 161
Tursiops truncatus, see Bottlenose porpoise
Twig rush (*Cladium marascoides*), Waquoit Bay NERR, 39
Two-toed amphiuma (*Amphiuma means*), ACE Basin NERR, 182
Tychoplankton, 194
Typha angustifolia
   Delaware NERR, 125
   Jacques Cousteau NERR, 67, 70
   Weeks Bay NERR, 220
Typha domingensis, Tijuana River NERR, 241
Typha glauca, Jacques Cousteau NERR, 70
Typha latifolia
   Delaware NERR, 125
   Waquoit Bay NERR, 38
Typha spp., Delaware NERR, 124–125, 152
Tyto alba, ACE Basin NERR, 188

U

Uca crenulata, Tijuana River NERR, 239, 244
Uca minax, Delaware NERR, 138
Uca pugnax, Delaware NERR, 138
Uca spp.
   ACE Basin NERR, 193
   Delaware NERR, 136, 138
Ulva lactuca, see Sea lettuce
Umbra pygmaea, Jacques Cousteau NERR, 92, 93
Upland chorus frog (*Pseudacris triseriata*), ACE Basin NERR, 182
Urochordates, Waquoit Bay NERR, 45
Urocyon cinereoargenteus, see Gray fox
Urophycis regia, ACE Basin NERR, 199
Urophycis spp., Jacques Cousteau NERR, 109
Urophycis tenuis, Waquoit Bay NERR, 46
Urosalpinx cinera, Delaware NERR, 162
U.S. Fish and Wildlife Service
   Jacques Cousteau NERR, 60
   Tijuana River NERR, 237
U.S. Fish and Wildlife Service Coastal Program, xiii
Uta stansburiana, Tijuana River NERR, 242, 243
Utricularia olivacea, Jacques Cousteau NERR, 72
Utricularia spp., Jacques Cousteau NERR, 72

Index **295**

V

Vaccinium angustifolium, Waquoit Bay NERR, 37
Vaccinium corymbosum, Jacques Cousteau NERR, 72, 82
Vaccinium spp., Jacques Cousteau NERR, 80, 82
Valiela, I.K., 43, 49, 51
Van Dolah, R.F., 193–194
Vernberg, F.J., 190
Viburnum prunifolium, Delaware NERR, 121
Viola pedata, Waquoit Bay NERR, 38
Vireo belli pusilius, Tijuana River NERR, 237
Vireo griseus, *see* White-eyed vireo
Vireo olivaceus, *see* Red-eyed vireo
Vireos, ACE Basin NERR, 187, 188, 189
Virginia rails (*Rallus limicola*), Jacques Cousteau NERR, 91
Virginia valeriae, ACE Basin NERR, 181
Virginia willow (*Itea virginica*), Weeks Bay NERR, 220
Viscido, S.V., 100
Volvox spp., Delaware NERR, 133, 157
Vulpes vulpes, *see* Red fox

W

Waquoit Bay NERR
  anthropogenic impacts, 46, 48–51
  benthic organisms, 43–46
  chemical contaminants, 49
  description, 35–36
  designation, 36
  eelgrass decline, 43, 45, 50–51
  environment, 43
  eutrophication, 48–51
  finfish, 46, 47–48
  fish nursery area, 43, 46
  habitats/species
    beaches/dunes, 41–42
    freshwater wetlands, 38–39
    mudflats/sandflats, 39–41
    riparian habitats, 38
    saltmarshes, 39
    sandplain grasslands, 37–38
    upland pitch pine/oak forests, 36–37
    vernal pools/coastal plain pond shores, 38
  invertebrates list, 44–45
  "kills," 49
  macroalgal mats, 43, 50–51
  map, 35
  motorboats
    chemical contamination, 49
    problems from propellers, 49, 51
  nitrogen inputs, 49–50
  nutrient loading, 43, 48–51
  overview, 51–52
  PAR (photosynthetically active radiation), 51
  pathogens (from septic systems), 49
  salt marsh plants list, 40
  septic system problems, 49, 50
  subwatersheds list, 36
  subwatersheds map, 37
  tidal creeks and channels/species, 42–43
  transition zone for fish assemblages, 46
Warblers, ACE Basin NERR, 187, 189
Washington clam (*Saxidomus nuttalli*), Tijuana River NERR, 247
Wasson, K., 25
Water celery (*Vallisneria americana*), Jacques Cousteau NERR, 67
Water chemistry controls, summary analysis, 14
Water degradation, anthropogenic impact types, xi
Waterfowl
  ACE Basin NERR, 202
  Jacques Cousteau NERR, 88
  Waquoit Bay NERR, 41
Water hemp (*Amaranthus cannabinus*), Jacques Cousteau NERR, 70
Water quality data, *see also specific sites*
  analysis of 22 sites, 13–14
  analysis over Phase I, 14–15
Watershed land use
  classes list, 19
  mapping, 19
Water temperature, summary analysis, 13–14
Watling, L., 133
Watson's salt bush (*Atriplex watsonii*), Tijuana River NERR, 239, 242
Wax myrtle (*Myrica cerifera*), ACE Basin NERR, 173, 174
Weakfish (*Cynoscion regalis*)
  ACE Basin NERR, 199
  Delaware NERR, 140, 162
  Jacques Cousteau NERR, 94, 109
Weeks Bay NERR
  anthropogenic impacts, 227–228
  best management practices (BMPs), 227, 228
  commercially important species, 227
  description, 217–219
  designation, 217
  development, 227, 228
  estuary communities (animals)
    benthic fauna, 225–226
    fish, 226–227
    zooplankton, 225
  estuary communities (plants), 223–225
  land protection, 227
  map, 218
  overview, 217, 228–229

physical description, 217–219
  salinity, 218–219, 226
  sediment, 219
  sediment distribution map, 219
  water depths, 219
watershed communities (animals)
  birds, 223
  herpetofauna/list, 220–223
  mammals/list, 223, 224
watershed communities (plants), 220
Wells NERR, restoration, 24
Wenner, C.A., 201
Wenner, E.L., 13, 191, 197
Western harvest mouse (*Reithrodontomys megalotis*), Tijuana River NERR, 243
Western meadowlark (*Sturnella neglecta*), Tijuana River NERR, 240
Western sandpiper (*Calidris mauri*)
  Jacques Cousteau NERR, 87
  Tijuana River NERR, 244
Western snowy plover (*Charadrius alexandrinus nivosus*), Tijuana River NERR, 240, 243, 244
West Indian manatee (*Trichechus manatus*), ACE Basin NERR, 202
Whales, ACE Basin NERR, 202
Whelks (*Busycon canaliculatum*), Waquoit Bay NERR, 45
Whelks (*Busycon carica*), Waquoit Bay NERR, 45
Whimbrel (*Numenius phaeopus*), Tijuana River NERR, 244
White catfish (*Ameiurus catus*), Delaware NERR, 140
White catfish (*Ictalurus catus*), ACE Basin NERR, 199
White croaker (*Genyonemus lineatus*), Tijuana River NERR, 252
White-crowned sparrow (*Zonotrichia leucophrys*), Tijuana River NERR, 240
White-eyed vireo (*Vireo griseus*)
  Delaware NERR, 145, 161
  Jacques Cousteau NERR, 91
White-footed mouse (*Peromyscus leucopus*)
  Delaware NERR, 145, 161
  Jacques Cousteau NERR, 86
White hake (*Urophycis tenuis*), Waquoit Bay NERR, 46
White mullet (*Mugil curema*), ACE Basin NERR, 201
White oak (*Quercus alba*)
  ACE Basin NERR, 174
  Delaware NERR, 121, 151–152
  Jacques Cousteau NERR, 80, 81
  Weeks Bay NERR, 220
White perch (*Morone americana*)
  Delaware NERR, 138, 140, 160, 162
  Jacques Cousteau NERR, 94, 106
  Waquoit Bay NERR, 38
White-rumped sandpiper (*Calidris fuscicollis*), Jacques Cousteau NERR, 87
White sand clam (*Macoma secta*), Tijuana River NERR, 247
White sucker (*Catostomus commersoni*)
  Jacques Cousteau NERR, 94
  Waquoit Bay NERR, 38
White-tailed deer (*Odocoileus virginianus*)
  ACE Basin NERR, 183
  Delaware NERR, 151, 161
  Jacques Cousteau NERR, 86
White-throated sparrow (*Zonotrichia albicollis*), Delaware NERR, 145
White water lilies (*Nymphaea odorata*), Jacques Cousteau NERR, 72
Widgeon grass (*Ruppia maritima*)
  Jacques Cousteau NERR, 67, 94
  Tijuana River NERR, 240
Wildcat Landfill Superfund site, 132, 133
Wild rice (*Zinzania aquatica*)
  Delaware NERR, 152
  Jacques Cousteau NERR, 70
Willet (*Catoptrophorus semipalmatus*)
  Delaware NERR, 161
  Jacques Cousteau NERR, 87
  Tijuana River NERR, 239, 241, 244
Willow oak (*Quercus phellos*), Jacques Cousteau NERR, 82
Willows (*Salix* spp.)
  Delaware NERR, 121, 152
  Tijuana River NERR, 241
  Waquoit Bay NERR, 38
Wilson's plovers (*Charadrius wilsonia*), ACE Basin NERR, 188
Wilson's storm petrels (*Oceanites oceanicus*), Jacques Cousteau NERR, 87
Windowpane (*Scophthalmus aquosus*)
  Delaware NERR, 140
  Jacques Cousteau NERR, 106, 109
Winterberry (*Ilex verticillata*)
  Delaware NERR, 123
  Jacques Cousteau NERR, 72
Winter flounder (*Pseudopleuronectes americanus*)
  Jacques Cousteau NERR, 100, 106, 109
  Waquoit Bay NERR, 42, 46
Wintergreen (*Gaultheria procumbens*), Jacques Cousteau NERR, 71, 72, 81

Winter wren (*Troglodytes troglodytes*), ACE Basin NERR, 188
Wolf, P.L., 190
Woodchuck (*Marmota monax*), Jacques Cousteau NERR, 86
Wood duck (*Aix sponsa*), Delaware NERR, 161, 163
Wood frog (*Rana sylvatica*)
    Delaware NERR, 140, 160
    Jacques Cousteau NERR, 82
Woodpeckers, ACE Basin NERR, 187, 188, 189
Wood stork (*Mycteria americana*), ACE Basin NERR, 173, 202
Wood thrush (*Hylocichla mustilina*)
    Delaware NERR, 145, 161
    Jacques Cousteau NERR, 92
Wood turtle (*Clemmys insculpta*), Jacques Cousteau NERR, 85
Wootton, L., 95
World Meteorological Organization, 22
Wrens, ACE Basin NERR, 187, 188, 189
Wright's panic grass (*Panicum wrightianum*), Jacques Cousteau NERR, 72

Yellow perch (*Perca flavescens*), Jacques Cousteau NERR, 92, 94
Yellow pondweed (*Nuphar lutea*), Delaware NERR, 152
Yellow rail (*Coturnicops noveboracensis*), Delaware NERR, 144
Yellow-rumped warbler (*Dendroica coronata*), Delaware NERR, 145, 161
Yellow shore crab (*Hemigrapsus oregonensis*), Tijuana River NERR, 239, 244
Yellow-spotted salamander (*Ambystoma maculatum*), Waquoit Bay NERR, 38
Yellowthroat (*Geothlypis trichas*)
    ACE Basin NERR, 188
    Delaware NERR, 144, 160
    Jacques Cousteau NERR, 91–92
Yellow warbler (*Dendroica petechia*)
    ACE Basin NERR, 188
    Delaware NERR, 145, 161
    Jacques Cousteau NERR, 91
Yemassee Wastewater Treatment Facility, 173
Yerba reuma (*Frankenia palmeri*), Tijuana River NERR, 242

X

Xerophyllum asphodeloides, Jacques Cousteau NERR, 71, 72

Y

Yellowbelly slider (*Trachemys scripta*), ACE Basin NERR, 182
Yellow bullhead (*Ameiurus natalis*), Jacques Cousteau NERR, 92
Yellow clam (*Florimentis obesa*), Tijuana River NERR, 247
Yellow-crowned night heron (*Nycticorax violaceus*), Jacques Cousteau NERR, 88
Yellow-green algae, Jacques Cousteau NERR, 72

Z

Zampella, R.A., 63, 92
Zanniuchellia palustris, Jacques Cousteau NERR, 67
Zapus hudsonius, *see* Meadow jumping mouse
Zebra mussel (*Dreissena polymorpha*), as invasive, 25
Zedler, J.B., 238, 239, 242, 245, 248
Zenaida macroura, *see* Mourning dove
Zimmerman, L., 183, 186
Zimmerman, R.W., 95
Zinzania aquatica, *see* Wild rice
Zonotrichia albicollus, Delaware NERR, 145
Zonotrichia leucophrys, Tijuana River NERR, 240
Zooplankton, categorization, 194
Zostera marina, *see* Eelgrass